AF527848

# Varieties and Hybrids of Vegetables

# Varieties and Hybrids of Vegetables

by

**D.N. Singh**
**Vishal Nath**

**2011**

**SATISH SERIAL PUBLISHING HOUSE**
403, Express Tower, Commercial Complex, Azadpur, Delhi-110033 (India)
Phone : 011-27672852, Fax : 91-11-27672046
E-mail :info@satishserial.com, hkjain1975@yahoo.com
Website : www.satishserial.com

*Published by :*

**SATISH SERIAL PUBLISHING HOUSE**
403, Express Tower, Commercial Complex, Azadpur, Delhi-110033 (INDIA)
Phone : 011-27672852 Fax : 91-11-27672046
E-mail : info@satishserial.com, hkjain1975@yahoo.com



ISBN 81-89304-88-7



Typeset at: **Harminder** *for* **Laxmi Art Creations** # 98 11 48 23 28
Printed at: Salasar Imaging Systems, Delhi-35

# Preface

Vegetable occupy an important place in diversification of agriculture and have played a pivotal role in food and nutritional security of ever growing population of our country. India is largely a vegetarian society, solely depending on vegetables for bulk of their nutritional requirement. Long life of vegetarians is not only due to the fact that vegetables provide much needed vitamins, minerals and other nutrients but also the green leafy vegetables, carrot, squash, melons and tomatoes act as antioxidant. Antioxidants are known to protect as by destroying the free radicals produced in our body. The antioxidants present in vegetables act as antidote for radicals and protect considerably from heart ailments. In nutshell, richness of antioxidants in vegetable has catapulted their significance in human health considerably. With the green revolution, we have been able to achieve self-sufficiency in food supply quantitatively but qualitatively, still much has to be done. Vegetables are very valuable for adding and supplementing the quality component to the food, as they are rich source of minerals, vitamins and other nutrients. India, bestowed with variable soil and climatic conditions, has the excellent potential for growing both the tropical and temperate vegetables.

India has emerged as the second largest vegetable producing country in the world. Presently, it occupies 7.49 million hectare area with the annual production of 116.03 million of tones, contributing 15% of the world production of vegetable. However, per caput availability of vegetables is far below the required at of 285g / caput / day for a balanced diet.

For successful vegetable production, better seeds, improved varieties, improved cultural practices and better plant protection methods are necessary. Due to rapid expansion of research and development (R & D) in the ICAR institute, State Agricultural University, private and public seed companies, number of varieties and hybrids have been released for commercial cultivation in the country. Production of hybrid vegetable is gaining momentum and it is apprehended that the next decade will be the decade of hybrids for sustainable development of horticulture as well as for higher economic returns to the farmers and vegetable industry. But due to lack of awareness and wide spread publicity about the characters of these varieties and hybrids, the beneficial effect of the research is yet to reach the ultimate users i.e. the farmers.

So we have made an endeavor to bring the facts about the varieties and hybrids evolved by various organization for the large scale cultivation. The characteristics of different varieties and hybrids have been presented in a lucid but informative ways. It is hoped that the book will serve as a guide and reference material to all those related to vegetable sector including students, research workers, agribusiness companies, NGOs and extension functionaries. Despite our best efforts, we realize that there is still scope for further improvement in this direction. We always welcome constructive suggestions of the readers for inclusion in next edition.

We may like to thank the publishers and authors of all the publications for information mentioned in the text. Last but not the least, we express our indebtness to all those who have contributed in publication of this book including composer, publishers, wellwishers and family members.

Date : 12.08.2010 **D.N. Singh**
Place : Bhubaneswar **Vishal Nath**

# Contents


*Preface* ........ *v*

**Chapter-1 Introduction** ........ **1**

**Chapter-2 Cole Crops** ........ **5**
Broccoli ........ 5
Brussels sprouts ........ 8
Cabbage ........ 10
Cauliflower ........ 28
Chinese cabbage ........ 41
Kale ........ 43
Knol-Khol ........ 44

**Chapter-3 Root Crops** ........ **49**
Beet root ........ 49
Carrot ........ 53
Chinese radish ........ 57
Horse radish ........ 58
Radish ........ 59
Rutabaga ........ 65
Turnip ........ 66

**Chapter-4 Bulbs and Tubers** ..... **69**
Amorphophallus ..... 69
Arrowroot ..... 71
Bunching onion ..... 72
Chinese potato ..... 72
Chive ..... 73
Colocasia ..... 74
Garlic ..... 78
Giant taro ..... 81
Great headed garlic ..... 81
Leek ..... 82
Multiplier onion ..... 84
Onion ..... 85
Potato ..... 93
Rhubarb ..... 103
Sweet potato ..... 104
Tannia ..... 109
Tapioca ..... 110
Welsh onion ..... 116
Yams ..... 117
Yam bean ..... 120

**Chapter-5 Solanaceous Vegetables** ..... **123**
Brinjal ..... 123
Capsicum ..... 147
Chilli ..... 153
Tomato ..... 181

**Chapter-6 Legume Vegetables** ..... **219**
Broad bean ..... 219
Cow pea ..... 221
Cluster bean ..... 228
French bean ..... 230
Hyacinth bean ..... 234
Jack bean ..... 240
Pea ..... 241
Runner bean ..... 250

| | | |
|---|---|---|
| | Sword bean | 251 |
| | Vegetable soybean | 252 |
| | Winged bean | 254 |
| **Chapter-7** | **Cucurbitaceous Vegetables** | **257** |
| | Ash gourd | 257 |
| | Bitter gourd | 259 |
| | Bottle gourd | 268 |
| | Cantaloupe | 276 |
| | Chow-Chow | 277 |
| | Cucumber | 278 |
| | Kokrol (Sweet gourd) | 288 |
| | Musk melon | 289 |
| | Pointed gourd | 298 |
| | Pumpkin | 302 |
| | Ridge gourd | 306 |
| | Round melon / Round gourd | 312 |
| | Small gourd | 313 |
| | Snake gourd | 315 |
| | Snap melon | 317 |
| | Spine gourd | 319 |
| | Sponge gourd | 320 |
| | Summer squash | 323 |
| | Teasel gourd | 325 |
| | Water melon | 326 |
| | Winter squash | 339 |
| **Chapter-8** | **Leafy Vegetables** | **341** |
| | Alternanthera | 341 |
| | Amaranth | 341 |
| | Basella | 346 |
| | Bathua | 347 |
| | Beet leaf (Palak) | 348 |
| | Celery | 351 |
| | Coleus | 353 |
| | Coriander | 353 |
| | Curry leaf | 357 |

| | | |
|---|---|---|
| | Fenugreek | 358 |
| | Kang kong | 362 |
| | Kurrat | 363 |
| | Lettuce | 363 |
| | Mustard | 366 |
| | Orach | 367 |
| | Parsley | 367 |
| | Parsnip | 369 |
| | Purslane | 369 |
| | Rakkyo | 370 |
| | Spinach | 371 |
| | Water leaf | 374 |
| **Chapter-9** | **Perennial Vegetables** | **375** |
| | Agathi | 375 |
| | Asparagus | 376 |
| | Bread fruit | 378 |
| | Chekkurmanis | 378 |
| | Drum stick | 379 |
| | Globe artichoke | 383 |
| | Jerusalem Artichoke | 384 |
| | Jhumpuri | 386 |
| | Pisonia | 389 |
| **Chapter-10** | **Other Vegetables** | **391** |
| | Baby corn | 391 |
| | Okra | 392 |
| | *Colour plates* | *405* |
| | *References and suggested readings* | *411* |

# 1

# Introduction

Indian Horticulture has evolved emensely over the past few years. After green revolution, it is now believed that horticulture is a better option to achieve evergreen revolution as because it is more suitable to our topography and agro-climates. This venture is attracting the faming community since it has proved to be intellectually satisfying and economically rewarding. Vegetable farming plays an important role in modern horticulture. It produces higher biomass resulting into better utilization of natural resources and provides food security to a large section of the society. Vegetable sector generates more employment opportunity with higher income per unit area. Being a rich source of vitamins, minerals and anti-oxidants, it plays a vital role in nutritional-security of our country. Vegetables cover about 3% of the total area under cultivation in India. The state-wise area, production and productivity of vegetable is given in Table-1.

Although we are second largest producer of vegetables in the world, yet our production is not sufficient to meet the ever increasing demand due to geometrical growth in population. There could be two ways to fulfill the gap between demand and supply of vegetables.

One way is to bring more area under vegetable cultivation and the other is to increase productivity. Due to rapid urbanization, industrialization, fragmented land holdings and depleting natural resources, the first option seems to be difficult whereas the latter to produce more from less is the only viable alternative in this resource starve situation. The future of vegetable industry therefore, lies in the enhanced productivity.

**Table 1 :** State wise area, production and productivity of vegetables in the country

| State/Uts | Area (000ha) | Production (000MT) | Productivity (T/ha) |
|---|---|---|---|
| Andaman & Nicobar | 4.3 | 32.6 | 7.56 |
| Andhra Pradesh | 264.9 | 4355.9 | 16.44 |
| Arunachal Pradesh | 23.7 | 110.0 | 4.64 |
| Assam | 331.4 | 4449.5 | 13.43 |
| Bihar | 824.3 | 13612.8 | 16.51 |
| Chandigarh | 0.1 | 1.7 | 17.00 |
| Chattisgarh | 277.0 | 2801.9 | 10.12 |
| D & N Haveli | 1.5 | 13.5 | 9.00 |
| Daman & Diu | 0.2 | 0.2 | 1.00 |
| Delhi | 39.7 | 672.0 | 16.93 |
| Goa | 8.2 | 84.3 | 10.28 |
| Gujarat | 366.1 | 6062.6 | 16.56 |
| Haryana | 280.9 | 3366.8 | 11.99 |
| Himachal Pradesh | 63.7 | 1150.7 | 18.06 |
| Jammu & Kashmir | 56.8 | 1247.5 | 21.96 |
| Jharkhand | 223.6 | 3394.9 | 15.18 |
| Karnataka | 393.6 | 4834.5 | 12.28 |
| Kerala | 163.5 | 3504.2 | 21.43 |
| Lashadweep | 0.4 | 14.0 | 35.00 |
| Madhya Pradesh | 201.4 | 2814.0 | 13.97 |
| Maharashtra | 433.9 | 6148.1 | 14.17 |
| Manipur | 10.3 | 91.8 | 8.91 |
| Meghalaya | 42.0 | 345.4 | 8.22 |
| Mizoram | 1.7 | 31.3 | 18.41 |
| Nagaland | 9.4 | 44.6 | 4.74 |
| Orissa | 660.1 | 8200.4 | 12.42 |
| Pondicherry | 2.7 | 54.7 | 20.25 |
| Punjab | 156.7 | 2501.9 | 15.97 |
| Rajasthan | 124.1 | 778.5 | 6.27 |
| Sikkim | 17.8 | 80.8 | 4.54 |
| Tamil Nadu | 252.9 | 7070.6 | 27.95 |

Contd...

| | | | |
|---|---|---|---|
| Tripura | 31.8 | 415.9 | 13.08 |
| Uttar Pradesh | 922.2 | 17555.3 | 19.04 |
| Uttaranchal | 77.1 | 995.3 | 12.91 |
| West Bengal | 1224.2 | 19194.6 | 15.68 |
| **Total** | **7492.2** | **116032.7** | **15.49** |

*Source :* NHB Data Book, 2006-2007

Quality seeds of desired variety suitable to a particular agro-climate are the key to uplift the productivity larvae. Therefore, extensive knowledge about the behavior of varieties and hybrids is a pre-requisite in healthy farm management system. Although different organizations are developing a number of varieties and hybrids, all may not prove worthy to a particular agro-ecological set-up. Hence, the choice of suitable cultivar determines the yield optimization.

As we know, India has rich diversity in genetic resources of a number of vegetable crops including brinjal, pointed gourd, cucumber, ridge gourd, sponge gourd, snake gourd, *momordica* spp., *Amaranthus* spp., *Chenopodium* spp., *Beta vulgaris* var. *bengalensis*, ash gourd etc. A number of vegetable crops known for secondary centre of diversity include okra, chillies and several species of cucurbitaceous. Due to seed propagation and selection by farmers, rich diversity has also been generated in tomato, cowpea, *Brassica* spp., amaranthus, french bean, bitter gourd, bottle gourd, coccinia and ash gourd. The lesser known indigenous vegetable need special emphasis for improvement. Hence, the study of the characteristics of the varieties is of great importance to the students, researchers and scientists to strengthen the gene pool of vegetables.

Tremendous efforts have been made in both public and private sectors for developing hybrids in number of vegetable crops. The share of hybrid varieties in several vegetable crops has gradually been increasing. Hence to update the informations about the varieties and hybrids available in the vegetable seed industry (market) is the need of the hour. The details descriptions about the varieties and hybrids in the succeeding chapters will help a lot in understanding the varietal wealth of vegetable crops.

□□□

# 2

# Cole Crops

## BROCCOLI

Broccoli/Sprouting Broccoli (*Brassica oleracea* var. *italica* L.); 2n = 18, is a member of family Brassicaceae is new in acerage and value. Broccoli is an Italian word from Latin Brachium meaning an arm or branch. In broccoli a main head bearing a cluster of flower buds is produced terminally on a fleshy branching elongated stem.

### Varieties and Hybrids

There are several types of crop varieties: Green broccoli, Purple broccoli, White broccoli, Pale broccoli and Dark green broccoli. The main difference between the varieties are colour of the bud. Purple and green types are more common of which green types are the most commonly cultivated. Purple types are hardier and withstand lower temperatures. On the basis of maturity early, mid and late types are available.

*Green types* are more commonly cultivated; these are early, medium and late type as per their maturity. The cultivars having solid, firm head with small green buds and few buds and few lateral shoots are preferred.

***Early cultivars*** are ready for harvest after 40-45 days of transplanting. De Cicco, Green bud and Spartan Early are common early cultivars. Early cultivars do not perform well under low temperatures. Waltham 29, Green Mountain and Coastal Atlantic are suitable for early and late planting.

***Some mid and late planting cultivars*** are Green Sprouting Medium and Green Sprouting Late, biennial in nature, matured in 100 days in winter.

***Purple types*** are more hardy and can withstand lower temperatures. Most of them are grown in European countries and harvested during winter. Colour development is adversely affected when grown under glass house or polythene house, hence, these types are suitable for protected cultivation.

Recently some varieties have been developed from IARI i.e. Pusa Broccoli 'KTS1', 'KTS 2', 'KTS 3', 'Palam Samridhi', and Palam Vichitra.

It has been reported that several forms of broccoli within the Italic group, which details are given below:

***Purple sprouting broccoli***: Over wintering N.W. European Biennials, branched, purple spears, variable, early and late forms overlapping.

***Purple Cape broccoli***: Overwintering N.W. European biennials, single purple heads variable.

***White sprouting broccoli*** : Over wintering N.W. European biennials, branched, white spears. Distinct early and late maturing forms of which the late type more strictly within the botrytis group .

***Purple Sicilian broccoli*** (*Cavolfiore violetto di Sicilia*) : Single heading, pale purple heads, also known in horticulture as "purple cauliflower".

***Couve broccoli Roxo de Cabeca*** : Tall purple heading/sprouting type from Portugal.

***Calabrese*** **(*Cavolo broccolo ramosa verde di Calabria*)** : Green sprouting broccoli from calabrian region. Now a worldwide crop and virtually single heading and hybrid cultivars.

***Black broccoli* (*broccolo neri*):** Highly branched, annual with dark-green spears and sickle-shaped leaves; from Rome area.

***Hybrids* :** Large number of $F_1$ hybrids cultivars are being marketed by different seed companies of Japan, USA and Europe. These are:

***Extra early* :** Southern Comet, Premium Crop, Clipper, Laser.

***Mid season* :** Corsair, Excalibur, Cruiser, Emerald, Corona.

***Late* :** Late Corona, stiff, Kayak, Green Surf belong to calabrese group. These are superior in yield, uniform in size and maturity have gained popularity and commonly grown in developed countries.

Some Japanese and European seed companies are also marketing hybrid cultivars in India; like-Packman, Pirate, Brigadier, Cap oven, Zues Hi Cream, Bacclus, Mars, Green Lofi, Tandon, Land Star, Green Mail, Green Dome and Shigmeri.

Some hybrids have been developed and marketed by Indian Seeds Companies. These are :

**Century Seeds**; CBH-1, CBH-2. **Bejo Sheetal Seeds**; Lucky, Fiesta. **Golden Seeds;** Rapid, Green Baby. **Namdhari Seeds**; NS 50. **Nath Sluis**; Nath special off-1. **Rallis India**; Prativa. **Sandoz seeds**; Everest. **Clause International**; Citation. **Indo American export**; Packmen and Piret.

**Detail characteristics of varieties and hybrids are given below :**

***Pusa Broccoli KTS 1* :** Its plants are medium tall (65-70cm long), foliage waxy and dark green with slightly wavy margins. Head solid green with small buds and slightly raised at the center. The main head size and weight are about 6cm and 15.4 cm (height x diameter) and 350-450g. It is ready 125-140 days after sowing and 90-105 days after transplanting in hills and reduced by 5-10 days in plains. It is medium maturing variety. It is resistant to mild frost and snowfall.

***Palam Samridhi* :** It has medium plant type with medium sized, dark green leaves. It is a green sprouting type with large, compact terminal, heads borne on long stalk. Terminal heads weigh 300-400 g. The subsequent tender sprouts arising from leaf axils. Its terminal heads become ready for harvesting in 85-90 days after transplanting. Heads remain free from yellow eyes and bracts. Its average yield is 150-200 q/ha.

***Palam Kanchan* :** It has long, broad, bluish-green upright entire margin leaves with prominent white midrib and veins. The head is large, compact, attractive and yellowish-green in colour. The heads attain marketable maturity 140-145 days after transplanting. The average yield potential is about 250-275 q/ha.

*Palam Vichitra :* It has medium, open, dark green leaves with redish tinge. Purple streaks are prominent on stems at seedling as well as adult plant stage. The head is medium, compact, and purple coloured and attractive. It attains marketable maturity 115-125 days after transplanting with an average yield potential is 225-250 q/ha.

*Punjab Broccoli 1 :* It is developed through selection by Punjab Agricultural University, Ludhiana . Foliage dark green with smooth leaf surface and wavy margin with bluish-tinge. Plant branched, bears large number of sprouts in the axil of leaves. It takes about 65-70 days from transplanting to harvest. Head and spears are dark-green with bluish-tinge. Average yield is 700 q/ha.

*Lucky :* It matures in approximately 60-65 days after transplanting. Head weight is 600-800 g and gives compact curd with fine buds. It can planted in winter season in the plains.

*Fiesta :* It matures in 65-70 days after transplanting. Head weight is 800-1000g. It is very compact and has fine buds and very high yielder, can be planted in winter season in the plains.

## BRUSSELS SPROUTS

Brussels sprouts (*Brassica oleracea* var. Gemmifera DC.); 2n = 18; belongs to the family Brassicaceae, got its name from Brussels in Belgium where it is grown since 13$^{th}$ century. Edible parts of Brussels sprouts is the miniature (5-8 cm diameter) cabbage like portions borne in axils of leaves, all along and around main stem.

### Varieties and Hybrids

There are two distinct type of cultivars, dwarf and tall.

*Dwarf cultivars :* They have short stem less than 50 cm in length and sprouts are small and crowded on the stem. These are mostly early cropper and most suitable for the areas where growing season is short. Common cultivars are Improved Long Island, Early Morn, Dwarf Improved, Frontier Zwerg and Kvik.

***Tall cultivars :*** Tall cultivars are grown in areas where growing season is longer especially in England, other parts of Europe and north Indian hills. Strains of cultivar Eveshan Rapid (Sweden) and Wilhelmsburg (Germany) are early in maturity but do not withstand hard winters.

***Late and hardier cultivars :*** Red Vein, Hilds Ideal, De Rosny Polarstjernen and Amager.

***Other varieties :*** High yielding dwarf strains are Bollo, Cambridge, Cambridge No.1, Cambridge No.2, Cambridge No.3, Cambridge No.4, Cambridge No.5, Dalington Early, Favourite, Goliath, Fill basket, Ovmskirth Giant, Rouslench late, Wroxton.

***Aphid resistant varieties :*** Uremo Inter, Irish Elegans. Roodnef.

***Powdery mildew resistant varieties :*** Bin Dee, Engard 2077, Frigo Star, Groeningse late, Seven Hills, Solid line, Triumph.

***Hybrids :*** The hybrids marketed by different seed companies of Japanese and Europe are Jade Cross, Avon Cross (first $F_1$ hybrids by National Vegetable Research Station at Wellesbourne), Doreman, Alkazar, Merlon, Poster, Rampart, Fortress, Predora, Rovoka, Sonara, Ladosa, Pearl Crystal, Oliver, Valiant, Rogor, Rider, Boxer, Richard, Royal Marvel, Asgard, Kundry, Rasmunda and Stephen.

***Hilds Ideal :*** It is an open pollinated cultivar recommended for growing in the hills and plains of north India from IARI, Regional Station, Katrain, Kullu, Himachal Pradesh. Its plant grows up to 60 cm and produces 45-55 sprouts, each weighing 7-8 g and the sprouts are compact and possess good flavour. A negligible percentage of loose sprouts at the base of the plants is also marketable. It takes about 115 days for first picking. The yield per plant is 250-400 g in four pickings at an interval of 10 days. Delay in picking will lower the sprout quality.

***Jade Cross (F) :*** It is 90 days hybrid cultivar. Sprouts are firm, dark green, closely packed on long stems, can be grown under wide range of climatic conditions.

***Diablo :*** It mature in 120-150 days after transplanting , during winter season. In order to get uniform sprouts, cut the top of the plant as soon as sprouts at the bottom start to develop. It can be boiled eaten with cheese spread. Plant height is 90 cm, winter variety, gives round solid green sprouts, which are well spaced on the stem. It was developed by Bejo Sheetal Seeds.

## CABBAGE

Cabbage (*Brassica oleracea* L. convar. *capitata* (L.). Alef. var. *capitata*); 2n = 18; syn. Band Gobhi, karam Kalla, Kobee, Patgobhi, Muttai Kose, Patta Kobi, Bandha Kobi, Bandha Kapi, Gobee, Yelekosu, Cabbage, Aku Gobhi; is a typical cool season cole crop grown for its tender heads, originated from a wild non-heading type, colewort (*Brassica sylvestris*). The genus *Brassica* includes about 100 species. Majority of which are native to the Mediterranean region.

Cabbage is one of the most popular temperate vegetable and occupies fourth position among vegetables in India as well as in the world. Area under cabbage cultivation in India is 2.58.000ha. with production of 56.78 lakh tonnes (2001-02).

Cabbage head is reach in Vitamin 'A', B2 and minerals. it possesses many medicinal properties. In ancestral time, it was used against gout, diarrzoea, stomach and coeliac troubles. Due to presence of Indol-3 Carbinal, it protects against bowel cancer. Cabbage juice is used as a remedy against-poisonous mushrooms.

Cultivated cabbage is a biennial crop for seed production and treated as an annual for vegetable production. The edible part, head is formed by thickening of edible bud with tightly packed overlapping leaves. The earlier formed leaves will be normal and after initial vegetative growth and on receipt of low temperature,further leaves will not open. Cultivated types exhibit variation in size, shape and colour of leaves and texture of head.

Bailey (1930) classified the cultivated *oleracea* into 7 divisions.

I. The leaf - races grown for the open foliage
   1. Branching shrubby kales - *B. oleracea* var. *fruticasa* ·
   2. Common biennial mostly unbranching kales - *B. oleracea* var. *acephala*
   3. Low cabbage like or Protuguese kales - *B. oleracea* var. *tronchuda*

II. The head races grown for the compact leaf - head or buds.
   4. Edible product a terminal head, the cabbages - *B. oleracea* var. *Capitata*
   5. Edible product lateral small heads or buds, the *Brussels sprouts* *B. oleracea* var. *gemmifera*

III. The inflorescence races grown for the modified thickened flower - shoots or malformed flowers.

6. Flowers only partially if at all modified - *B. oleracea* var. *italica*
7. Flowers thickened and malformed into a head or several heads - *B. oleracea* var. *botrytis*

Nieuwhof (1969) reported three cultivated forms of cabbage;

1. *Brassica oleracea* var. *Capitata* L. *f. alba* - White cabbage
2. *Brassica oleracea* var. *capitata* L. *f. rubra* - Red cabbage
3. *Brassica oleracea* L. *f. sabauda* - Savoy cabbage

White cabbage is widely grown in India while red and savoy cabbages are not of much importance in India.

Cabbage is highly cross-pollinated and degree of cross-pollination is as high as 73%. Cross pollination is mainly due to sporophytic self-incompatibility, determined by upto 10 alleles at the 'S' locus. Protogyny is very conspicuous in cabbage and stigma remains receptive for five days before anthesis. Cytoplasmic and genetic male sterility are also noticed in certain lines. Anthesis occurs 8.00-10 hr.; pollen fertility is maximum on the day of anthesis. Stigma is receptive 2-3 days before to the day of anthesis. Pod is a siliqua and will be ready for harvest 50-90 days after flowering depending on temperature.

## Varieties and Hybrids

Most of earlier cabbage varieties were European introduction. Varieties differ in colour, size and shape of head, duration of crop, ability to with stand low or high temperature etc. White Cabbage is widely grown in India and in this group three shapes, viz., round, conical and flat exist, of which round head is more common.

The presently grown cultivars in the country are classified into three groups according to planting time and maturity.

1. ***Early varieties*** : The varieties of this group are usually ready within 60-70 days. The varieties are - Golden Acre, Pride of India, Copenhagen Market, Early Drumhead, Sel 8, Express,
2. ***Mid season varieties*** : The mid season varieties become ready for harvesting in 80-90 days. The varieties are - All Head Early, Wisconsin All Green, September.

3. ***Late varieties*** : These varieties require long winter season for a good crop. The heads are ready for harvesting in 90 to 120 days. The improved cultivars are - Pusa Drumhead, Indian Eclips, Danish Ballhead, Late Flat Dutch, Surehead *etc.*

## Hybrids

*I.A.H.S* : Ganga, Jamuna, Kaveri, IAHS-5, IAHS-7, IAHS-9, Krishna, Green Express,Tropeakan, Snowpak, Constenja.

*Mahyco Seeds* : Wonder Boy (MHCB-261), Navkranti (MHCB-600), MHCB-640, Hari Rani, Kalyani, Kamdhenu, Haripriya, MCV-1, Hybrid No.23, Shree Ganesh Gol.

*Century* : Meenakshi, Konark, Green Gold, Pashan, Geetanjali, Kapilash, Kanyakumari, Vardan, Keertiman, CH-21, CH-55, CH-65, CH-33, Raj-2, Rajrani.

*Rallis India* : Bajrang, Sujata, Suvarna.

*Sreemaa Seeds* : Gold Star, Priya, Green Samrat, Green Chalanger, Rajar Bal, Green Diamond (Imp.).

*Heritage Seeds* : Green state, Hari kalyan, Surya, Lord Murugan

*Royal Slouis* : Deepa, Sona,Abha, Namratha, Royal Express, Diamond Express, Dish, Indu,Glory, Kunwari, Geetanjali,Vignesh, Meghana, Maya, Moti.

*Syngenta* : Coimbatore, Equatoria, Green Globe, Quisto, Viswas, Pluton, Greendiar (OP)- Mosketiar, Combinar, BC-26, BC-34, BC-35.

*Takai Seed Co.* : Green cornet, Emerland Cross, T-621.

*Nath Seeds* : Nath 401, Nath Lakshmi, Nath Indu, NL-138, Nath Puja, Sumati, Gloria, Namrata

*Bejo Sheetal Seeds* : Bajarang, Suvarna, Sudha, Swapna, Sonalee, Sadhana, Sumithra, Surekha, Siddhi, Suchithra, BSS-44. Sujata, Swati, Sarita, Shanti, Sonali, Benson, General, Fresco, Governor, Rotonda, Fieldrocket, Gonzales, Morris, Fieldman, Invento, Gideon, Ramada , Bejo 1857 and Bronco.

*N.S.C.* : Green Boy, Green Express.

*Sandoz Seeds* : Quisto, Vishwas.

*Pro-Agro Seeds* : Bahar, Pragati.

*Sungro Seeds :* Hero, Super Hero, S-101, S-103, S-107, Hybrid 123, Hybrid 124, Hybrid 125, Hybrid 126, Hybrid 133, Improved Savitri (OP).

*Unicorn Seeds (Choice) :* Varun, Hema II, Haragola Vaibhav, Evergreen, Super Cross, Harini, Ratna, Red Queen.

*Novartis Seeds :* Combinor, Equatoria, Green Globe, Quisto, KS-Cross, Summer Queen, Vishwas.

*Tropica Seeds (Green CO.) :* BC 38, BC 35, CJN 12, Asia Cross, Orient, Tara, Tropical BC-34, Red Ball.

*Namdhari Seeds :* NS 21, NS 22, NS 25, NS 26, NS 27, Jeevan.

*Nunhems Seeds :* Bahar, Mansi, Bharat (45), Pragati, Unnati, Ria (105), Greenball.

*Golden Seeds :* Quick Ball-45, Golden Ball, Super Ball (50), Gray Ball-60, Goody Ball, Big Ball-75, Sumithra, Bobby, Bhima, Shristi, Red Baby.

*Tokita Seeds :* Blue Joys, Shatabdi, Hare-Krishna, Green Kid, Tokita-9651, Delhi Ball-65.

*Ankur Seeds :* Mahima, Simran.

*Annapurna Beej :* All Season, Shatabdi Express, Green Star, Shaktimaan, Maharani-70, Chakadola, Kohinoor.

*Biostadt Seeds :* Chase 301, Chase 302.

**Details Characteristics of Varieties and hybrids are given below:**

***Golden Acre :*** A selection from the ' copenhagen market', the heads are small, round, weighing 1.0 to 1.5 kg. It matures in 60-70 days after transplanting. The outer leaves are few and cup-shaped. The harvesting should not be delayed beyond 10 days of head-formation, otherwise the heads have a tendency to burst open.

***Pride of India :*** The plant type of this variety is similar to golden Acre but it is about a week later in maturity. It is available about a week later than the 'Golden Acre'. Head size 1.5 to 2.0 kg. slightly bigger frame than 'Golden Acre'. An introduction recommended by YSPUHF, Solan.

***Sel 8 :*** A selection from an intervarietal cross EC 10109 X EC 24855 made at Katrain. The variety is characterized by a short-stalk, medium-sized leaves, light green with wavy margins and slightly puckered leaf-blade. Heads slightly flattish round weighing 1.5 to 2.0 kg. also available a week later than the 'Golden Acre'. It is resistant to black-rot.

***Copenhagen Market :*** In most of the areas it has been replaced by 'Golden Acre'. But it is still grown on a limited scale in some pockets. Head size is 2.0 to 2.5 kg. heavy outer frame. Shape of leaves and outer whorls are same as in 'Golden Acre'.

***Express :*** 'Golden Acre' is marketed by some of the seed companies under this name. It resembles 'Golden Acre' in all respects except it is said to be slightly earlier in maturity.

***Sri Ganesh Gol :*** Being marketed by Maharashtra Hybrid Seed Company (Mahyco). Its plant type is like that of 'Golden Acre' but it matures a week early. Its head size and shape also resembles that of 'Golden Acre'. It has got a much better staying ability.

***Hari Rani :*** It is also marketed by Mahyco and is very popular in the Southern States, especially the Nilgiri hills. It has a medium-sized frame with light green foliage. The heads are flattish round weighing 2.5 to 3.0 Kg.

***Meenakshi :*** Marketed by century seeds. The medium-sized, round heads possess a good staying power.

***Green Boy :*** Marketed by National Seeds Corporation. An introduction from Japan. Early in habit, produces medium sized round heads.

***Konark :*** Extremely good material for North East region, West Bengal and Orissa. A mid-season hybrid with flattish-round solid head weighing 2.5-3.0 kg with large frame and strong outer leaves. Have good holding ability. More adapted for sowing during rainy season in these regions. Head is crisp, tasty and excellent for fresh market.

***Raj-2 :*** A dark-bluish green ball headed hybrid suitable for fresh market and long trans-shipment. Maturity 80-85 days, medium large head grows upto 4.0 kg depending on field conditions. Leaves are thick and core short. Tolerant to black rot. This hybrid has been bred for a wide adaptability.

***CH-21 :*** Extra early dark green ball headed hybrid with solid texture. Heads weigh 750 g to 1.0 kg with very short core.Inner leaves are very dense and outer excellent wrappers. Excellent field holding ability enables harvest at will to catch best market. Tolerant to cabbage yellow and Tip burn.

*Kalimpong Eclipse Drumhead :* This variety is developed at the Horticultural Research Station, Krishna Nagar and at the Vegetable Research Station, Kalimpong,Darjeeling. This variety is suitable for West Bengal. It is comparatively late variety, heads become ready for harvesting in 100 to 110 days.The heads are large, flat and compact.

*Kalimpong Eclipse Ball :* This variety is also developed at the HRS, Krishnanagar and the VRS, Kalimpong. This is an early variety. Heads are ready for harvesting 60 to 80 days. Heads are oval in shape, big sized, and compact.

*Red cabbage :* All the red cabbage cultivars are tolerant to diamond back moth insect which is serious pest of cabbage. It has distinct coat of wax and produces a head of 1-2 kg, taking about 90 days from transplanting to head formations.

*MCV-1 :* Heads are ready after 80 days from transplanting. It has round shape, bigger head (1.5-2.5 kg) and compact. The foliage is dark green with waxy margins.

*Hari Rani Gole :* It is a good hybrid for medium late maturity. It produces medium sized, ball shaped dark blakish green, solid heads with good wrapper leaves. Each head weigh 1.5-2.0 kg. It can be harvested after 95-100 days of transplanting.

*Quisto :* It is a high yielder hybrid and has ability to stand over severe hot conditions. Heads can stand in the field conditions upto 70 days after maturity. Heads are compact round and very solid with dark blackish colour. It is good for tropical climate.

*Kranti :* It is ready for harvesting within 60 days after transplanting. It is good for close spacing planting. Heads are 0.8-1.2 kg weight, round and compact.

*Mitra :* An early hybrid with excellent head to plant ratio. Heads are fresh green uniform, compact, ball shaped with weight around 0.8 to 1.2 kg each. Good field retention ability.

*Early Drum Head :* An early variety, foliage light green, heads flat, leaves medium to large curving inward enclosing the head loosely. Leaves forming the head fold over each other at the centre.

*Jersey Wakefield* : A pointed head cultivar is preferred by some growers but it has very limited market. It takes about 55 to 60 days for head formation. The outer leaves are few and slightly crinkled. Average weight is 1.0-1.5 kg. It has got better staying power. This is also preferred for raw eating due to better taste.

*Savoy group* : This group is also not popular in India. All the cultivars have blistered leaves. The group has all the shapes like pointed, round and flat. The commonly catalogued cultivar is Chieftain. It has fairly light green foliage and large, rather smooth, flat ground head.

*Pusa Sambandha* : An early maturity variety with wide adaptability. Heads round and compact. Suitable for close planting.

*Green Express* : Marketed by the National Seeds Corporation (NSC). Even before the NSC started marketing the variety, it had become popular in West Bengal, where its seeds had been smuggled from Bangla Desh. Medium sized plants with few outer leaves. The head size depends upon the time of planting. Those planting earlier can attain head size up to 3 kg while in the subsequent plantings the head size decreases.

*Pusa Drum Head* : Selection made at Katrain, and an introduction from Japan (EC 6774), the variety is the earliest in this group. The heads are flat, slightly loose weighing 3-5 kg. The outer leaves are light green with prominent midrib. It has field-resistance to black-leg disease (phoma lingam). Heads are ready in 80-90 days after transplanting.

*September* : An introduction from the German Democratic Republic and a most popular open pollinated variety in the 'Nilgiri hills. It is characterized by a large frame, bluish-green leaves with serrated margins. It has a long stalk and the plants fall on one side after the head-formation has completed. The heads are flattish oblong, very tight and weigh 4 to 6 kg. Highly susceptible to black-rot. It takes 96 to 100 days from planting to head formation.

*Late Large Drum Head* : Marketed by different seed companies. It matures in 110 to 115 days. The heads are large flat, somewhat loose.

*Drumhead Savoy* : It has a dark green foliage which is much blistered or wrinkled. The quality is superior but it is not commercially acceptable in our country.

***Pusa Ageti :*** It has been developed from TKCBH-28 ($F_1$hybrid) procured from Taiwan. Selection were made $F_2$ generation onwards by using simple recurrent selection. It is the first heat tolerant variety which can produce seed in subtropical climate. It takes about 70-90 days from transplanting to harvesting. Plant has medium sized frame, small stalk, upright growing habit and wax coated grey-green foliage broad lamina with prominent mid rib and wavy margin. It has flattish-round, compact, non-cracking heads with large wrapper leaves. Head weight varies from 600-1200 g yielding 110-380 q/ha depending upon the time of transplanting. Optimum temp for its growth and head development is 18-20$^0$C but it can grow and form head when the mean temp. is between 15-30$^0$C and day temp. should not be more than 35$^0$C, the crop can be planted in march, August and October.

***Pusa Mukta :*** It is a cross between EC 24855 and EC 10109, both exotic parental lines. It has a short stalk, medium sized puckered light green leaves with wavy margin. Heads are flattish-round, weighing 1.5-2.0 kg. It is ready for harvest 65-70 days after transplanting, a week later than 'Golden Acre'. The flat and compact headed 'Pusa Mukta' yields 58.41 tonnes/ha. It is a field resistant to black-rot and is almost free from insect-pest attack.

***Pusa Synthetic :*** Plants with short stalk, cup shaped leaves, early maturity in 60-70 days after transplanting. Heads are very compact and good quality. Average weight about 1.0 kg and it is a high yielding.

***Bajrang (BSS-50) :*** It has been developed by Beejo Sheetal Seeds Pvt. Ltd, Jalna. Heads are dark green, smooth, highly compact, average head weight 1 kg ; outer leaves are upright; mature in 65 days but can stand on the field for 45 days after maturity; tolerate high temp upto 36$^0$C; resistance to fusarium wilt. Female parent (BH-8,20)- Early in maturity and has compact round head. Male parent (NORT-14)- Early maturity and has highly compact, round head.

***Suvarna (BSS-32) :*** It has been developed by Beejo Sheetal Seeds Pvt. Ltd. Jalna. Heads are round, average head weight 3 Kg; strong smooth outer leaves; matures in 110 days. Female Parent (ET-2-2) - Medium late in maturity blue leaves, vigorous growth; round and medium heads. Male parents (BOL-13) : Late in maturity blue leaves; flattish round and big heads.

***Sudha (BSS-115) :*** It has been developed by Bejo Sheetal Seeds Pvt. Ltd. Jalna. Heads are highly, compact, good interior structure, matures in 75 days and can be grown round the year; average weight of head is 2-3 kg; have resistance to *Fusarium* wilt. Female parent (FRICO-11): Medium late maturity; leaves are bluish and plant structure is good; has flattish round and medium heads. Male parents (LAWIC): Early in maturity; bluish leaves, compact, round and medium heads.

***BSS-44*** **:** It has been developed by Bejo Sheetal Seeds Pvt. Ltd. Jalna. Heads are roundish, average weight 2.5 Kg; matures in 75 days but can stand for 20 days in field after maturity; grows well both in cold and hot weather; resistance to *Fusarium* wilt. Female parent (BOL-20): Medium early in maturity, bluish leaves; compact, round and big heads. Male parent (BI X 12)- Early in maturity, bluish leaves; round and compact heads.

***Varun*** **:** A medium maturing hybrid, medium compact globe shaped head is medium green in colour and weigh 1.0-1.5 kg.; excellent field holding ability is also adaptable for high density planting and long distance shipping.

***Hema II*** **:** An outstanding hybrid produces flattened globe shaped heads of weight 1.5-2.5 kg, good tolerance to bursting gives an extended harveting time.

***Hara gola*** **:** This early maturing hybrid produces globe shaped head with extra fine quality crispy sweet flesh. Fowits weigh around 1.0-1.25 kg.

***Vaibhav*** **:** Early maturing hybrid with excellent field holding ability makes it best choice. Globe shaped compact heads are green and weigh 1.0-1.5 kg., very easy to grow.

***Evergreen*** **:** Very promising and early maturing hybrid with relatively very good field standing ability. Head is round, firm, compact and weigh about 1.0-1.2 kg.

***Super cross*** **:** Very promising, early maturing hybrid with good field holding ability. Head is semi-flat, very attractive and weighs about 1.3-1.7 kg., easy to grow in hot and wet conditions.

***Harini*** **:** Early maturing hybrid with excellent adaptability for upland / high altitude areas. Head is semi flat, compact and weighs around 1.2 kg. This variety has good field holding ability and is suitable for long distance shipping.

***Ratna :*** An outstanding hybrid with medium late maturity. Big size heads are dark green in colour and very compact. Heads weigh approximately 1.5-2.5 kg.

***Red Green :*** This newly introduced hybrid cabbage produces red colour round shaped head which are firm and compact. Slightly upstanding in habit. Matures in 70 days from transplanting. Highly tolerant to wet conditions in the field. It is suitable for fresh market and processing.

***NS 21 (F1) :*** Very vigorous hybrid with an excellent plant stand. Foliage is dark bluish green in colour with strong uniform bloom cover giving tolerance to insect pests. This hybrid has very uniform heads which are flat in shape with a weight of 1.5–2.0 kgs. Produces very compact dense heads of very good firmness and short core. It matures in 75-80days.

***NS 22 (F1) :*** Tropical type hybrid, vigorous plants, uniform growth and maturity. Heads are round with very good firmness and density. The heads weigh 1.5-2.0 kgs each with good field holding capacity and is suitable for spring and summer sowings, matures in 70 days.

***NS 25 (F1) :*** This early hybrid has upright habit, uniform compact heads and good wrapper leaves. The heads are round and weigh 1.0-1.25 kg each. It matures in 55 days. This hybrid is suitable for close planting because of small compact heads. It has very good field holding capacity.

***NS 26 (F1) :*** Very good hybrid with vigorous plants and a green foliage with good wrapper leaves. The round heads are very compact with good density and firmness. They weigh 1.5-2.0 kgs each with good field holding capacity. This hybrid is suitable for spring and summer sowings with a maturity period of 65-70 days.

***Quick Ball- 45 :*** Extremely early maturing hybrid, ready for harvest after 45 days of establishment of transplanted seedling. The plants are medium sized and heads are round, compact, dark green and weigh 0.8-1.0 kg., suitable for planting from the second week of July and onwards and up to December in the northern parts of the country and throughout the year in southern parts. Suitable for high density planting. Tolerant to black rot and tip and burn diseases.

***Golden Ball :*** Early maturing hybrid, ready for harvest after 50-55 days of transplantation. The heads are uniformly round, compact, green weigh 1.0-1.2 kg with dark green wrapper leaves. Field holding capacity is very good. The hybrid has very good field tolerance to major fungal diseases and black rot.

***Super Ball -50 :*** Strong and vigorous early maturing hybrid with excellent head and plant ratio. Heads are compact, round with gray green outer waxy leaves and weigh 1.0-1.5kg . Outer waxy leaves provide good protection against the pest diamond back moth. Excellent field holding ability and wide adaptability makes this hybrid quite popular. Tolerant to black rot, black leg, tip burn and diamond back moth.

***Gray Ball- 60 :*** Early maturing vigorous hybrid ready for harvest in about 55-58 days after transplantation. Heads are very compact, round, gray green and weigh 1.0-1.2kg. Good field holding capacity and tolerant to black rot and tip burn diseases.

***Goody Ball :*** An excellent hybrid with bluish green leaves and compact round heads, weighing 0.8-1.2kg. Field holding capacity is excellent and heads do not burst for a long time after maturity. Suitable for planting from July to January in northern India and throughout the year in southern parts.

***Big Ball–75 :*** A relatively late maturing hybrid, ready for harvest in 70-75 days after transplantation. The heads are very compact, dark green, predominently globe-shaped and weigh 1.5 -2.0 kg. The hybrid has excellent field holding capacity and very good field tolerance to black rot and fungal diseases.

***Sumithra :*** Early maturing hybrid and heads can be harvested in about 58–60 days after transplantation. Campact heads are round, dark green and weigh 0.9–1.2 kg., tolerant to black rot disease.

***Bobby :*** A medium early maturing vigorous hybrid. Heads can be harvested in about 70–75 days after transplantation. Heads are round, compact with grey green leaves and weigh 2.0–3.0 kg. Its versatile nature makes it ideal for long distance transportation. Tolerant to black rot, tip burn and diamond back moth.

***Bhima :*** A versatile late maturing hybrid ready for harvest in about 75–80 days after transplantation. Heads are very compact and solid, short cored, flattened globe in shape, weigh 2.5–4.0 kg. and covered with dark green wrapper leaves. Tolerant to black rot disease. Excellent field holding capacity allows flexible harvest. Ideally suited for all season cultivation in southern India, especially Karnataka.

*Shristi* : Late maturing hybrid, extremely suitable for planting in southern plains. Heads are relatively large, compact, semi-globe to flat shape with attractive outer dark green leaves and weigh 2.5 to 3.5 kg. Heads mature in 80 -85 days and have excellent field holding capacity. Can be grown throughout the year in the plains of southern parts of the country. Tolerant to black rot and black leg diseases.

*Red baby* : A medium maturity hybrid, compact heads can be harvested in 65–70 days after transplantation. Heads are round, uniformly dark red with strong waxy outer leaves and weigh 1.0–1.2 kg. Inside colour uniformly dark red and layers finely arranged with a short core. Possess good storage ability. Tolerant to black rot disease and head bursting.

*Mahima (F1)* : Categorised under round early group of cabbage. Good head firmness. Round shaped, light green coloured and compact heads weighing 0.8 to 1.0 kg. approximately. Head maturity in 60-65 days after planting with a field holdng capacity of 8-10 days.

*Simran (F1)* : Categorised under round early group of cabbage. Good standing ability. Round shaped, green coloured and compact heads weighing 1 to 2 kg. approximately. Head maturity in 60-65 days after planting with a field holding capacity of 12-15 days.

*Priya (F1)* : A tropical type hybrid, medium vigorous plant with very uniform growth and maturity, round and very compact. Heads are of 1-1.5 kg. in size. Ready to mature in 60-60 days after transplanting.

*Green Dimond (Imp.)* : An extra early variety maturing in 50-55 days after transplanting, white coloured, suitable for early spring, sowing in western and southern region of India. Summer sowing in northern region of India.

*Green-Samrat (F1)* : A mid season hybrid weighing 2.5-3.0 kg of medium large, round and compact headed. Ready for harvesting 70-75 days after transplantation. An excellent material for strong and distance transportation. Tolerant to black-rot, fusarium wilt.

*Gold-Star (F1)* : An excellent hybrid of cabbage. Heads are very solid and don't burst even after several days of maturity. Dark green coloured leaves have a protected layer. Maturity 60-65 days after transplanting. Average head weight is 1.5 kg.

***Meenakshi :*** It is a deep green and head size is up to 5.0 kg and can kept in field for more days without any damage. Generally head size varies from 2.5 to 3.0 kg. Mature in 70-80 days.

***Konark :*** It is cultivated throughout the year and give more profit. Head weight vary from 2.5 to 3.0 kg. It is more attractive and has high consumer preference. Mature in 75-80 days.

***Green Gold :*** It is green type and head mature within 55 to 65 days after planting. Head weight varies from 1.5 to 2.0 kg.

***Pashan :*** It is early type and head mature within 55 to 65 days . It is resistant to root rot and tolerate rain, so it can be cultivated in rainy seasons.

***Geetanjali :*** It mature within 80 to 90 days after planting. Head weight vary from 2.5 to 3.0 kg. It is highly suitable for cultivation in Orissa and Bengal.

***Green Diamond :*** It is round type and early variety. It mature in 60 days after planting. Head are compact and uniform in maturity and size. It is highly beneficial for farmers and greenish type head with weight from 1.0 kg to 1.5 kg.

***Kopen Hegen Market Improrted :*** It is cultivated in more acerage in Orissa . Its heads are medium in size and varies from 1.0 to 3.0 kg ; oval, green and compact head and seed are sown in nursery during August and Sept. It mature within 75 days after planting.

***Glory :*** It is late type and seeds are sown in nursery during Sept – Oct. Heads mature 90 days after sowing of seeds. Heads are green, round and big size. Weight vary from 2.5 kg to 3.5 kg. Late planted crop give small size heads.

***Priya (F1) :*** A tropical type hybrid , medium vigorous plant with very uniform growth and maturity. This hybrid mature within 60 to 70 days after planting of 25 days old seedlings. Heads are round and dark green. Average head weight is 1.5 kg. It yields 700-1050 q/ha.

***Gold Star (F1) :*** It mature within 60 days after planting of seedlings. It is dark green and compact head type. Head are very solid and do not burst after several days of maturity. Average head weight is 1.5 kg. It yields, 200 to 250 q head/Acre.

***Green Samrat (F1) :*** A mid seasons hybrid, has the head weight of 2.5 kg to 3.0 kg with bigger size. It has dark green and compact medium large round head. Heads mature within 80–90 days after planting. Tolerant to black rot and fusarium wilt. An excellent hybrid for sowing and distance transportation. It yields 700-1000 q/ha.

***Primero*** **:** It is a red cabbage. Mature in 65-70days after transplanting. Attractive round head, very compact. Its standing ability is 30-40 days after maturity. Head weight 1.0-1.2 kg and used for salad decoration. It gives attractive dark red colour heads.

***Benson*** **:** This is an early hybrid and good for intercropping in sugarcane. Crop mature in 45-50 days after transplanting. Head round with less outer leaves, medium compact. It can stand one week in the field after maturity. Average head weight is 800 g and suitable for August to December sowing.

***Fresco*** **:** It produce uniform round, compact head, mature in 70-75 days after transplanting. It can stand 25 days in the field after maturity. Head weight vary from 2.0-3.0 kg., it can be planted in all three season and gives good heads even in the summer season. It gives good taste and an attractive head.

***Field rocket*** **:** It gives flattish round, compact head, with well covered leaves, mature in 85 to 90 days, after transplanting. It can stand 25 days in the field after maturity. Head weight is 2.5-3.0 kg. Resistant to fusarium wilt. It can be sown round the year except very hot summer. It is good for transport.

***Gonzales*** **:** It produces very attractive, round, compact head weighing 0.8-1.0 kg each, which have less outer leaves. It matures in 50 days after transplanting and has standing ability 15-20 days after maturity. Suitable for August-December sowing in plains.

***Morris*** **:** It is 65-70 days maturing, round shape and compact hybrid. Head weight is 1.0 to 1.5 kg and is recommended specially for sowing in summer season.

***Fieldman*** **:** It is very durable hybrid, can give head in all seasons. Head is round with less leaves and very compact with green leaves. Head weight is 1.0 to 1.2 kg and has standing ability 30 days after maturity. Crop mature in 60-65 days after transplanting. Suitable for kharif, rabi and summer crop and good for long transport.

***Invento*** **:** It produces attractive round, compact with short core head, which mature in 60-65 days after transplanting. Average head weight is 1.0 kg and has standing abilityof 30 days after maturity. It is suitable for August to December sowing in the plains. It is good for transport and resistant to fusarium wilt.

*Gideon :* It gives very compact, bluish green, round head with pointed upward leaves. Maturity 75-80 days after transplanting and standing ability is 35 days after maturity. Average head weight is 2.0 kg. Good for round the year production and suitable for long storage. It is resistant to fusarium and xanthomonas.

*Ramada :* It is suitable for long storage and transport. It has good vigour and suitable for round the year planting. It produces flattish round, very compact head with 2.0-3.0 kg weight. It matures in 80-85 days after transplanting and can stand 25-30 days in the field after maturity. It is resistant to fusarium wilt.

*Bronco :* This is a very uniform, high yielding hybrid which can be grown round the year and specially recommended for sowing in summer and rainy season. It produces very compact, round and firm head weighing 1.5-2.5 kg each. It mature in 70-75 days after transplanting and can stand 30 days in the field without bursting. It is resistant to fusarium and highly tolerant to xanthomonas.

*Melisa :* It matures in 75-80 days after transplanting. It has wrinkled leaves with dark green outer leaves and a dense interior structure and has good taste. It is healthy with very good disease tolerance. It is suitable for August-November sowing.

*General :* This is a 60 days Cabbage with 1 ½ kg. head weight. It is medium compact with 15 days standing ability and tolerant to fusarium.

*Bejo 1857 :* It is a 85-90 days maturing hybrid. The head is flattish round and weighs 1 ½ to 2 kgs, with attractive green colour. It is very compact and can stand for 30 days in the field without bursting. It has a long storage ability i.e. it can be stored at room temperature after harvesting for 25 days. It is recommended for round the year sowing, as it has good tolerance to fusarium.

*Bahar :* Medium early maturing hybrid. Medium sized heads, round and compact with short core and bluish outer leaves. Excellent performance under low land summer fall cropping. Tolerant to black rot, soft rot, and least damage from diamond back moth.

***Governor :*** This is a 75 days maturing, round shape hybrids. Average head weight is 1 ½ kg. and it can stand in the field for 30 days after maturity. Having disease tolerance and can be sown round the year.

***Rotonda :*** It is round, very compact hybrid, which matures in 75 days after transplanting. Average head weight 1 ½ kg. with good field standing ability. It has good tolerance to fusarium and can be grown round the year.

***Pragati :*** An early maturing hybrid with compact round heads and green outer leaves. Late bursting and excellent performance under spring and fall season and partial resistant to fusarium yellow.

***Green Ball :*** An early maturing hybrid, compact ,round heads with green leaves, wide adaptability. Tolerant to black rot and bursting. Average head weight is 600-700gms.

***601 :*** A medium- early maturing hybrid is suitable for fresh market and long transportation, Ball shaped heads are 1-1.5 kg in weight, very firm, short cored with attractive bluish green colour and excellent field holding ability.

***Unnati :*** A late maturing hybrid (75-85 days ), deep green leaves with large plant frame, flattish heads are compact, good for storage and long distance transportation. Excellent field holding ability. Average head weight 3.0-4.0 kg.

***Mansi :*** An early maturing hybrid has compact, round heads with green leaves. Wide adaptability. Intermediate level of partial resistance to black rot, bursting and high temperature. Aaverage head weight is 900-1250 g.

***Bharat (45) :*** This promising hybrid is round in shape and early maturing (55-60 days). Medium sized heads are bluish green in colour, partial tolerant to diamond back moth, high temperature and weighing about 1.25-1.50 kg.

***Ria (105) :*** Medium maturing hybrid (65-70 days). Round heads are compact, dark green in colour with good taste. Good field holding ability. Average head weights 1.5 to 2.0 kg.

***Summer Queen (F1) :*** Plants are strong, compact with 10-12 auxiliary leaves. Heads are attractive green, semi-flat, weighing about 2.0 to 2.5 kg. Heads will be ready in 55-60 days after transplanting. It is suitable for growing round the year, especially for summer. This is a good hybrid for hot and humid climate. It is tolerant to soft-rot and cracking.

*Vishwas (F1)* : It is very early maturing hybrid. Leaves deep green, fleshy-broad, open. Heads are medium size, moderately compact, round and uniform green coloured, weighing 0.8 to 1.0 kg each. Head is ready in 45-50 days after transplanting. Uniform maturity which make single harvest possible. Matured head does not split up to one week. It is suitable for close planting (70-80 thousand plants/ha), could be plant densely. Yield 50-60 mt/ha under optimal conditions. Tolerant to fusarium wilt.

*Combinor (F1)* : Compact, vigorous, strong plant with 8-10 auxiliary leaves. Heads round, green, compact, weighing approximately 1.0-1.25 kg each. Ready for harvest 60-65 days after transplanting. Most suitable for cool and dry season. Tolerant to head bursting. Excellent standing ability of 45-60 days after the crop is ready.

*Equatoria (F1)* : Strong plant with vigorous growth, 8-10 auxiliary leaves. Head flattened attractive green coloured, globe-shaped head with height 14-16 cm, diameter 20-22 cm and weight 2.5-3.0 kg., head is ready within 70-75 days after transplanting. Tolerant to head bursting and low temperature.

*Green Globe (F1)* : Plant vigorous, compact with 8-10 auxiliary leaves. Head is round, attractive green, weight is 1.0-1.25 kg. Extremely early maturing. Head is ready within 50-55 days after transplanting. Suitable for growing round the year. Tolerant to cracking, makes the harvest period more adjustable.

*Quists (F1)* : Plant strong, self supporting, sturdy, erect compact and well built. Bluish-green, fleshy, broad, leaves with serrated collars. 10-15 auxilliary leaves. Head compact, dark green, heavy, round and uniform. Ready for harvest in 75-80 days of planting. Short, tender core with fine internal structure. Head weigh 1.0-3.0 kg. Matured head does not split for 90 days. Very testy and good for table, cooking or for process-industry. Tolerant to fusarium wilt. Yield 60-100 mt/ha under optimal conditions.

*KS-Cross (F1)* : Plants strong, sturdy vigorous with 10-12 auxilliary leaves. Head is green, semi-flat, 13-15 cm in height and 19-21 cm in diameter, weighing about 1.0-1.5 kg with attractive appearance. Very early maturing and ready within 56-60 days after transplanting. Tolerant to bacterial soft rot, black rot and suitable for growing round the year and specially for hot and humid climate.

***Vijay :*** Medium maturing hybrid (80-85days). Semi round, heads are compact, dark green in colour with good taste. Excellent field holding ability. Ideal for processing and fresh use. Tolerant to yellow and black rot.

***Dublin :*** This is a new medium sized variety with a fresh medium green colour. The head is filled quickly and it generates a firm fair sized ball, which is well closed at the base.

***Hero :*** It mature in 80-85 days after transplanting, head flat, bluish green, firmness excellent, average weight 2.0 kg, excellent field retention ability and very strong against pests and diseases.

***Super Hero :*** It mature in 65-75 days after transplanting, head semi-flat, green, firmness excellent, average weight 2.0 kg, tolerant against pests and diseases.

***S-101 :*** It come to maturity in 55 days after transplanting, head ball shaped, glossy green, very compact, average weight 1.2 kg, and extra early hybrid with excellent taste, tolerant to diseases.

***S-103 :*** It matures in 80-85 days after transplanting, head ball shaped bluish green, excellent firm, average weight 2.5 kg, suitable for long distance transportation, possesses outstanding field holding ability.

***S-107 :*** It comes to maturity in 70-75 days after transplanting, head flattened globe, green, very good firm, average weight 2.0 kg, excellent holding capacity.

***Hybrid 123 :*** It matures in 45 days after transplanting, head ball type, deep green, solid excellent firm, average weight 1.2 kg, good field holding capacity.

***Hybrid 124 :*** Plants come to maturity 50-60 days after transplanting, head flattish round, light green, good solid, average weight 1.2 kg, moderate field holding capacity.

***Hybrid 125 :*** Plants mature in 65 days after transplanting, heads round, light green, good solid, average weight 1.4 kg, good field holding capacity.

*Hybrid 126* : Plants come to maturity 65 days after transplanting, heads round, green, good solid, average weight 1.5 kg, very good field holding capacity.

*Hybrid 133* : Plants mature in 65 days after transplanting, heads flattish round, bluish green, good solid, average weight 1.5 kg, excellent field holding capacity.

*Improved Savitri (OP)* : Plants mature in 65 days after transplanting, heads round, green, solid round, average weight 1.2 kg, suitable for tropical conditions.

## CAULIFLOWER

Cauliflower [*Brasoica oleracea* L. convar. *botrytis* (L.) Alef var. botrytis (Syn. *B. botrytis, B. florida botrytis, B. cauliflora*)]; 2n = 18; syn heading broccoli, Fulgobhi, Phool kobi, Fulwar, Fulkapi, Fule Kobi, Poogbi, Gospoovu, Malakobi, Kosuguddae, Fulvar, Phool Gobhi; belong to the family Brassicaceae, is a cool season vegetable grown for its white and tender curd formed by the shortened undifferentiated flower parts. The word cauliflower is derived from the Latin Caulis (Cabbage), floris (flower), variety botrytis (Meaning budding) and originated from wild cabbage coleworts (*Brassica sylvestris*). It is also reported that cauliflower originated from a related species *B.cretica*.

Cauliflower is a diploid with 2n = 2x = 18. The original temperate cauliflower was biennial. In India there are two separate groups of cauliflowers viz., Tropical annual Indian cauliflower and temperate annuals commonly known as Snowball types. In cauliflower, stem is short and thicker to the some extent as that of cabbage. Leaves are large and oblong, younger leaves are sessile. Unlike in other cole crops, buds do not arise in leaf axiles. Edible parts 'curd' is a compressed hypertrophid pre-floral meristematic growth, which terminates main stem of plant.

Cauliflower was first introduced in India, from England in 1822 by the British. Imported seeds were tested in different parts of the country during May-July (hot months) which corresponds to growing period in England. This resulted in adaptation of some introduced varieties to the warmer and humid conditions in the country. Farmers made selections for early types in the adapted types for their own use.

Local strains thus produced were better adapted and earlier in maturity for curd production during November-December compared to crops raised from imported seeds. This marked the development of an entirely new cauliflower types known as Indian cauliflower. This developed in the last 170 years in India. Sutton seeds, India produced four early Indian varieties in 1929.

The differences between Indian and European cauliflowers are given below :

| Indian types | European types |
|---|---|
| Tolerant to heat | Not tolerant to heat |
| curd formation at and above 20°C | Curd formation at 5-20°C |
| Annual | Biennial |
| Yellow curds with loose strong flavour | Snow white curds with mild or no flavour |
| Early | Late |
| Less jurenile phase | Long jurenile phase |
| No need of vernalization but needs cold treatment at 10-13°C for 6 weeks | Need vernalization at 7°C for 8-10 weeks |

Gill (1993)

Development of Indian cauliflower helped to extend cauliflower cultivation from temperate countries to tropics and sub-tropics. Genes for tolerance to high temperature and rainfall of Indian cauliflower were also utilized for developing improved varieties like Pua kea in Hawaii, Campinus in Brazil, Improved Japanese and D96 in Israel and Extra Early in Taiwan.

## Varieties and Hybrids

The cauliflower varieties are very responsive to temperature and photoperiod, so require specific temperature for their curd initiation and development. Varieties are classified into different maturity groups based on temperature requirement and harvesting time of curds under North Indian Plains. According to temperature requirement varieties have been grouped into following -

| Group | Cultivars | Planting time | | Temperature requirement |
|---|---|---|---|---|
| | | D/S | D/P | |
| 1. Extra early | Early kunwari,<br>Patna selection,<br>Early Market<br>Pusa Early synthic | March-April | Late April-<br>Early May | 20-27°C |
| 2. Early group | Pusa Deepali<br>Pusa kartiki<br>Katki<br>Arka Kanti | Late June-<br>Early July | Late July-<br>Early Aug. | 20-25°C |
| 3. Medium group | Improved Japanese<br>Pant subhra<br>Pusa Himjyoti<br>Pusa Hybrid-2<br>Pusa Sharad | End of July<br>to early Aug. | Early to<br>late Aug. | 16-20°C |
| 4. Mid late group | Pusa Synthetic<br>Marghashira<br>Pusa subhra. | End of Aug.<br>to early Sept | End of Sept.<br>to Early Oct. | 12-16°C |
| 5. Late group | Dania<br>Pusa snowball-1,2,16<br>Pusa snowball K-1<br>Punjab Giant-26<br>Punjab Giant-35 | Late Nov. to<br>Early Dec. | Dec.-Jan | 10-16°C |

Great Care should be taken to select suitable varieties to each month of planting for getting curds in appropriate mouths.

***Hybrids :*** There are so many hybrids developed by different seeds company and Agriculture University for cultivation during different periods. Most important are given below.

*I.A.R.I.* : Pusha Hybrid-2, Swarna Hybrids

*I.A.H.S* : Himlata, Himani, Himgiri, Swati, Guardian, White Centessa, White Baren.

*Century Seeds* : Early Himdev, Early Himlata, Early Himangini, Aghrani, Himdhara, Paush Himyug, Maghi Himyug, Pathar, Neelachal, Himageti, Himyug, Hindan, Himangini, Ageti Himkaran, Ageti Rapids, Ageti Himpaat.

*Rallis India* : Priya, Poonam, Preeti, Prema, Bajarang, Sujata, Suvarna.

*Nath Sluis* : Nath Sweta, Nath Suparna, Nath Ujjala.

*Clause international* : Solo, Tuing, Parla, Malini, Maharaja, Maharani.

*Haritage seeds* : Dhangari, Aagehari, Himkartika.

*Sandoz India* : Serano, Pawas, Himgiri.

*Syngenta* : Super Express (OP), Pawas (Hybrid), Sarano(Hybrid)

*Pahuja Seeds* : ES-77 Hybrid, Snow baby, Agahani, Pusi, Maghi, Early Snowball, Late Snowball, Hybrid No.71, Hybrid Snow Misteek, Hybrid Snow Crown.

*Royal Slouis* : Sushma Hybrid, Subhra, Swetta Hybrid, Ujjala Hybrid, Kintore Hybrids, Sukanya Hybrid, Reema Hybrid.

*Sungro Seeds* : Basanti (OP), Chandramukhi (OP), Hybrid - 218, Hybrid - 212, Hybrid Kunwari, Summer Special, First Early, Improved Early, Sungro ES-67, Hybrid Summer, Early Synthetic, Kuwari hybrid, Hybrid Ceco, Improved Kartik, Sungro Aghani, Sungro Pusi, Sungro Maghi, Sungro Agati Snowball, Sungro Snowball-16, Sungro Holland Special, Sungro Super Snowball, Sungro late snowball, Snowgress, Winter Queen.

*Bejo Sheetal Seeds* : Goodman, Fastman, Cassius, Somersat, Amazing, Lateman.

*Mahyco* : No.310, No.311, Himani Hybrid, Swati Hybrid.

*Novartis Seeds* : Pawas, Serrano, Super Express.

*Tokita Seeds* : TSX-0128, TSX-734.

*Nunhems Seeds* : Himgauri, Shweta, Sarita (NP 1000), Tushar, Gaurav (97002), Parvati, Himpushpa (NP 1801).

*Golden Seeds* : Garima, Summer Special-10, Rimzim, Shravan, Sharad Safedhi-70, Mahima-80, Sharmilee-85.

*Namdhari Seeds* : Ati Shighra (NS 1351), NS 60N, Basant (NS 245), NS 60, NS 66, NS 76, NS 75, NS 84, NS 90, Shivangi.

*Tropica Seeds (Green CO.)* : Palma, Hoggar, White Snow.

*Shreemaa Seeds* : Early Selection (H.Y.), Anarkali, Parijat (Imp.), Puspanjali (Imp.), Puspagiri, Himsagar.

*Annapurna Beej* : Himkanya, Kaveri, Cycllone, Rangila, Sukh Sagar.

*Biostadt Seeds* : Captain 401, Captain 402.

**Details Characteristics of important varieties and hybrids are given below:**

***Early Kunwari :*** It is the earliest cultivar released by the Punjab Agricultural university. It has bluish-green leaves with waxy blooms. Early maturing, semi-erect, partial blanching habit curd, colour creamish. Field tolerant to alternaria blight and black rot. Curds are not so compact and sensitive to riceyners.

***Pusa Early synthetic :*** It is developed from I.A.R.I., New Delhi, curd is yellowish white, medium sized and moderately compact. Suitable for planting in the end of June or first week of July. Average curd yield 110 q/ha. Suitable for sub-Humid Sutlej-Ganga alluvial plains and humid to semi arid western Ghats, Karnataka plateau and Orissa.

***Arka Kanti :*** It is developed from IIHR, Bangalore. It is a early variety with medium plant height. Heads compact white and mature in 60 days. Average curd yield is 25 tonnes / ha.

***Pant Gobhi-2 :*** Early maturing, compact curds, composite; creamy white curds, ready for harvest in 125 days after planting.

***Pant Gobhi-3 :*** Early maturing; curds medium sized, solid white, nursery sowing upto 15$^{th}$ May. Curds ready for harvest in 110 days after planting.

***Dania Kalimpong :*** More commonly grown in Eastern parts of India. Plants sturdy, wavy leaves; curds medium deep; it is less sensitive to fluctuations of environment.

***Pant Gobhi-4 :*** It has medium long stem, sparse, semi erect leaves, hemispherical creamish-white, medium, compact, non-ricey curds.This is a November maturity group, takes 115 days for marketable curds.

***Pusa snowball-2 :*** It has upright leaves covering head tightly and slightly puckered at the margins. The curds are solid, white with very good staying power and slightly raised in the center. It takes 110-135 days from sowing to curd harvesting. It yielded 150-120 q/ha in hills and 400-500 q/ha in plains.

***Swarna :*** It can be grown from September to December in plain and round the year in the hills. Curds are white, compact and ready within 80-85 days after transplanting. It yielded 1.0 to 2.5 kg on an average curd weight.

***Pawas :*** It is most suitable early hybrid, which can be grown from middle May to August. It produces 700-800 g average curd weight and ready within 60 days after transplanting.

***Summer King :*** It produce 400-500 g on an average curd weight. Whitish round shaped curd and ready for harvesting with 65 days after transplanting. It can be grown during summer season in north plain.

***Hisar 1 :*** It is suitable for mid late season under Haryana conditions. Medium to large sized white heads. It yields 250 q/ha.

***Punjab Giant-26 :*** Main season variety, foliage green, plant height 57 cm with a spread of 75 cm. Curds solid, white, medium in size, developed by P.A.U., Ludhiana.

***Punjab Giant-35 :*** The leaves erect and taller than Punjab Giant-26.The curds are snow white in colour, compact and large sized. The yield potential is higher (about 325 q/ha).

***Snowball-16 :*** The leaves broad, semi-erect, darkgreen with waxy coating. The curds are snow-white compact, medium to large sized. Its average yield of curds is about 200 quintals per hectare.

***Pusa Hybrid-2 :*** It is a combination of a self-incompatible line (female parent) and sel. 1-3-18-19 (Male parent). Plants are semi erect with bluish-green upright leaves, resistant to downy mildew. The curd is a creamy white, very compact, weighing on an average 907 g. It matures form mid Nov to mid-Dec. the curd- plant ratio is 10 per cent higher because the curds are very compact and with lesser number of leaves compared with "Improved Japanese". The yield is 23 tonnes/ ha. Temperature between 16-20$^0$C is ideal for curd formation. Optimum sowing time is late July.

***Pusa Katiki :*** One of the early releases from the IARI, New Delhi, having medium-sized plants and bluish-green leaves, medium sized curds.

***Pusa Deepali :*** Released by IARI, New Delhi, developed through inbreeding from the local collection. Early season, crop in 100-120 days. It is a hot weather cauliflower and temperature requirement is 20-25$^0$C during curd initiation and development. The plants are medium tall, erect, leaves a waxy green with rounded apex and inner leaves covering the curd moderately in early stages of development. The curds are somewhat round, very compact, medium sized, weighing between

500 and 700 g and white to creamy-white colour. Due to its erect habit and narrower frame (45-50 cm), it is possible to accommodate 35-40% more plant per unit area. It is highly tolerant to riceyness, fuzzyness when grown in proper climate.

***Pusa Himjyoti :*** A product of selection at the IARI, Regional Station, Katrain. It has erect bluish-green leaves with a waxy coating. It is pure line selection from 'MG 2-4' (Maghi Sri Ganga nagar Selection), give stable yield in both the sowing (April and June). It has erect bluish green leaves with waxy coating. The leaves cover the head tightly in the initial stages. Curds, are pure white and they retain their colour even after exposure. The average curd weight is 500-600g. It takes 62-75 days from transplanting to curd formation. A good crop may yield up to 150-170 q/ha.

***Pusa Sharad :*** It has semierect and open type plants with small stalk. Leaves are glabrous, bluish-green, long, petiolate oblong with narrow apex, wavy margin, prominent midrib and ear like lobes at the base of the lamina. Its marketable curd becomes ready 85 days after transplanting. It is about two weeks early in maturity over the Improved Japanese. Curds are retentive white, knobby, very compact, semidome shaped and about 750-1000g in weight. Average yield is 26 tonnes/ha. The optimum temp. eratuse for its curd initiation and development is 16-20$^0$C. Seed should be sown by the end of July to early August depending upon the locality and climatic conditions.

***Improved Japanese :*** It is an introduction from Israel with long, erect, narrow leaves and medium-sized curds. A mid season variety, crop in about 90-95 days.Stem long. Curds medium to large, compact and white. The average curd yield is 145 q/ha.

***Pant Shubra :*** Developed at Pantnagar. Outer leaves semi-erect, inner leaves partially cover the curd, compact, slightly conical, non-ricey, slightly creamish white curds. The average curd yield varies from 185-225 q/ha.

***Pusa Synthetic :*** Developed at the IARI, New Delhi, synthesized from the lines with good general combining ability. Plants are erect, very vigorous, 70-90 cm tall, bluish-green to green, leaves, numbering 22-26 at the time of curding; curds some what self balanched, colour varies from white to creamy white, fine-textured and compact. It takes about 130 days from sowing to 50 percent maturity. The most ideal temperature for curding is 12-16$^0$C. Sowing and transplanting time are end of August and end of September.

This is the first-synthetic variety, recommended so for in the vegetable crops. It can set seeds in plains without sacrificing yields. The cauliflower is tolerant to curd-blight and farmer can produce their own seed for 3-4 generations, vigour and yield are hardly affected.

***Pusa Shubra :*** A selection form a triple cross (MGS 2-3', '15-1-1' and D 96') developed at the IARI, New Delhi. Plants erect with somewhat long stalk and light bluish-green leaves. Light bluish-green twisting back wards from the middle. Curds are of medium size flat, compact, white, each weighing 700-800 g. Once ready, they should be harvested without delay. They have a tendency to loose, yet least prone to ricyness. The best temp. for curd formation is 12-16$^0$C. In plains sowing in late august and transplanting by late September are consider ideal. The curd yield is about 25 tonnes/ha.

***Dania :*** Variety developed at Kalimpong, successful in the eastern region, more particularly the eastern hills. It has very sturdy plants with wavy leaves and medium-deep curds. It is sensitive to fluctuating environment.

***Pusa Snowball 1 :*** Selection made at Katrain from an inter-varietal cross EC 12012 X EC 12013. Straight, upright leaves, tightly covering the curd. Curds are compact, medium in size and snow-white in colour. Suitable time for sowing in northern India is the first fortnight of October.

***Pusa Snowball K-1 :*** Field resistance to black-rot. Selection made at Katrain from an exotic material. It has upright, wavy, light green leaves, self-blanched, snow white, hi-dome curds. Becoming increasingly popular with the growers because of its curd quality and late maturity. Period of fresh availability is extended by at least 15 days.

***Romanesco :*** It is a fractal in structure, with conical head formed of small peaks of buds. It has slightly waxy, large, grey green leaves, bigger in curd size (0.6-0.75g each) complex, symmetrical head features whorls of pointy, vivid green florets, self blanching type. A good source of vitamin-C (for antiageing, wound healing, decreasing blood cholesterol. It takes 120-125 days for head formation after transplanting.

***Pusa Snowball K-25 :*** It has been developed by mass pedigree method of selection from cross of heading broccoli EC 103576 and Pusa Snowball 1. Its curds are very solid, compact, medium, sized, retentive shining white with very good staying power. Inner leaves cover the curd very tightly. Leaves are waxy, slightly bending towards inner sides with puckered margins. It takes 175-180 days from transplanting to first harvesting at katrain, and 105 to 130 days in different parts of the country. It is a very late variety, average yield being 175-300 q/ha.

***Pusa Meghna :*** It has been developed from heterozygous material collected from Hazipur (Bihar), using recurrent selection method. Belonging to early maturity (September) group of cauliflower. It has semi-spreading plants with 5-7 cm long stalks and 45-48 cm height. It has medium sized, light green, smooth, glabrous, ovate, petiolate and non-blistered leaves with entire margin waving outward. It produces retentive white, compact, medium-sized curd weighing 350-400 g. It takes about 90-100 days for maturity and temperature range for curd development is 20-27°C. It is ready for harvesting, from early September.

***Goodman :*** Mature in 65 days after transplanting, head weight 2.0 kg and compact, snowwhite curd. Seeds sown in Sept.-Oct. and tolerant to fusarium, self blanching leaves structure. The leaves are spreading and cover the head very nicely.

***Fastman :*** it mature in 60-65 days after transplanting. Very compact and snowwhite curds with weight 1.0 to 1.5 kg. Suitable for cultivation during winter season in plains. Resistant to fusarium. It is an early maturing cauliflower which gives nice white, non-hairy curds.

***Cassius :*** This is vigorous growing hybrid, which gives a strong and healthy plant. It is having very nice self-blanching leaves which gives snowwhite curods. Mature in 65-75 days, good comapct head (curd) with average weight of 2.0 kg. Recommended for August to December sowing. Tolerant to *Fusarium* and other diseases.

***Somersat :*** It has got a better heat tolerance, matures in 65-70 days with curd weight of 2.0 kg each. Curds are very well protected with leaves. Recommended for August to December sowing. High leavel tolerant to fusarium wilt.

***Amazing :*** It is an uniform mid early cauliflower which gives non hairy white curd, mature in 70-80 days after transplanting. Compact with self blanching leaves, which gives snow white curd with weight of 1.0 to 1.5 kg. Recommended for August to November sowing in the plains and tolerant to fusarium wilt.

***Lateman*** **:** It mature in 80-90 days after transplanting. Very compact self blanching white head, with upright growing leaves. It gives good heads having attractive white colour with average weight of 2.0 kg. Recommended for September to November sowing.

***Ati shighra (F1) (NS 1351)*** **:** Early summer cauliflower with open plant habit. Curds are medium white in colour and semi dome in shape with good firmness. Each curd weighs 0.5 kg and matures in 50 days. Comes up well in North and North East India (22-30 degree latitude).

***NS 60N (F1)*** **:** A typical summer cauliflower with early maturity. The hybrid has open canopy with medium vigour. Each curd weighs 0.75-1kg with good white colour and firmness. The curds are dome shaped with maturity of 55-60 days. It has wide adaptibility.

***Basant (F1) (NS 245)*** **:** A good hybrid with open type plants. The hybrid has good white curd colour with semi dome shape and good firmness. Each curd weighs 1.0-1.5 kg with a maturity of 55-60 days. It is suitable for late summers and can perform well in the Southern states (latitude range of 13-26 degree).

***NS 60 (F1)*** **:** It is a very popular hybrid with wide adaptability. It is a mild season tropical cauliflower with semi blanch. The hybrid has very good attractive curds with compact dome shape and good firmness. Curds are milky white in colour and weigh 1.25-1.5 kg maturing in 60-65 days.

***NS 66 (F1)*** **:** Mid late season cauliflower with excellent uniformity and curd quality. Plants are very vigorous with medium blanch. Curds are compact and smooth. They are dome shaped and have excellent firmness and density. Each curd weighs 1.5-1.75 kg with excellent attractive pure white colour and mature in 70-75 days.

***NS 76 (F1)*** **:** This cauliflower has very vigorous plants with good blanch. Curds are compact and dome shaped with excellent milky colour, firmness and density. The curds weigh 1.5-2.0 kg each. It is mild season cauliflower maturing in 75-80 days.

***NS 75 (F1)*** **:** This hybrid cauliflower is very much suitable for early winter season. It has very vigorous plants with dark coloured foliage and an excellent blanch covering the curds. Each curd weighs 1.5-2.0 kg, which are dome shaped and admirable snow white colour. Curds are compact with good firmness and density. Matures in 80-85 days.

*NS 84 (F1)* : This hybrid is of winter type late cauliflower with vigorous plant habit and an excellent bloom. Plants are uniform with compact curds and superior blanch. Each curd weights 2.0 kgs and is of a superb pure white colour and deep dome shape. Matures in 85-90 days.

*NS 90 (F1)* : It is a late season cauliflower with large heads bred for winter season sowing in areas with long winters. It has very vigorous plants with excellent blanch giving good tolerance to cold. Curds are compact, deep dome shaped with excellent pure white colour. Each curd weights 2.0-2.5kgs and matures in 90 days.

*Shivangi (F1)* : Snow ball type cauliflower with very vigorous plant habit. It is suited very well for late season sowings. This hybrid has very good blanch giving very good protection to the curds. Curds are very compact and dome shaped with excellent firmness. Each curd weighs 1.5-2.0 kgs and is of milky white colour. Matures in 90-95 days.

*NS 50* : A mild maturity hybrid. Heads are compact, uniform and deep dome shaped. Holds shape very well even when harvesting is delayed. Uniform heads with tolerant to downy mildew and black rot. Suitable for processing, fresh market and home garden.

*Garima* : Early maturing vigorous hybrid. curds can be harvested approximately 45-50 days after transplantation. Leaves are bluish green coloured. Curds are dome shaped, white and weigh 700-800 g., suitable for planting from middle of February to end of May. Tolerant to black rot and heat.

*Summer Special - 10* : Early maturing hybrid suitable for sowing after mid may to July in northern India and during February to April in southern region. Curds mature in 45-50 days after transplantation. The plants are erect with medium frame. Curds are white, dome-shaped, compact and weigh 800 to 900g and have good self blanching habit.

*Rimzim* : Early maturing hybrid. Curds can be harvested approximately in about 55-60days after transplantation. Curds are well protected by inner folding leaves, white in colour, semi-dome shaped, and weigh 700-1000g.; excellent adaptability to different agroclimatic regions. Suitable for rainy season cultivation. It is recommended to plant at wider spacing to obtain uniform curds of larger size.

***Sharavan :*** Early maturing hybrid, maturing in about 60 days after transplantation. Curds are pure white and of excellent quality and dome shaped, weigh 800-1000g. Upright leaves of the plant provide good protection to the curds. Tolerant to soft rot *(Erwinia* sp.) and suitable for rainy season cultivation.

***Sharad safedhi-70 :*** Top quality hybrid in early snowball group. Curds are snow white, dome-shaped and relatively large weighing 1200to 1500g., upright leaves provide excellent protection to the developing curds. The hybrid has flexible adaptability to different cliamatic conditions.

***Mahima-80 :*** Late maturing hybrid of mid snowball group. Curds ready for harvest in about 80-85 days from transplantation. The plants are vigorous, upright and have excellent self blanching habit. Curds are dome-shaped, pure white and weigh 2000 to 2500g and fairly tolerant to black rot disease. The bluish green coloured waxy leaves provide some protection from insect pests also. This hybrid is recommended for winter season cultivation.

***Sharmilee-85 :*** Vigorous growing hybrid of late snowball group. Curds can be harvested in about 85-90 days after transplantation. Curds are snow white, high dome-shaped and weigh 1750 to 2000g. This hybrid is recommended for winter cultivation in the plants and for summer in the hills.

***Anarkali :*** Upright plants. Sown in September-October, average weight 1.5 kg. curds are being harvested within 65-70days, curds are white, dome shaped.

***Parijat :*** An extra early variety, maturing in 50-60 days after transplanting snow white curd. Suitable for early spring in western and southern region. Summer sowing in northern region.

***Puspanjali :*** An early-mid variety, leaves are medium sized upright, greenish blue colour, curds are solid dome shaped and fine stincture, weight 0.8 kg., mature in 70 days after transplanting.

***Puspagiri (F1) :*** Suitable for western and southern region of india, maturing in 55-60 days after transplanting, curds are very compact medium sized and of snow white colour.

***Himsagar (F1) :*** Suitable for western and southern region of India. Maturing in 65-70 days after transplanting, curds are very compact, medium sized and of snow white colour.

***Early selection :*** It is a early cauliflower. Seeds are sown in nursery during. May–June then after 25 days old seedings are transplanted in main field. It mature within 70-75 days after transplanting.

***Marghashira :*** It has pure white and big size curd. Seeds are sown in nursery during September. 25 days old seedings are transplanted in the main field. It matures within 80–90 days after transplanting.

***Amarabati :*** It is highly profitable variety and mature within 65 to 70 days. Curds are white and compact. Seed sowing should be done in September. It is medium group cauliflower.

***Aparajita :*** It has big white and compact curd. Seeds are sown in nursery during September and curds are harvested in December. It is medium type cauliflower.

***Himdev :*** It is extra early type grown in summer and rainy seasons. It mature within 35-40 days after planting. It requires low spacing for high return.Curd size is medum and size varies from 250g to 300g.

***Himdahra :*** It is extra early type and cultivated in rainy season. It matures within 65-75 days after planting. Curd size bigger i.e., 700–800 g., curd uniform and compact, required narrow spacing for higher return.

***Himlata :*** It is tolerant to high temperature and grown in summer. It matures within 50-55 days after planting. It has highly compact curd and has permanent white colour. It is popular and well known to through out country.

***Himyug :*** It is late type and seed is sown during late Nov.-Dec. in nursery and 25 days old seedling are transplanted in main field. It matures within 70-80 days after planting. Curds are bigger in size i.e. 1100g. Leaves are dark green and curds are more white and compact.

***Himangini :*** It is extra early type and cultivated in summer and rainy seasons.It matures within 55-60 days after planting. Curds are more compact and can be procured for more days in the field.

***Shweta :*** A medium maturing hybrid (70-75 days after transplanting) suitable for tropical and subtropical regions. Plants have thick leaves which provide protection to the curd. Solid and tender curds are white in colour, dome shaped weighing approxmately 500 - 600 gms.

***Basanti (OP) :*** Plant erect and vigorous, maturity 85-90 days after transplanting, curd set at temperature 20°C, shape dome, snow white, average weight 2.0 kg, most suitable for late planting in North India.

***Chandramukhi (OP) :*** Plant vigorous and upright, first picking 75-80 days after transplanting, curd set at temperature 15°-17°C, dome shaped, snow white, average weight 1.5 kg, typical self branching type and keeps the curd white for longer period.

***Hybrid - 212 :*** Plant green, maturity 45 days after transplanting, curd white, flattish dome, maximum curd weight 900 g, suitable for early sowing (tropical condition).

***Hybrid 218 :*** Plant deep green, maturity 60 days after transplanting, curd white, flattish dome, average weight 650 g, suitable for areas having milder climate.

A few $F_1$ hybrids of 3 - 5 months maturity have also been developed by Australian Company. Some of these hybrids are Flanca, Sebal, Gija and Nedcha.

## CHINESE CABBAGE

Chinese Cabbage (*B. Chinensis, B. pikinensis*); 2n = 20 belongs to the family Brassicaceae is an annual and has few characteristics of common cabbage. It is grown as vegetable to a limited extent. Two more or less distinct species are grown: pe - tsai (*Brassica pekinensis*) and packchoi (*B. chinensis*). The pe - tsai resembles a cos lettuce but produces a much larger head which is elongated and compact, the leaves are slightly wrinkled, green, thin and much veined. The midrib is broad and light in colour. The pakchoi varieties resemble Swiss chard in habit of growth. The leaves are long, dark green, and oblong or oval. This type does not form a solid head. The most popular varieties are Chilli and Wong Bok.

Chinese cabbage is a good source of protein, vitamin, carbohydrate and minerals. It is used as pot herb and consumed fresh in salad or cooked as vegetable like other green vegetables. It can also be used as a substitute for lettuce and celery.

### Varieties and Hybrids

There are different types of chinese cabbage: (a) Loose leaved variety (var. *dissoluta* Li). (b) Semi - heading variety (var in fraction Li). (c) Fluffy topped heading variety (var. laxa Tse et Lee) and (d) heading variety (var. *cephalata* Tse et Lee). This variety is further classified into three types: (i) Ovate type (*f. ovata* Li), (ii) Flat topped type, (*f, depressa* Li) and (iii) Cylindrical (*f. cylindrica* Li).

Chinese cabbage varieties have also been classified into three groups according to growing seasons : (i) Spring varieties (January to April). (ii) Tropical varieties (April to July). (iii) Autumn or main season varieties (August to September).

**Characteristics of some important varieties are given below**

***Tianjin Qingmaye*** **:** It is grown in Tianjin area of China since very long time. It has good quality, low cellulose content, disease resistance, yield storage and transport quality.

***Naebyung 60 Days*** **:** It is a early cultivar, has significantly more vitamin C (36.8 mg/100 g fresh weight) than other cultivars.

***Raseosamgye and Olympic*** **:** These two cultivars have highest percentage of marketable heads (85 %) at harvest in comparison to others.

***Nepos, Nestor, Nerva and Spectrum*** **:** These cultivars perform well and do not produce bolter when planting is done in cold glass houses from second week of March and harvesting is done in third week of May.

***Chiifu, Kaga, Hotoren, Aichi and Kenshin*** **:** These cultivars are popular in China having good morphological character and perform good in their growing region.

***Peking*** **:** It is a fairly late Japanese variety, released in late 1979. It has a tight head weighing about 3.0 kg and few leaves.

***Chihili*** **:** It produces a tall narrow and cylindrical head.

***Wong Bok*** **:** It is short and very chunky in shape.

In India, there are not many varieties recommended for cultivation under local conditions. Punjab Agricultural University has released Chinese Sarson No. 1 in 1987 for general cultivation in the State.

***Chinese sarson No. 1*** **:** It is non - heading type with light - green, broad leaves and semi - erect plant habit. Its mid - rib is white, succulent and tender. It takes about 30 days for the first cutting after transplanting. It is quite rejuvenating, giving 6-8 cuttings. Its average yield is 4 tonnes/ha. It possesses field resistance to alternaria leaf spot.

***Hybrids*** **:** Some hybrids are also in cultivation, which are described below:

***Hybrid*** **59 :** It is an early maturing, higher yielding and more heat tolerant than others.

***Yuan Pao*** **2 :** It has been developed for commercial cultivation

**65 *Days* :** It produces more than 80 tonnes per hectare yield.

***Michihli* :** It is heading type give yield of 54.2 t/ha. Cylindrical head and tightly folded dark green leaves, medium maturity and uniform head. White midribs very crispy and tender.

***Manha*** *(Cheongbang x Yungching 30 Days)* **:** It is a good performer and has moderate resistance to *Erwinia carotovora* and turnip mosaic virus.

***Sambok*** *(Yungching 60 Days x Ta Feng Feng Lu)* **:** This hybrid shows moderate resistance to *Erwinia carotovora* and turnip mosaic virus and also has good performance.

***TaoyuanYashu*** **2 :** This $F_1$ hybrid is responsive to nitrogen fertilizer and high temperature and it is high yielder.

***Sumo* :** It has dark green foliage, leaves hairless and shiny. Mature in 55 to 60 days after sowing. Head oval shape, weighing 1.8 to 2.0 kg. The crisp head has a pleasant flavour and is ideal for salad and good substitute of head lettuce. Resistant to virus, soft rot and downy mildew.

***Tropic Early* :** It produces marketable heads within 40 days after transplanting in tropical low land areas. The head is very firm and the leaves are hairless, thick and very juicy. It is suitable for fresh consumption, frying and processing. It is tolerant to bacterial soft rot and mosaic virus.

***Rex (CCA 24)* :** Rex is adopted to be cultivated in the tropical highlands. It mature in 65 days after sowing. The head is oval shape and light green in colour with an internal orange yellow colour and very attractive. It is resistant to club root disease and shows good tolerance to soft rot and downy mildew.

## KALE

Kale (*Brassica oleracea* var. *acephala* DC.); 2n = 18; belongs to the family Brassicaceae, also known as Karam-sag is an underutilized vegetable. It is not yet commercialized in any part of India. It is a biennial but cultivated as an annual. Rosette of leaves and young shoots at the top of the stem which resemble cabbage prior to heading. It does not form a head like cabbage.

Kale is temperate and hardiest vegetable crop and can withstand temperature as low as 10° to 15°C. The quality produce is obtained under cool temperature. A few varieties are less hardy, can be grown in sub - tropical climate. It prefer full sun. The leaves of Kale are cooked similar to cabbage greens or in stir fries. Kale is used as winter vegetable when other vegetables are scare. The leaves are cooked as single or mixed with other vegetables.

### Varieties and Hybrids

**The followings are some important varieties of Kale are :**

*Dwarf varieties* : Dwarf green curled scotch, dwarf moss curled.

***Medium tall varieties:*** Moss curled, Hamburgen-Market, Karam saag (grown in Jammu and Kashmir).

*Tall varieties:* These are comparatively less productive. Hybrids mature earlier and tend to be more productive. Scotish and Siberian varieties are more suited for temperate regions.

## KNOL - KHOL

Knol-khol (syn Kohlrabi) *Brassica oleracea* var. *Gengyloides* L. (*B. Oleracea* var. *Caulorapa; B. Caulorapa* L.); 2n = 18, is also known as Ganth gobhi, Nool khol, Oalkabi, Old Kapi, Navilukosu, Gaddagobhi, Nawalkol, Knol Khol,. Kohlrabi is the german name for cabbage, turnip, since it resembles an above ground turnip. It is a herbaceous plant of which swollen stem just above the ground is the edible portion. All cultivars grown in India are exotic introductions. However, they are being maintained and multiplied by different organizations.

### Varieties and Hybrids

The popular varieties of knol - khol are Early White Vienna, White Vienna, Early Purple Vienna, Purple Vienna, Kyote No. 3 King of North and Large Green.

Some introductions from foreign countries are :

***Szentesi Faber :*** It is a summer cultivar, has a thin, pale green skin and crisp flesh, early bearing, produce a higher marketable yield and show greater resistance to alternaria leaf spot.

***Roggli's*** : This hybrid is out door white and is medium type, resistant to bolting and tuber do not become fibrous soon.

***Goliath :*** It is a late cultivar maturing between 70 - 100 days after planting. Tubers are white and blue type.

***European varieties :*** Wiesmoor White Forcing, Gaugels Forcing White are quick growing (4 - 6 week of Planting) and resistant to bolting.

***Short duration and resistant to low temperature varieties*** : Stuicka, Polni Ranar, Czekes, Blanr speak, Weiver blan, Prasake bele, Rogglis, Soroksar White, Knaufs Frushweis, Prasfer weisser Treib, Trero, Primerera.

***Varieties suitable for summer cultivation*** : Blaso, Pollux, Reliktess Bleu. These are most suitable cultivars for summer cultivation in temperate regions.

***Variety moderate resistant to Fusarium wilt***: Vienna White 1350.

***Variety Resistant to Palasmodiophora brassicae*** : K 240 from Bulgaria

***Variety Resistant to Xanthomonas campestris*** : Szentesi Tartos Kek

***Cultivars resistant to cold and free from bolting***: Wiesmoor White Forcing, Kuauf's Early White and Moravia.

***Varieties with short foliage*** : Wiesmoor White Forcing and purple variety Knauf's Ideal.

***Some promising varieties available in India through seed dealers are*** : 'Suttons' Earliest white (Suttons): Green Glory (Verma); Golith white (V.Sadashiv); Earliest Purpole, Early White, King of Market, Green Pearl, Petsri, Michihli (Pocha).

***Hybrids*** : The first hybrid was developed by the firm Roggli. These hybrids were Rowel, Roblau and Rosta. These are early bearer. There are several hybrids developed by seed companies. **Rallis India:** Lakshmi; **Tropica Seeds** (Green CO.) : Champion, Dakini; **Annapurna**

**Beej** : Winner No.4 Imp., Raja, Kirtiman, Bishal ; **Bejo Sheetal Seeds:** Korist, Kossak and Kolibri.

**Detail characteristics of varieties and hybrids are given below:**

***Early White Vienna*** : An early variety takes 50-55 days for the development of knobs after transplanting. Knobs are globular-round and green with fewer leaves. Flesh is cream-white, tender and crisp.

***White Vienna*** : An early variety having dwarf plant. Stems (tubers) globular, light green and very smooth. The flesh creamy white, tender with delicate flavour. It takes 50 to 60 days for knob formation.

***Early purple Vienna*** : An early variety which takes 55-60 days for knob formation. Foliage purple, knobs globular round, large with purple skin and light green flesh.

***Purple Vienna*** : Purplish-blue leaves, knobs are globular round, large in size with bluish - purple tinge, flesh light green. It takes 55 to 65 days from transplanting to knob - formation.

***King of North*** : Plant height 20-30 cm, dark green foliage with broad leaf - blade. Knobs are flattish round and dark green in colour. Leaf - sheath is large and well - spread over the knob. Mature in 65-70 days.

***Large Green*** : It is a late variety which takes 60-70 days for harvest. Plants are vigorous with dark green foliage, knobs, flattish round and dark green, adapted well to hilly regions.

***Palam Tenderknob*** : It is an improved variety of knol khol, bred through mass selection at Department of Vegetable Science, HPAU, Palampur. It has small light green foliage. It forms turnip-shaped knobs, which remain split free. These are round, flat, stringless, light green, sweet and fleshy. These knobs attain marketable size of 5-7 cm diameter weighing 250-300 g in about 80 days. Its average yield potential is 250-300 q/ha.

***Korist*** : It matures in 55-60 days after sowing, The approximate weight is 150 g and has good standing ability up to 30 days after maturity. It keeps growing even after 65 days and can grow up to 300 g. in size, without cracking and without fibre development.

***Kolibri*** : It is attractive red coloured knol khol with more sweet taste. It matures in 55-60 days after sowing.Approximate weight is 150 g and has good standing ability up to 30 days after maturity. It is tolerant to cracking and there is no fibre.

***Kossak*** : It is a big sized knol - khol, which matures in 65-70 days after sowing. The approximat weight of tuber is 1.0 kg.

***Champion (F1)*** : This hybrid variety grown in dry and cool climates. It gives high yield and very uniform, green bulb of globe shape, weight about 600 - 700 g at maturity. Very good bulb quality, maturity 60-65 days. Erect growing habit and very good shelf life.

***Dakini (F1)*** : Plants are erect medium and tuber mature early in about 57 days after sowing. Heavy yielder, pale green bulb of semi - globe shape. Weight 500-600 g at maturity. Very good shelf life.

***EWV Improved*** : It is developed by Tokita Seed, India. Pale skin, white flesh, crisp knobs are flat round to roundish. Ready for harvest after 55 days of transplanting. Average knob weight about 200-250 g. Knobs are with less foliage and smooth surface.

□□□

# 3

# Root Crops

## BEET ROOT

Beet root or garden beet (*Beta vulgaris* L.); 2n = 2x = 18; family Chenopodiaceae is known as Chukandar in hindi. It is a popular root vegetable grown in home and market garden in the USA and Europe. The name beta came from Romans. The species is thought to be derived from *Beta maritima,* known as sea beet, indigenous to southern Europe. Though, it is grown in almost all the states of India, it is not as common as radish, carrot and turnip.

*Beta vulgaris* ssp. vulgaris contains many cultivars including sugar beet, beet root, mangolds and fodder beet. Beet root is a biennial, producing the enlarged hypocotyl and so - called roots and a rosette of leaves in first year, and flowers and seeds in second year. The fleshy enlarged roots are edible.

### Varieties and Hybrids

**Beets are generally classified into two major groups on the basis of the edible parts.**

(i) The beets grown mainly for their fleshy roots belong to the crassa, or garden, group and include yellow beets and the common table, or red beets.

(ii) This group is grown as leafy vegetables belong to the cicla group and include leafy beets and Swis Chard.

**Varieties of beet root has been grouped on the basis of shape of roots :**

***Flat :*** Flat Egyptian

***Short top shaped :*** Flattened at top and bottom with rounded sides and conical or tapered base. var. Crosby Egyptian, Early Wonder, Asgrow Wonder.

***Round or globular :*** Roots are round or globular in shape, var. Detroit Dark Red, Crimson globe.

***Half long :*** Length is shorter than long types, var. Half Long Blood, Winter Keeper.

***Long :*** Roots are long, may grow as much as 40 cm, quite popular in Europe, var. Long Dark Blood.

***Hybrids :*** MAHYCO - Mahyco Lal; Ralish India – Dayana; Clause International - Warrior.

*Bejo Sheetal Seeds :* Action,Pablo.

*Golden Seeds :* Lalima, Nunhems Seeds : Bolivar.

**Table 1.** Varieties, hybrids and their source

| Variety | Source |
|---|---|
| Crimson Globe | IARI, New Delhi |
| Detroit Dark red | IARI, New Delhi |
| Father Ball | USA |
| Egyptian Blood | USA |
| Edmond | USA |
| Eclipse | USA |
| Crosbby Egyptian | USA |
| Early Wonder | USA |
| Ooty - 1 | ICAR |
| Mahyco Lal ($F_1$) | MAHYCO |
| Action ($F_1$) | BSS |
| Fablo ($F_1$) | BSS |
| Lalima ($F_1$) | Golden Seeds |
| Bolivar ($F_1$) | Nunhems |
| Ruby Queen | Peto Seeds |
| Ruby Queen Improved | Tokita Seeds |
| Bassano | - |
| *Corumba* - | - |
| Dayana | Ralish India |
| Warrior | Clause International |

**Detail characteristics of varieties and hybrids are given below:**

***Crimson globe :*** The roots are globular to flattened globe, medium red with little shoulders. The flesh is medium dark red with indistinct zones. The top is medium to tall with large, bright green leaves having maroon shade. It is heavy yielder.

***Detroit Dark Red :*** Roots are perfectly round with smooth, uniform, deep red skin. Flesh dark blood red with light red zoning, tender and fine grained, top small, leaves dark green, tinged with maroon. It is a heavy yielding cultivar, maturing in 80-100 days.

***Crosby Egyptian :*** Roots are flat globe with a small top root and a smooth exterior. The internal colour is dark purplish red with some indistinct zoning. The top is medium tall, green with red veins. This cultivar reaches maturity in 55-60 days after sowing and shows pronounced white zoning when grown in warm weather.

***Early Wonder :*** The roots are flattened globe with rounded shoulders with a smooth, dark red skin. The interior is dark red with some lighter red zoning. The top is heavy, green with red veins. This cultivar take 55-60 days after sowing to reach edible maturity.

***Ooty 1 :*** It is a selection from the local type. It is a high yielding, variety with potential of 31-45 tonnes yield per hectare. The skin is thin and the flesh colour is blood - red. The duration from sowing to harvest is 120-130 days. It is suitable for growing in south Indian hills. It is released by Tamil Nadu Agriculture University.

***Edmond :*** Moderate size, handsome, rounded roots with, smooth, deep, red , good grain and flavour.

***Eclipse :*** Uniformly globular, bright red, fine grained and sweet. One of the best quick growing early beets.

***Egyptian Blood :*** Top quite small, roots fair size, rich deep red, a standard early variety.

***Mahyco Lal :*** Top is medium with erect leaf crown, maturity 65-70 days. Root globe with close tail, deep red, crisp flesh. Uniform with smooth texture, retain orginal quality without ringing. Root 125-130 g each. Yield potential 40-45 t/ha and resistant to premature bolting.

***Action :*** This is an early maturing hybrid, which become ready in 50 - 55 days after sowing. The average beet weights 100-150 g. It can stand in the field for 30 days after maturity, without cracking or losing colour and taste. The size keeps increasing upto 400 g.

***Pablo* :** It matures in 55-60 days from sowing and weigh 120-150 g each. On maturity, if not harvested, keeps growing upto 450 g without cracking or losing taste and colour and does not develop fibre. It can be grown very well in all three seasons.

***Lalima* :** It is a vigorous growing early maturing hybrid. Uniform roots can be harvested in about 60 days after planting. Tops are erect with bright green and glossy foliage. Roots are top to globe shaped, smooth, dark red, very sweet and free from zones. The hybrid has good tolerance to *Cercospora* leaf spot, *Rhizoctonia* root rot, downy and powdery mildews and drought. Yield potential is 45-50 t/ha. Suitable for both fresh market and processing industry, especially for pigment extraction.

***Ruby Queen* :** It is a vigorous growing early maturing variety. Uniform smooth shouldered beet with small tap root which holds its shape well even when crowded. The interior colour is deep red. Roots can be harvested in about 50-55 days after planting. Roots are crisp, sweet, round and dark red. Yield potential is 40-45 t/ha.

***Bolivar* :** Very strong plant type produces round uniform beet, with smooth skin and deep red internal colour. Upright leaves facilitate early harvest. Very slow bolting. Good for fresh market and processing. Roots dark red, 200-250 g each, diameter 12-14 cm with round shape. First harvest within 45 days with good keeping quality.

***Ruby Queen Improved* :** Plant vigorous, light green leaves with red vines petiole. Roots crisp, sweet, round, dark red and ready for harvest after 55 days of sowing. Slices are very dark red and free from white rings.

***Corumba* :** It has very uniform, medium long cylindrical roots. Excellent internal and external colour with good shelf life.

***Other favourite varieties* :** Golden Tankard, Golden Yellow, Mammoth, Mammoth Long Red, Seneca Detroit, King Red, Detroit Red, Detroit Perfect, Rubia, Formanova, Little Ball, Kogel and Red Ball, which are common in temperate countries.

## CARROT

Carrot (*Daucus carota* L.); 2n = 2x = 18; syn. Gajar, Gajor, Gajjar, Gajargadda, Kaaret, Kempu, Mulangi, Mormuj, Bulmuj. Carottee, Cullive, Gemeiner, Mohre, Karoth,Jazar, Zardak, Garijara, belong to the family umbelliferae (Apiaceae), is a cool - season root vegetable grown round the world in temperate climates during spring, summer and autumn seasons and in subtropical climates during winter. It has a significant place as an ingredient in soups and sauces and in dietary compositions and also as a salad. Carrot is an important source of carotene. Temperature is the single environmental factor affecting root shape and colour in the crop.

The European carotene carrot (syn. Wastern carotene carrot, temperate carrot) has been derived from the Asiatic anthocyanin containing forms of carrot (syn. Eastern anthocyanin carrot, tropical carrot). The yellow coloured carotene – rich forms were also thought as a mutants of anthocyanin carrots.

The genus *Daucus* has 2 sub - genera i.e. *Eucarota* and *Gummiferea*. The species under *Eucarota* are *D. carota* L., *D. maritimus, D. major, D. sativus, D. maximum*. The *Gummiferea* consists of mainly perennials. The species under *Gummiferea* are : *D. cumutalis, D. hispanicus, D. fontanesii, D. boeconiel* and *D. gummiferea*. The species *D. carota* and *D. sativus* are only cultivated species. There is another view that all the above are sub-species of *D. carota* as *D. carota* spp. *Carota* and *D. carota* spp. *Maritimus*.

### Varieties and Hybrids

Varieties/cultures with long, orange coloured and smooth roots are preferred in India. Many varieties, both indigenous and exotic, differing in temperature requirement, length, size, shape and colour of roots and duration of crops are grown in India. The carrot cultivars may be classified on the basis of shapes of their roots or on temperature response.

Classification based on shape of roots :

1. ***Long rooted*** : Roots may be 25 cm or more in length, generally tapering, these perform best in comparatively light soil.
2. ***Half long rooted*** : Root length does not usually exceed 20 cm.
   (a) Roots cylindrical with straight or sloppy shoulder e.g. Nantes.

(b) Roots tapering with blunt or semi blunt tip. e.g. Chantenay, Imperator.

3. ***Short stump rooted*** : These cultivars are suitable for growing in heavy soils.
   (a) Heart shaped e.g. Oxheart
   (b) Oval e.g. Early scarlet Horn.
   (c) Round e.g. French Forcing.

**Classification based on Temperature Response to Flowering**

**1. Temperate or european or biennial types**

The European carrot is cold tolerant, orange coloured and rich in carotene. These cultivars are biennial in characters and require low temperature(4.8-10.0 C) treatments for flowering to occur. These do not produce seeds in the plains of India, e.g. Nantes, Chantenay, Half long, Imperator, Early Nantes, Pusa Yamadagini and Ooty 1.

**2. Tropical, asiatic, oriental or annual types**

The Asiatic carrots are heat tolerant, deep red or purple coloured, high-yielding and low in carotene content. These cultivars do not require any low temperature treatments for flowering, they form seed freely in the plains of India e.g., Pusa Kesar, Pusa Meghali, Hissar Gairic.

## Hybrids

***Mahyco*** : Hybrid No. - 1

***Bejo Sheetal Seeds*** : Samson 196, Bangor, Carson, Oranza.

***Novartis India Seeds*** : Premia

***Nunhems Seeds*** : Cerumba.

***Namdhari Seeds*** : NS 820 (OP), Namdhari Kuroda (OP).

***Sungro Seeds*** : Desi Red.

***Tokita Seeds*** : Kuroda Improved, Samurai Red.

**Detail characteristics of varieties and hybrids are given below:**

***Nantes :*** It is an European variety grown in plains for root production. It is half long, thin, well - shaped with stump and forming a small thin tail. It has delicious flavour, fine grain texture and self - coloured core with orange - scarlet flesh colour.

***Chantenay*** : It is also an European type having attractive roots with deep reddish orange - colour.The root is smooth, half long with thick shoulders, gradually tapering towards a distinctly stumped end. The flesh is beautiful rich orange, tender, sweet and fine textured with distinct core.

***Pusa Kesar*** : This variety is selected from a cross between Local Red and Nantes, half long.The leaf top is markedly shorter than that of the Local Red.The roots develop a narrow Central Core and the core is also sufficiently red coloured.The root, stay a month longer in the field without showing any sign of bolting. It seeds freely in the plains but the seed crop matures about 15 days later. It has higher (38 mg) carotene.

***Desi Red*** : It has medium foliage, mature in 80-90 days after sowing. Roots tapering, 15-20 cm long, appealing Red, Uniform Red, internal colour, core small and self coloured. High yielder.

***Corumba*** : It has very uniform, medium long cylindrical roots. Excellent internal and external colour with good shelf life. Good for cold storage.

***NS 820 (OP)*** : It is high yielding, early vigorous plants very uniform long, straight and cylindrical roots. Very small self coloured core. Crisp flesh with good sweetness and excellent quality. It has very good storage and transport qualities.

***Namdhari Kuroda (OP)*** : An early vigorous, high yielder with uniform long cylindrical roots. Roots are smooth, deep orange with very small self coloured core. Flesh is very sweet, crisp and endowed with good storage and transport qualities.

***Nantes Half Long*** : The variety commonly known as Nantes is evolved at IARI Regional Station, Katrain; roots are small, slim, rough, sweet, cylindrical and stumpy with abrupt tail; the core is small and self coloured; yield is 15-20 t/ha in 90-110 days.

***Early Nantes*** : Roots almost cylindrical terminating abruptly in small thin tail, 12-15 cm long, orange flesh with self coloured core; duration 90-100 days.

***Chaman*** : Developed at SKUA&T, Srinagar, roots long, cylindrical and semi blunt; tolerant to cracking; yields 25-27 t/ha.

***Pusa yamadagini*** : Developed at IARI, Regional Station, Katrain, roots 15-16 cm long, orange with self coloured core, slightly tapering and semi-stumpy with medium top; yield 9-10 t/ha in 90-100 days.

***Ooty - 1 :*** Developed at Horticultural Research Station (TNAU), Udhagamandalam; roots are 25 cm long with deep orange colour, yield 45-50 t/ha, in 100-110 days.

***Zino*** : Introduced from Germany in the Nilgiris hills of Tamil Nadu by the state Department of Horticulture, roots deep orange with self coloured core, duration 110-120 days.

***Danvers*** : It is imported variety, suitable for fresh market and processing. Roots long tapering or slightly rounded end. This is a mid season variety, with large foliage. The cortex is average and yellowish core.

***Imperator (Nantes x Chantenay)*** : It is suitable for fresh market. Root long, short tapered end. The colour of cortex is orange with slightly less pigmented core. It is a mid to late maturing variety. These above cultivars are European type.

***Sel - 233 :*** A derivative of a cross of Nantes x No.29 and released by PAU, Ludhiana in 1978. A desi type with all good qualities of Nantes. Harvesting can be delayed without bolting and impairing its edible qualities. It takes 90-100 days for root formation.

***Gajar No.29 :*** A selection from local material of Punjab. A desi variety with long tapering light red and thin roots. Released by Punjab Agricultural University.

***Kuroda Improved :*** A very uniform foliage and root variety. Deep range colour, self coloured core, 18-20 cm long, round shoulder with stump roots, ready for harvest after 80-90 days of sowing. Ideal for summer sowing to harvest in fall and winter.

***Samurai Red :*** Medium sized vigorous plant with green leaves,. Roots are 30-35 cm long, dark red with self coloured core, ready for harvest after 90 days of sowing. Roots are very uniform, smooth, sweet and juicy.

***Pusa Meghali :*** It is an advance generation selection , selection 5A from a cross between ' Pusa Kesar ' and 'Nantes '. It has a short top, smooth roots, orange flesh, self-coloured core and stumpy to slightly tapering roots. The tropical 'Pusa meghali' contains 11,571 mg of carotene per 100g on fresh weight basis. It is suitable for sowing as an early (Aug-Sept ) as well as late (Oct.-Nov.) crop. It is preferred for early-sowing and can be marketed in 110-120 days. It produces 25-28 tonnes /ha of carrots. The variety is capable of setting seeds in the plains.

***Hisar Gairic (H C1) :*** It has been developed through mass selection from the germplarm collected from various parts of the country. It matures in 120 days, the leaves are light-green, roots long, tapering, attractive, light brick red thin self core, less fibre and forking. It gives an average yield of 28.8 tonnes/ha and carotene content is 96.2 mg / 100g of fresh weight. The roots are 18.1 cm long and 3.9 cm across.The high yielding 'Hisar Gairic ' is for early sowing.

***Hybrid No - 1 :*** It has been developed at Maharashtra hybrid seed Co. Ltd., Jalna. Roots are deep orange, uniform, length 7-9 inches, matures early ; semi table for processing and fresh marketing.

***Samsoon 196*** : It is Nantes type, very high yielding, having soft edible core. It gives long carrots with vigorous upright growing leaves, having good taste. Maturity mid early and tolerant to powdery mildew.

***Bangor*** : It matures in approximately in 100 days. It is Borlikum type carrot, tolerant to powdery mildew and very high yielding.

***Oranza*** : It is having edible core and is good for slicing, having good yield, dry matter content is 9 to 12 %. Maturity mid early and Nantes type.

***Carson*** : It matures in approximately 110 days from sowing. Shape is Chantenay type, having healthy foliage and gives attractive coloured carrots.

***Premia ($F_1$)*** : Early hybrid and nantes type for early fresh market. Short foliage and produces long (15-18 cm), slender cylindrical carrots with rounded shoulders. Its external and internal colour is an attractive orange red and its core is fine. Resistance to cracking is very good. Normal yield ranges between 45-49mt/ha under optimum growing conditions.

## CHINESE RADISH

Chinese radish (*Raphanus sativus* Linn. var. niger); 2n=18; belongs to the family Brassicaceae; is a favourite crop of the home gardens because it is easily grown and is ready for use in very short period

(3-6 weeks) after seed sowing. It is a fast growing erect annual herb for root production. It is probably originated from the old world, cultivated long before Greeks and Romans came to know of it as rhaphan is and radix respectively.

Chinese radish is biennial for seed production but cultivated as annual for root production. The top of the plant is heavy. It is a large and heavy rooted, broad shouldered, thick skinned. The roots have an attractive pink colour and weighing on an average 1-2 kg each. Its roots are mildly pengent and an excellent vegetable for cooking. During severe winters of Ladakh, when no fresh vegetables are available, chinese radish is available in the market. The edible portion contains: protein 1.0 g; Calcium 32 mg; iron 1.4 mg; vitamin C 26 mg; potasium 400 mg and energy 30 kcl (per 100g of edible portion). The roots become ready for harvest in 65-70 days, if harvesting is delayed, remain unaffected from any pithyness, pungency etc. due to its thick skin.

### Varieties and Hybrids

There is no improved varieties of this crop, available. Only local genotype is grown in area of cultivation as per availability.

Some hybrids have been developed by Tropica Green Co. Ltd. i.e., Tropical Seed Company. These hybrids are Long White (F1) and Longo (F1).

***Long White (F1) :*** It can be grown in both high land and lowland. Excellent, uniform appearance with white colour skin with upright and dark green foliage. Root weight 600-800 g, highly uniform, 22-26 cm long with firm and delicious flesh. It is tolerant to heat and drought, suitable for spring, summer cultivation. Best suited to good draining soil. Resistant to Downy mildew, fusarium yellow and other common radish diseases. It matures in 45-50 days. Good for fresh market.

***Longo (F1) :*** It is a slow bolting hybrid with excellent root enlargement at low temperatures. Very smooth root skin and soft flesh. Resistant to fusarium wilt disease. Highly uniform roots of 35-40 cm long, 7-8 cm in diameter and 1.0-1.2 kg in weight. It is very popular for fresh market.

## HORSE RADISH

Horse radish (*Armoracia rusticana* Gaertn; syn. *Cochlearia aromoracia* L.; *Armoracia lapathifolia* Gilib); 2n=32; belongs to the family cruciferae;

is a stout glabrous perennial herb, native of south-eastern Europe, with long fleshy cylindrical roots, large oblong serrate leaves and small white flowers. The grated roots are used as a condiment, particularly with roast beef. The pungent taste is due to the glucoside sinigrin which is broken down in water by enzyme action.

It grows a height of 60-100 cm. The leaves are, dark green, wavy indented and resembles spinach. The tap roots in a napiform in shape and tapers downwards ending in a slender long root (25-30 cm) and 25-30 cm thickness having smaller roots and root hairs all round. On the basis of 100 g edible portion, it contains protein 3.2%, fat 0.2%, ash 1.8%, total carbohydrate 21.4 and fibre 2.4%. Besides, roots are rich sources of vitamin C. The fresh material contains on an average 302 mg of ascorbic acid per 100 g edible portion. The boiled extract of its green leaves is medicinally useful to cure and reduce obesity in human beings if taken regularly with honey every morning.

### Varieties and Hybrids

Similar to many under utilized vegetables, horse radish is also having local plant types but for obtaining higher yield and making it commercial crop, varietal improvement is urgently needed.

## RADISH

Radish (*Raphanus sativus* L.); 2n = 2x - 18, syn. Mula, Mullangi, Mulo, Mullanki, Mullamgi, Mooli, Muli, belongs to the family Cruciferae (Brassicaceae, is popular root crop grown for its tender tuberous roots which are eaten raw as salad or as cooked vegetable. It has unique pungent flavour. It is also preserved by storage, pickling, canning or drying. Cultivated radishes are put under 3 different species based on plant morphology and ecological requirements.

(1) ***Raphanus indicus*** : The Indian types which are frost tolerant, pungent and mostly used for pickle making.

(2) ***Raphanus raphamistroides*** : The Japanese type which are intermediate to Indian and European types, growing in the coastal regions of Japan and neighbouring regions and the turnip shaped giant radishes cultivated in the Sakurjuna Island of Japan.

(3) ***Raphanus caudatus* L.** : The non - tuber forming rat - tail radish, is closely related to the Indian type of radishes. The Indian type

of radishes got evolved in the Near Eastern and Indo - Burma centre of Origin. Its presumed ancestor has been lost.

The European types originated directly or through hybridization from a few wild species like *R. maritimus, R. landra* and *R. rostratus*. The Japanese types are derived from *R. sativus f. raphanistroides* (Syn. *R. raphanistroides*).

## Varieties and Hybrids

Radish varieties/cultivars or hybrids differ in shape, size and skin colour of roots and duration of crops.The varieties can be divided into three groups.

1. ***European or temperate types*** : These types are small in size, with excellent quality and mainly used for salad purpose, var. Pusa Himani, White Icicle, Scarlet Globe, Scarlet long, Kashi Sweta.

2. ***Asiatic or tropical types*** : Tropical types are more pungent than temperatre types and have large roots. Var. : Pusa Deshi, Pusa Chetki, Pusa Rashmi, Japanase white, Punjab Safed, Punjab Pasand,Punjab Agethi, Kalyanpur No.1, Arka Nishant, Co - 1, Long greens, Kalyani White.

3. ***Indian type*** : These are usually white with conical shape, attaining 25 - 40 cm in length and are said to be more pungent. An indigenous type; Newari or Jaunpuri Giant, cultivated around Jaunpur in Uttar Pradesh, has roots of 75-90 cm length, 50-60 cm girth and weight upto 5-15 kg.

A Baramasi is also an indigenous type which can be grown throughout the year.

***Others Cultivars*** : Burpee's white, Burpee's red giant, sparkler, scarlet globe, bombay red, Chinese rose, contai, sakurjuna ( enormous size from Japan).

***Cultivars For Plastic Culture*** : Jygeva 169, Drienbrunnen, pinkie and comet, Rubin Harkovskij, CO1 radish, Kalianpur type 1.

## Hybrids

***Mahyco*** : Hybrid Radish No.11, Hybrid Radish No.841.

***Century Seeds*** : Shravani, Lalima, Gulnar.

***Bejo Sheetal Seeds*** : Radish - Koren, Radish Reggae.

*Unicorn Seeds* : Shogunmino.

*Annapurna beej* : Kalam Kathi, Ause Mula.

***Sungro Seeds :*** Improved Chetki, Improved Japanese white, Summer Queen, Komal, Early long white, Hill Queen, Himani, Long Red, Chinese Pink.

***Tropica Green :*** Long white, Longo.

*Tokita Seeds* : Mino Early Long White.

**Detail Characteristics of Varieties and hybrids are given below:**

**Temperate types**

***White Icicle :*** It is a medium - short European variety which matures in 30 days. The skin is pure white, thin and tender whereas the flesh is icy white. Crisp, juicy, mild and of sweet flavour with just enough pungency to appeal the appetite. The root is solid icicle - shaped straight and tapered. It has given as good harvest in plains.

*Rapid red white Tip* : It is a very early European variety which matures in 25 days. It has a short top with small bright red roots with white tip. The flesh is pure white and crisp with mild pungent flavour.

*Pusa Himani* : The roots of Pusa Himani are 30 - 35 cm long and 10 to 12 cm in girth with green stem end. They are semi - stumped to tapering with short stalks.The skin is pure white and the flesh is crisp and sweet flavoured with mild pungency. It does well in both hilly and plain areas.

*Scarlet Globe* : Roots bright red, round, small, flesh white, takes 25-30 days from sowing to harvest.

*Scarlet Long* : Roots long, tapering with red skin and white flesh.

*Kashi sweta* : A selection made from Chetki population at IIVR, Varanasi; roots 25-30 cm long, straight and tapering with pointed tip; yield 45 t/ha in 30-35 days.

## ASIATIC TYPES

*Japanese white* : It is Asiatic variety which matures in 45-50 days. The root is cylindrical and about 20-30 cm long.The skin is pure white, smooth and the flesh is snow white, crisp, solid and mildly pungent with top medium - large with deeply cut leaves.

*Arka Nishant* : It matures within 45 - 55 days. Roots are medium sized, marble - white, crisp texture, pleasant aroma, free from early bolting, pithiness, splitting and forking, each root weighing 300-400 g.Root and shoot length ratio is 1 : 1. Pungency mild. It yields 200-300 q/ha.

*Pusa Chetki* : It has been developed at IARI, New Delhi.Roots are medium long, white, good textured, weighing 300-400 g each.It yields about 200-250 q/ha in 50-60 days.

*Pusa Deshi* : Roots white with green shoulders, 30-35 cm long, medium thick, tapering, pungent, tap medium, dark green cut leaves. Maturity 50-55 days. Suitable for sowing from mid August to September.

*Punjab Safed* : Developed at Ludhiana; roots are pure white, tapering, smooth, mild in taste, 30-40 cm long and 3-5 cm thick.

*Pusa Safed* : Developed at Ludhiana; roots are pure white, tapering, smooth, mild in taste, 30-40 cm long and 3-5 cm thick.

*Punjab Pasand* : Roots are long white and semi stumped. Suitable for sowing in October - Novemebr.

*Punjab Agethi* : Roots medium long, red skinned at top and white skinned at lower half, the most suitable for sowing from April to August.

*Kalyanpur No. 1* : Roots 22-23 cm long, white with green shoulder and tapering.

*Co - 1* : Roots milky white, 23 cm long and cylindrical, yield 9-10 t/ha in 45 days.

*Shravani* : A hybrid inspired by the popular open-pollinated Barkha Bahar. The development ensures no premature bolting, thus giving flexibility of earlier sowing. The leaves being glabrous make this hybrid an excellent cooking material. Roots have excellent white skin with no neck - colour, tender and juicy texture with mellowed test. Approximately 35-40 cm long and 4.5-5.0 cm diameter. Quite tolerant to pithiness and forking. Uniform early tapering roots. Maturity in 40-45 days.

*Lalima* : As the name suggests this is a red skinned hybrid with white core, fine texture and high pungency. Highly tolerant to premature bolting. Roots on an average are 30 cm long with white root end and 4.5 cm diameter. Maturity 50-55 days from sowing.

***Gulnar*** : A hybrid developed in 'PINK' class. Almost pith free and does not bolt prematurely. Leaves smooth devoid of pickly hair. Very attractive smooth pastel - pink skin with pure white core, tender and succulent and mild in taste. Approximately 40-45 cm long and 5.5-6.0 cm diameter. Uniform medium maturing roots are ready for picking in 45-50 days of sowing.

***Mino Early Long White*** : Leaves are serrated, dark green, shiny with few hair. Roots are pure white, straight, grows upto 45-60 cm in length. Flesh is also pure white with pungency. Strong against yellow wilt and viruses. Best selling summer and fall radish.

***Long White (F1)*** : It can grow in both highland and low lands. Excellent quality, uniform appearance with white colour skin with upright and dark green foliage. Root 600-800 g each, highly uniform and 22-26 cm long with firm and delicious flesh. Tolerant to heat and drought, suitable for spring, summer culture, best suited for good draining soil. Resistant to Downy Mildew, Fusarium yellow and other common radish diseases. Mature in 45-50 days and good for fresh market.

***Longo (F1)*** : Slow bolting hybrid with excellent root enlargement at low temperatures. Very smooth root skin and soft flesh. Highly uniform roots of 35-40 cm long, 7-8 cm in diameter and 1.0-1.2 kg in weight. Resistant to fusarium wilt disease. Very popular in fresh market.

***Mino Early Long White*** : Leaves are serrated, dark green, shiny with few hairs. Roots are pure white, straight, grows up to 45-60 cm in length. Flesh is also pure white with pungency. Strong against yellow wilt and viruses. Best selling summer and fall radish.

***Rapid red white topped or French Breakfast*** : This is a globular variety which becomes ready in 26 days after sowing, most suitable for kitchen garden. The globular portion is red and tip is white. It is temperate type and is good for winter season sowing.

***Hisar Muli No. 1*** : It is developed at H.A.U., Hisar and is more popular in Haryana. Its roots are straight, white and ready to harvest after 50-55 days of sowing. It is sown from September to October.

***Jaunpur White*** : It is more popular in Jaunpur district of Uttar Pradesh. Roots are sometimes 1.0 to 1.5 metre long, 15 to 20 cm in diameter, 15 - 20 kg in weight, white in colour, soft and mildly pungent.

It is more suitable for vegetable and pickles. The roots are ready to harvest at 75-85 days after sowing.

***Shogunmino*** : This F1 hybrid derived from the popular Mino Early variety. Roots have beautiful white skin, very crisp and juicy texture. Produces very uniform long roots about 30–40 cm with uniform maturity. Very slow in pithiness development, can grow well under high temperatures. Avoid growing in peak winter of north India.

***Improved Chetki*** : Crop mature in 40 - 45 days after sowing, roots straight and tapering, blunt at tip, colour pure white; no green neck, length 20 - 25 cm and late in pithiness. In North India, it is suitable for planting from mid-March to end of August. Avoid planting under low temperature.

***Improved Japanese White*** : It mature in 60 days after sowing, root shape straight, cylindrical and stump rooted, colour white, 30–35 cm long, very late pithiness, roots are very attractive, clean and with less fibre and suitable for planting in October – November.

***Summer Queen*** : Crop maturity is 55 days after transplanting, roots straight and tapering, white coloured, 30 cm long, late in pithiness. Performs best in summer and rainy season.

***Komal*** : It mature in 50 days after sowing, roots cylindrical and stump, white, upper part sometimes has greenish tinge, length 20–25 cm, very late in pithiness. Specially developed for areas where leaves are eaten, roots are less pungent, leaves soft and palatable, suitable for planting from February to May.

***Early Long White*** : It comes to maturity at 40-45 days after sowing roots are straight and tapering, pure white colour, 35-40 cm long, late in pithiness, suitable for growing in spring - summer and autumn - winter seasons and good heat tolerant.

***Hill Queen*** : It mature in 55–65 days after sowing, roots straight, slightly tapering, lustrous white, 40–45 cm long, very late in pithiness, suitable for planting in October – November. Roots have very little fibre, fetches premium price in the market.

***Himani*** : It comes to maturity in 55–65 days after sowing, root shape straight, tapering, colour pure white, length 25–30 cm, very late pithiness, suitable for late planting in North India.

***Long Red*** : It matures in 50–55 days after sowing, root shape tapering, colour red, length 25–30 cm, pithiness medium late, performs well under high temperature, high humid conditions suitable for sowing from May to August.

***Chinese Pink*** : It matures in 55 days after sowing, roots stump shape, colour pinkish red, 20–25 cm long, late in pithiness, suitable for planting from June to September.

***Long white (F1)*** : It can be grown in both high land and lowlands, excellent quality, uniform appearance with white colour skin with upright and dark green foliage. Root weight 600 - 800 g, highly uniform at 22-26 cm long with firm and delicious flesh. Tolerant to heat and drought, suitable for spring, summer cultivation. Mature in 45-50 days. Good for fresh market.

***Longo (F - 1)*** : Slow bolting hybrid with excellent root enlargement at low temperature. Very smooth root skin and soft flesh. Resistant to fusarium wilt. Highly uniform roots of 35-40 cm long, 70-8 cm in diameter and 1.0-1.2 kg in weight. Very popular in fresh market.

***Nadauni*** : This is another variety which is popular in Himachal Pradesh. The roots are long, tapering, smooth and light pink in colour. The tops are dark green with cut leaves. Developed from a local material.

***Chinese pink*** : Roots 12-15 cm long, semi stumpy to stumpy. Pink with white colour towards the tip; takes about 50 - 55 days from sowing to root formation. An introduction identified by Dr. Y. S. Parmar University of Horticulture and Forestry, Nauni (Solan). It is Asiatic type variety.

## RUTABAGA

Rutabaga (*Brassica napus* var. *napobrassica* Petem); 2n = 38, amphidiploid; belongs to the family Brassicaceae; is also known as Swede or swedge turnip or lapland turnip. They resemble giant turnips, except they have a long, leafy neck, smooth, bluish-green leaves and huge roots. Their roots often grow half way out of the ground. It takes a long time to mature. It is believed to be the result of a natural hybridization between turnip and cabbage. Rutabaga is biennial in habit but grown as an annual. Leaves are bluish, thicker and smooth. It produces a large, hard and swollen root, elongated or round in shape. The skin of root is yellow or white in colour.

Rutabagas are higher in total dry-matter and total digestible nutrients than turnips. Roots are good source of vitamins A and C and are very high in potassium content. Steamed roots are mash with butter. The enlarged and elongated roots are used in soups and stems. It is used as side vegetable as well as mixed.

## Varieties and Hybrids

There are white fleshed and yellow-fleshed varieties of rutabaga which are commercially grown in different regions.

1. *Yellow fleshed* : These types are usually preferred and commonly grown for commercial purpose.
2. *White fleshed* : These types are also grown but not so common. *Laurantiana* is one of the widely and commonly grown variety.

# TURNIP

Turnip; *Brassica rapa* L. (*B. Campestris* var. rapa); 2n = 2x = 20; syn. Salgom, Shalghum, Shalgam, Salgum,Gonglu, Shaljum, Thipperseema, Mullanki, Turnip, Turnip - Kannad, is a popular root vegetable in north India and is used as salad, cooked or pickled. It belongs to the family Brassicaceae; growth and flowering of turnips are similar to that of other cruciferous vegetables. Flowers are hermaphrodite and highly cross pollinated. It is herbaceous annual grown for the root production but biennial for the seed production.

## Varieties and Hybrids

There are two distinct groups of turnip (i) biennial or temperate or European type; (ii) annual, tropical or Asiatic type. The seeds of former group can be produced in the hills and plains

Turnip cultivars have also been classified into several groups on the basis of morphological characteristic i.e. shape and colour of the root and top.

***Classification:*** Turnip cultivars are classified as : (1) White Fleshed, (2) Yellow Fleshed and (3) Foliage Cultivars.

1. ***White Fleshed :*** It has three different types : (a) Purple topped, (b) Green topped, (c) White topped.

   (a) ***Purple topped :*** It has two types – (i) Flat type, var. purple top Milan, (ii) Globe type, var. Purple Top White Globe.

(b) ***Green topped :*** It has two types (i) Globe type, var. Green Top White, Green Globe, (ii) Long type : var. Cow Horn.

(c) ***White topped :*** It has four types :

(i) Flat type; var. White Milan, White Flat Dutch, Norfolk.
(ii) Globe type; var. White Stone, Quick Silver.
(iii) Half - Long type; var. White Egg, White Gem.
(iv) Long type; var. Lily White.

2. ***Yellow Fleshed*** : It has three different types (a) Purple topped, globe type: var. Aberdeen Purple Top. (b) Brodge or green topped, globe type : var. Aberdeen Green top, Amber Globe. (c) Yellow topped, globe type : var. Yellow Globe, Golden Ball, All Seasons.

3. ***Foliage cultivars*** : Flat Japan, Shogoin, Seven Top.

***Variety suitable for saline and sodic soil :*** Sel 66, Snow ball, and purple top white globe.

Cultivars of both temperate or European and Asiatic types are now available in India.

**Brief description of improved varieties are given below :**

## Temperate or European Varieties

***Pusa Chandrima (Asiatic x European)*** : It is early maturing with high yield and possess all the good qualities of the European type. Roots are medium to large in size, 8-9 cm in length and 9-10 cm in diameter. The tops are medium but not so deeply cut. The skin is smooth, pure white, fine grained with sweet and tender flesh. The root attains maturity in 55 - 60 days after sowing. It is temperate or biennial cultivar and does not produce seeds in the plain of India.

***Pusa Swarnima (Japanese white x golden ball) :*** It matures 15 days earlier than Golden Ball and yields 40% more. The roots are flattish round, 6-7 cm long and 7-8 cm diameter with creamy yellow skin, flesh amber coloured, fine textured and mild flavoured. Tops are medium and not very deeply cut. This gives more seed yield than Golden Ball. It matures in 65-70 days.

***Golden Ball*** : The roots are perfectly globe - shaped, medium sized and smooth. It has bright creamy yellow skin and pale, amber coloured flesh of fine texture and flavour. The top is small erect and with cut leaves.

***Snowball*** : This variety is an early temperate type, with medium sized small top. Leaves are small and light - green. Roots are round and medium sized, flesh light - yellow, finely grained, sweet and tender. Roots are ready to harvest after 60-65 days of sowing.

***Purple Top White Globe*** : It is a high yielding large rooted cultivar. The roots are nearly round, upper part purple, lower portion creamy, flesh is white, firm, crisp and mildly sweet - flavoured. The top is small, erect with cut leaves. It is adapted to cooler months.

***Early Milan Red Top*** : It is an extra early and high yielding cultivar reaching maturity in 45 days. The roots are deep flat with purplish - red tops and white underneath. The flesh is pure white, well - grained, crisp and mildly pungent. The tops are very small with 4-6 sessile leaves.

## Tropical or Asiatic Varieties

***Pusa Sweti*** : It is an Asiatic cultivar suited for sowing in October. It has attractive white roots. It produces seeds in the plains of India. It matures in 45-50 days after sowing and yields 20-30 tonnes/ha.

***Pusa Kanchan (local red round x golden ball)*** : It has good qualities of both the types. The skin is red, but the flesh is creamy yellow with excellent flavour and taste. The leaf top is shorter than the Local Red Round. It is earlier by 10 days than Local Red Round, but the roots can be kept longer in the field without being spongy. It sets seeds in the plain.

***Pubjab Safed –4*** : This variety is early maturing , commonly grown in Punjab and Haryana. The roots are attractive, medium size round and white. The taste is good.

***Turnip L1*** : The root is round, pure white, smooth, rattailed and crisp with mild flavour.It can be harvested within 45 to 60 days after sowing. It yields about 262 quintals/ha. It was developed at Ludhiana.

***White - 4 :*** It is an early maturing deshi type. Its roots are white, round and medium and developed in about 60 days. The average yield is about 200 q/ha. It was released by the Department of Agriculture, Punjab.

***Red - 4 :*** It is also an early maturing deshi type. Its roots are red and medium. The roots are ready to harvest after 55-60 days of seed sowing. It yield's about 200 q/ha, It was also released by Department of Agriculture, Punjab.

□□□

# 4

# Bulbs and Tuber Crops

## AMORPHOPHALLUS

Amorphophallus (*Amorphophallus paeoniifolius*, syn. *A. campanulatus* (Roxb.)Blume); 2n = 28; also known as Elephant foot yam, Suran, Zamikand, belongs to the family Araceae, is a remunerative and profitable stem tuber crop. The crop is gaining popularity due to its shade tolerance, easiness in cultivation, high productivity, less incidence of pests and diseases, steady demand and reasonably good price.Tubers are mainly used as vegetable after thorough cooking. Chips are made of starch rich tubers. Tender stem and leaves are also used for vegetable purpose.

Cultivated species *A. paeoniifolius* is a robust herbaceous plant 1.0-1.5 m tall. Large dissected tripartite leaves constitute the luxuriant outspreading crown like foliage, borne on a fairly thick single upright stem. The aerial pseudostem which is round with characteristics irregular blotches is botanically a leaf petiole.

Amorphophallus is a tropical/subtropical, perennial crop due to underground stem, it is treated as an annual with a duration of 8-9 months. It is propagated through corm.

The genus *Amorphophallus* comprises about 40 species of these, *A. campanulatus* (Roxb) Bl. is the species cultivated in India. Very little genetic variableity is available in the crop. It is cultivated as a rainfed crop as well as irrigated as practised in East and West Godavari District of Andhra Pradesh.

## Varieties and Hybrids

Elephant foot yam is two types.

(1) One has smooth corms propagated by cutting the corms into pieces- this variety causes irritation of the throat when eaten and yet is grown because of its heavier yield and lower price of seed.

(2) The other kind has daughter corms growing on the parent corm and is much superior to the former type in edible quality and is almost non-irritating if cooked properly. Its seed is more expensive and its yield is less but the produce fetches a better price in the market.

Few improved cultivars have been released for commercial cultivation in the country, which detail characteristics is given below:

***Gajendra (kovvur)*** : It is a high yielding cultivar with vigorous growth habit. Corms are smooth, without acridity and have better quality of flesh. The corms growth is uniform but it does not produce daughter corms. Corm yield exceeds 100 tonnes/ha., duration 210-260 days.

***Santragachi*** : Plants are vigorous in growth and produce corms having slight acridity. The corms are rough with light cream colour and bear several daughter corms which take about 3-4 growing season to attain the marketable size. The cultivar is suitable for eastern India and yielded about 50-70 tonnes per hectare, depending on the size of planting materials.

***Bidhan Kusum*** : This variety has been developed through selection of local germplasm from East Godavari district of Andhra Pradesh. It is a non-irritant, smooth corm type and can be grown successfully twice a year. It is mostly single corm bearing variety. Corm is globose with depression at the top, having 40-60 cm girth, weighing 5-8 kg depending upon seed corm size. Flesh colour is cream and bulking rate 400-800 kg/ha/day. It has out yielded Gajendra by 19.67 %. Green petioles having emerging folded leaves are highly delicious as vegetable, during green harvest corms, the pseudostems are also used as animal feed chopped with paddy straw.It is superior in corm yield (65 t/ha) excellent cooking quality and in build ability to tolerate major insects-pests and diseases.

***Sree Padma*** : It is developed at CTCRI, Thiruvananthapuram, yield 42 t/ha.

***Thiya Kanda*** : It is a local variety of Andhra Pradesh cultivated in East and West Godavari district as irrigated crop and give high yield.

***Dorada Kanda*** : It is upland variety of Andhra Pradesh, cultivated as rainfed crop and more acrid also give very high yield.

***Karuna Kilangu*** : A popular local genotype of south Arcot and Tanjore districts of Tamil Nadu also come under the species *A. campanulatus,* though the corm shape and size is much smaller.

***C 3*** : This culture (Accession) collected from Trivandrum and tested five years at College of Agriculture, Dholi (Bihar). It produced good quality corm and gave more yield of 38 t/ha.

***C4*** : This culture (Accessions) collected from Hyderabad and tested five years at College of Agriculture, Dholi (Bihar). It gave higher yield of 42.5 t/ha alongwith good quality corm.

Some accessions, such as AM 15, AM-27, AM-43 and AM-51 have been identified for good yield potential (40-60 t/ha) and culinary quality.

## ARROWROOT

Arrowroot (*Maranta arundinacea* L.); 2n = 48; belongs to the family Marantaceae; popularly known as West Indian arrowroot, is a lesser known under utilized promising tuber crop with immense applications in food, feed and industrial sector. It is a versatile crop, which can adopt relatively quality fit well in diverse cropping systems.It can produce high yield of acceptable products and generate economic returns. Arrowroot is mostly cultivated by tribal and marginal farmers for the rhizomatous edible tubers, which contain good quality starch. Its rhizomes are valued as food stuff, particularly for infants. It is also used for making various bakery products, especially glue and paste as a base for face powder, as an ice cream stabilizer and in carbon less paper used for computer printouts. Its starch possesses demulcent properties and is used in the treatment of intestinal disorders. Its rhizomes are also a good substitute for maize in broiler rations. The fibrous debris after starch extraction is used as a feed or manure.

Arrowroot does not set seeds and is normally propagated vegetatively from rhizome apices (bits); sometimes suckers are also used as planting material just after harvesting of the crop. The crop attain maturity 10-11 months after planting. Maturity is indicated by yellowing, wilting and drying up of leaves. Under favourable conditions rhizome yield of 20-25 tonnes/ha can be obtained. Intercropping arrowroot with coconut, arecanut and rubber are also profitable. The amount of starch obtained is 16-18 % of fresh weight of rhizomes.

### Varieties and Hybrids

Generally yellow coloured local cultivars are grown. However, cultivars having blue rhizome give higher yield of starch than yellow coloured cultivars.

## BUNCHING ONION

*Antilla Red (Allium fistulosum)* : It has strong flavour and well appreciated for cooking. It is resistant to the pink rot disease and Alternaria.

*DUE* : It has wide adaptability with a strong root system. F1DUE grows very vigorously under hot and cold conditions with its heavy rooting system. Dark green colour would remain even under harsh weather conditions such as cold temperatures. Very firm neck with about 40 cm glossy white part. Excellent for export. It can grow all the year round in tropical conditions.

## CHINESE POTATO

Chinese potato (*Solenostemon rotendifolius*; syn. *Coleus parviflorus*); 2n = 56, 64, also known as coleus; belong to the family Labiatae, is a seasonal crop cultivated for its edible tubers, which resemble miniature potato and has a characteristics flavour and are used as vegetable after cooking. Plant is a bushy herbaceous annual with succulent stem and around ovate petiolate leaves having characteristic aroma. Coleus is a vegetatively propagated crop so variability is limited.

## Varieties and Hybrids

There are usually two types of coleus. One is having small sizes tubers and the other one large-sized. The variety having small tubers but good flavour. Wehreas large tuber type produces higher yield.

***Shree Dhara*** : It is an improved variety. It yields about 20-25 t/ha over a 6 month period. It is released by the CTCRI, Thiruvananthapuram in Kerala. In other places local types are grown.

***Nidhi*** : This is developed by Kerala Agricultural University. Under Kerala condition crop comes up well during July to October and yield 27.9 t/ha in 120-130 days. Individual tubers are oblong shaped with an average length of 25 cm. Average tuber weight is 8.3 g.

## CHIVE

Chive (*Allium schoenoprasum* L.); 2n = 16; belongs to the family Liliaceae; is used as a spice since the early middle ages in Europe. The leaves of chives are used as appetizing condiment in soups, cooked meals, salads and also as sandwitch filling. The spicy flavour of chives is caused by the content of garlic oil. It contains relatively high amounts of vitamin C , carotin and calcium.It belongs to the permanent cultures and grows on practically all soils, which are not too poor or dry.

The genus *Allium* is recognized to have more than 600 species. Besides *Allium schoenoprasum,* there are other species important for human nutrition in the section *schoenoprasum,* like *Allium cepa* (Onion), *Allium fistulosum* (Welsh onion), and *Allium ascolanicum*(shallot). The natural habitats of *Allium schoenoprasum* are Asia, Europe and Northern America including the Arctic. Presumably, it come from central Asia or the Mediterranean and spread across Europe up to the 70$^{th}$ degree of latitude, across middle Eastern and Northern Asia as well as across Northern America up to the Arctic area as a wild plant. The cultivated form appears to have been domesticated on numerous occasions.

*Allium schoenoprasum* is a bulbous, self compatible, perennial herb. It is propagated by clone by the production of daughter bulbs along short rhizomes as well as sexually by the reproduction of seeds. The latter is preferred for commercial cultivation.

### Varieties and Hybrids

Chive can easily propagated by clone and also by seeds. Large numbers of cultivars/varieties have been developed by Europe and America and maximum number of accessions are maintained by Czechoslovakia, Germany, Hungary, India, Russia and USA.

## COLOCASIA

Colocasia/Toro is also known as Kachu, Kachalu, Saru, Alooka, Dasheen, Great-leaved caladium, Elephant ear, Chemadusmpalu, Sambu, Sapan, Kizhangu, Alu, Kanuad, Kesavina gedde, Alvi, Alavu, Arvi, Ghuiya, Ghuiyan; belongs to the family Araceae; is a herbaceous plant consisting of a main corm from which cormels, roots and shoots arise. Corm is botanically a stem and is covered with concentric rings of leaf scars and scales. Cormels arise from lateral buds present under each scale or leaf base. Morphologically main corm is the stem and stolons and carmels are lateral branches of stem. It is a cross pollinated due to protogyny. Colocasia and xanthosoma together are known as cocoyam. In colocasia, leaves are peltate and sterile appendage is at top of spadix. In xanthosoma, leaves are segitate and sterile appendage is absent.

*Colocasia esculenta* (Syn Dasheen), 2n = 28, 42 are two types :

*C. esculenta* var. *esculenta* : Produces a large main corm with a few cormels. This includes wild types and is grown under a wide range of moisture conditions from flooded to rainfed upland conditions. This group has 28 chromosomes and are diploids. Leaves are generally elongated. Sterile appendage of spadix is shorter than the male part. Eddoe types with cormels bigger than mother corm belong to this group. Naga Kaju of Nagaland, Khasi Bhunga of Meghalaya, Kovur of Andhra Pradesh, local cultivars like Panch Mukhi, Koni Kachu and White Gauriya belong to var. esculenta.

*C. esculenta* var. *antiquorum* : This includes both cultivated and wild forms. It has small or medium sized main corm and a larger number of side cormels which are edible. It is raised as an irrigated or rainfed upland crop. Plants have 42 chromosomes and are always triploids and do not set seeds. Leaves are almost round. Sterile appendage is larger than male part.

Several varieties/cultivars/cultures differing in colour, shape, size and acridity of tubers and colour, size and shape of lamina and petiole are found in India.

## Varieties and Hybrids

In Colocasia there are two varieties - *C. esculenta* var. *esculenta* and *C. esculenta* var. *antiquorum*. In variety antiquorum, the axillary buds develop into long slender areal or subterranean stolons which develop shoots at their tips. In variety *esculenta,* there are two types; the "Dasheen" and the "eddoe". In "Dasheen" type, the side tubers may be reduced or absent and the mother tuber alone stores the food and swells up. In "eddoe" type, the side tubers are as big as the mother tuber or little smaller and plenty in number and are used as propagation materials for the next season. In this group, normally the side tubers (cormels) are edible.

**Details Characteristics of the varieties and hybrids are given below:**

***Kandyam C7*** : It is a selection from local collection of Arunachal Pradesh. The tubers are long with more girth and more weight. Yield potential is 45 t /ha with average yield of 35.6 t /ha.

***Panch mukhi*** : The world panch mukhi means a type of colocasia with five mouths or five tubers attached together. It can be kept for more than a year after harvest. It is grown successfully in the hills above an elevation of 1,200 m and has very good cooking quality.The yield varies from 25 to 40 t/ha.

***Koni kachu*** : Colocasia producing many daughter corms. It is very good for currying and has a good cooking quality. Yield 12-25 tonns/ ha.

***Naga Kaju*** : Colocasia with bigger sized mother tubers and small daughter tubers. Yield 12 to 28 t/ha. It is a Negaland variety.

***Khasi Bugga*** : It contains more acridity as compared to other varieties. Yield 14-16 t/ha.

***Sahasra mukhi*** : A collection from Maharashtra was found to be superior in yield in evaluation studies conducted at Dholi (Bihar). Cormel yield of 18 t/ha was recorded.

***Telia*** : It is early and good for summer cultivation, duration 5-6 months, corms are black in colour.

***Muktakeshi*** : It has very good cooking quality and tolerant to leaf blight. Duration 5-7 months.

***Sree Rashmi (C-149)*** : The plants are tall (1-1.5m), erect with medium broad leaves, and purplish green petiole and sheath. The corms are big and cylindrical and cormels medium sized, yield being 15-20 t/ha. Duration 7-8 months. All plant parts-leaf, petiole, corm and cormels are non-acrid and edible. The cormels have good cooking quality and excellent taste. The starch content is about 15 % and protein 2.5 % on fresh weight basis.

***Sree Pallavi (C-266)*** : 'Sree Pallavi' is a variety identified from the germ plasm collection of indigenous origin. Plants are tall (1-1.5 m), green petioled and possess comparatively bigger sized corms and small cormels, plenty in number (20-25/ plants). Cormels have good cooking quality with bluish, starchy and tasty pulp. Duration 6½ - 7 months and produces 15-18 tonnes/ha of corms and cormels with 16-17 % starch and 2-3 % protein content on fresh weight.

***Jhankadi*** : It has good cooking quality, tolerant to leaf blight. Duration 5-7 months.

***Topi*** : It is tolerant to drought condition and possess good cooking quality. It also tolerant to leaf blight.

***Mana Saru*** : It is tolerant to leaf blight. High yielder. Duration 6-7 months. It is a low land variety.

***Dasheen* type**

***Kaka Kachu*** : It is promising variety which produces side corms which are fleshy with better taste. The yield varies form 22 to 40 tonnes per ha.

***Munshikuntia Kachu*** : It is just like wooden pegs in size and shape. It is very soft and highly palatable. It has calcium oxalate. Yield varies from 22 to 40 tonnes/ha.

***Shankar Saru*** : It has big size corms, white in colours good texture, good cooking quality and high yielder.

***Bilaspuri Arum*** : It is a dasheen type taro. Leave and petiole green. Mother corm smaller with more soft cormels of roundish and uniform size each weighing 50-100 g. The flesh corm is white with low acridity, can be easily got rid of by boiling. Yield up to 30 t/ha.

In Hawaii taro varieties are classified as low land taro and upland taro varieties.

***Low land taro*** : Lehua, Mali, Big lehua, Moi, Api, Maokea, Pikokea, Piialii and Bunlong.

***The upland taro*** : Varieties are Bun-long, Mana ulu, Mana opelu, Eleele, Makoko.

Colocasia or taro is also classified according to usage. Poitaro varieties are sticky type and poi, which is basically a paste is prepared from them Varieties lehua, Maoli, Big lehua, Api, Piko kea, Moi and Piialii. Table taro and chips taro varieties are grow principally for table use or cooking. For taro chips Bun-long is used. The corms can be sliced and fried immediately without pre cooking. Samoan taros are also used for cooking.

***Azad Arvi 1*** : The Azad Arvi 1, a new varietyof colocasia has been released. It mature 15 days earlier than local variety. Its edible corms become ready for picking in 135 days.

***Jhangri*** : This variety is identified suitable as rainfed crop in Orissa and it is cultivated locally.

***Thamarakannan (C-9)*** : It is a popular variety cultivated locally in Kerala for the mealy big cormels (20-40 g).

***Sree Kiran*** : This is a F1 hybrid with high yield and good cooking quality.

***Kannanchembu*** : It is a famous local variety with small tubers of excellent quality and has characteristics of both esculenta and antiquorum.

Some local genotypes are also used in different parts of the country.

***In North India*** : Sel. 3, Sel. 11, Goraiya, Faizabadi, Banshi, Ladhora, Deshi Banda.

***In Central India :*** Worm 1, Worm 2, Gyano Nos 12, 22, 36 and 40.

***In South India*** : Sarkachu, Gurri Kachu, Asu Kachu, Kalla Kachu, Char Kachu, Satmukhi, Shree Palvi and Sree Rashmi.

***ML1*** : It is a selection from local germplasm of Meghalaya. Its leaf colour is light green while petiole is purple green. It consists of medium sized corms (length 10-12 cm and diameter 5-6 cm) and small to medium-sized cormels (length 6-7 cm and diameter 2.5-3.0 cm). On an average, it gives 25-27 tonnes/ha. It is moderately resistant to leaf blight.

***ML2*** : A selection from local germplasm of Meghalaya, its plants grow 60 cm high. Leaf length is 25-30 cm while breadth 20-25 cm. It has medium-sized corms (length 10-12 cm and diameter 5.6 cm) and 7-8 small to medium sized cormels. On an average it gives yield of 20.0-22.5 t/ha. It is moderately resistant to leaf blight.

*ML 9* : It is also a selection from local germplasm of Meghalaya. Both leaves and petiols are light green. It has medium - sized corms (length 8-10 cm and diameter 6-7 cm) and small to medium-sized cormels (length 5.5-6.5 cm and diameter 3.0 -3.5 cm). It gives 20-22 t/ ha yield. It is comparatively tolerant to leaf blight.

***Vallabh Nikki*** : Its plants grow 38-42 cm high. The length of leaf is 29-30 cm and width 18-20 cm. The number of leaves is 4-6 while corms 24-26. Vallabh Nikki has been developed by selection from germplasm, maintained at SVBP Univ. of Agric. and Technology, Meerut. It is high yielding (281.05 q/ha) and resistant to leaf blight. It is grown successfully in plains as well as hills. This is superior to other varieties in yield and well accepted by farmers due to high yield and other desirable characters. Thus, Vallabh Nikhi is superior to Azad. Arvi 1 (control). It yielded better at Meerut, Saharanpur and Jhansi.

Varieties Bangali Banda, Desi Banda, Faizabadi, Bansi, Lathara are very common in North India. Three wild varieties, commonly known in Bengal as Kalla Kachu, Char Kachu and Ban Kachu are more valued for their leaves and stalks than for their tubers. Some recommended and local varieties of Colocasia are Guri Kachu, Asu Kachu, Sar Kachu (good for aquatic situations for Bengal). No.22,12,36,40 and warm 1 and 2 (Maharashtra), S-3 and S-11 (for Punjab).

## GARLIC

Garlic (*Allium sativum* L.), 2n = 2x = 16 ; syn. Lahsoon, Poodu, Lassan, Veluthull, Lasun, Rashun, Nohoyo, Vellulli,Lason, Shumandha, Bhutagna, Mahasuda; belongs to the family Alliaceae; is vegetatively propagated crop. Mostly local cloves are used on the basis of colour. The underground bulb consists of 10-50 segments called cloves and are enclosed in a thin sheath of white or pink colour.

### Varieties and Hybrids

Two types of varieties are available depending on their colour i.e., red and white. The white coloured varieties are more popular as they are less pungent. Red coloured varieties are mostly used for medicinal purposes. They are more pungent as compared to white varieties.

Amongst white varieties some produce small-sized bulbs while some of them produce big-sized bulbs. The different varieties of these groups are as below :

(1) ***Small sized varieties*** : T 56-4, Creole, Italian and Tabiti

(2) ***Big-sized varieties*** : Fawari, Rajalle Gaddi, Solan

Some others are Madrasi, Jamnagar Local, Ladwa, Aru, Sweta etc.

Varieties developed by different Institutions/Organizations are given below :

***MPKV, Rahuri*** : Godavari and Sweta

***CCSHAU, Hissar*** : HG-1 and HG 6.

***IARI, New Delhi*** : Pusa Sel.10.

***PAU, Ludhiana*** : LCC 1.

***VPKAS, Almora*** : ARU 52.

***NHRDF*** : Agrifound white (G-41), Yamuna Safed (G1), Yamuna Safed 2 (G 50), G 282 and Agrifound Parvati.

***Acclimatise varieties*** : Early White or Medican varieties, Creole and Tatule and Italian.

**Detail characteristics of varieties/hybrids are given below :**

***Fawari*** : It has big size bulb with white colours and compact cloves. It is grown in Bellary region of Karnataka and high yielder with good keeping quality.

***Rajalle Gaddi*** : It is grown in Karnataka and A.P. It gives high yield and has good keeping quality. It has medium size bulb with more pungency.

***G-6*** : It has been found to be tolerant against soil salinity.

***HG-7*** : It grows tall with high number of leaves/plant, best yielder with high number of cloves/bulb and weight of bulbs and cloves.

***Godawari*** : It is good yielder but premature sprouting in the field is very high.

***Madrasi*** : It is a desi/local variety and is very popular in Rajkot division of Gujarat and also a very high yielder.

***Creole*** : It is earlier than Italian, has a flatter broader leaf a greener top, a shorter stem, fewer wrapper leaves and larger cloves. It is a heavier yielder, but its keeping quality is somewhat inferior.

***Jahiti*** : Have the largest cloves and also are the darkest.

***Agrifound white (G-41)*** : The bulbs are compact, silvery-white skinned with creamy flesh. Diameter of bulb is 3.5-4.5 cm and number of cloves/bulb is 20-25. The variety is susceptible to purple blotch and stemphylium blight which are common in northern India. TSS 41 %, dry matter 43 % and good storer. Average yield is 73 tonnes/ha.

***Yamuna Safed (G-1)*** : The bulbs are compact, silvery white spinned with 25-30 sickle shaped cloves and creamy flesh. Diameter of bulb is 4.0-4.5 cm. It is tolerant to insect pests and diseases like thrips, purple blotch and stemphylium blight. Total sluble solids (TSS) 38 %, dry matter 39.5 %. The average yield is 15.0-17.5 tonnes/ha. It is recommended for cultivation all over India.

***Yamuna Safed-2 (G-50)*** : The bulbs are compact, attractive with 35-40 cloves/bulb and white creamy flesh. Bulb diameter is 3.5-4.0 cm; TSS 38-40 %; dry matter 40-41 %. The average yield is 15-20 tonnes/ha. The variety is recommended for northern parts of India.

***Yamuna Safed-3 (G-282)*** : The leaves are wider than other varieties. Bulbs are creamy-white and big size (5-6 cm diameter), having 15-16 cloves/bulb. Total soluble solid 38-42 %, dry-matter 39-43 %, medium storer, yield 17.5-20.0 tonnes/ha; suitable for export purposes. It performed very well in northern and central India.

***Agrifound Parvati (G -313)*** : A selection from the material collected from Hong-Kong market. It is a long day type and as such is suitable for cultivation in hill of northern states. Bulbs are of bigger size (5-6 cm diameter), creamy-white with Pinkish-tinge, 10-16 cloves/bulb, tolerant to common diseases. Average yield 17.5-22.5 tonnes/ha. It is medium storer and suitable for export.

***Ooty 1*** : Bulbs bigger in size with dull white in colour, 20-25 cloves/bulb, yield 17 t/ha in 125-130 days. Resistant to thrips.

***Pant Lohit*** : Moderately resistant to purple blotch. Cloves have pink tinge colour.

***T 56-41*** : This variety bear medium size bulbs and is white in colour.

***Jamnagar*** : It has bigger size white colour bulbs with compact cloves.

## GIANT TARO

Giant Taro (*Alocasia indica* (Roxb.); 2n = 28 ; belongs to the family Araceae; the fleshy aerial stems of the giant taro is the primary product. The stem tuber is cut into pieces and eaten as a vegetable after cooking, usually in curries or stews. Giant taros are sometimes used as a source of very white easily digested starch or flour. The underground stems are also used for food after thorough cooking. The leaves may be used for medicinal purposes in India and Pacific Islands. It is found naturally along river banks, but can not withstand water-logging. Suckers are commonly used as planting material, but shoot tips with a few inches of stem and rolled up young or sections of stem having two or three buds are also frequently used.

The species is essentially tropical and three species, *A. indica, A. macrorhiza* and *A. cucullata* are cultivated in Assam and West Bengal as food crop. Crop life is usually 12-18 months. The plant corm remains in the ground for about 3 months after reaching maturity without any deterioration.

### Varieties and Hybrids

Giant Taro shows a narrow genetic variability. A few numbers of local strains are popular with the growers. They are having the corms of different shape and size and used for cultivation. There is no named variety and locally available materials are used for seed purpose.

## GREAT HEADED GARLIC

Great-headed garlic (*Allium ampeloprasum* L. var. ampeloprasum); 2n = 16, belongs to the family Alliaceae, resemble robust types of *A. sativum,* but the exterior bulblets are never present in garlic. Scape round,solid, upto one metre or more in height, not curled or curved; inflorescence large and globular; spathe, which has a long beak, is shed at maturity. It is distinguished by the lage leek like umbel, lacking bulbils, and the small bulblets round the main bulb. The clones are relatively sterile hexaploid and are propagated vegetatively. The bulbs have a flavour intermediate between onion and garlic. It is grown to a limited extent in the Mediterranean region, Europe and America, but always as a home-garden plant.

### Varieties and Hybrids

A number of named clones are grown in different countries. Many clones named behind the name of regions and locality of production.

## LEEK

Leek (*Allium porrum* L.; syn. *A. ampeloprasum* var. *porum* L.); 2n=32; belongs to the family Alliaceae is anon-bulb farming crop. It is cultigen derived from the wild *A. ampeloprasum*. It is grown for its etiolated pseudostem formed by the leaf sheath. The economic part is blanched stem and leaves which are highly nutritive and rich in medicinal properties, especially for heart ailments. It can be consumed as salad, in soup and in various cooked foods and can be quite relished for its onion flavour and nutritive value. Since Indians are fond of green onions, therefore, leek thought to be good alternative leaving grreen onions to form bulbs.

The leek has greater cold resistance than the onion; bulbs are not formed. It does not have a rest period, and it is not influenced by photoperiod. Plant is tall hardy biennial with white narrowly ovoid bulbs and broad leaves. Leek is highly cross-pollinated due to heteromorphism and male sterility. It has milder and more delicate flavour than onion, though of coarser texture. In medicine the bulbs are used as arenal stimulant.

### Varieties and Hybrids

In leek many high yielding and promising varieties have been evolved. 'Borde Pearl' is one such high yielding line resistant to bolting. 'Bluevetia Early' is suitable for both summer and early autumn culture. 'Copenhagen Market' and 'Elephant' are short-necked type and 'Gennivillies' and 'St.George' are long-necked types with resistant to frost.

Some other have been developed by pedigree methods are :

***'Miner'*** : An autumn or early winter variety derived from crosses and family selection in the variety 'St. George', has long stem and gives high yield.

***'Splendid'*** : Cultivated throughout the season.

***'Renova'*** : Derived by selection from the variety 'Regions', has darker leaves, longer stalk and less tendency to form bulb and is more uniform in size and shape.

***'Special 050'*** : It is suitable for forcing.

***American Seed men varieties*** : London Flag, American Flag, Elephant or Monstrous Carentan, Giant Musselburgh. Lyon.

***Belgium varieties*** : White Alberta, Blizzard, and Carina are for winter. Acadia, Derrick, and Electra are suitable for autumn cultivation.

***Canadan varieties*** : Longa, Odin, Kilima, Goliath, Siberia, and Artico.

***High yielding variety*** : Carina and Alberta.

***Others*** : Regions Large Roven, Conqueror and Prize Taker.

***Tropica Seeds (Green Co.)*** : France, Large Long Summer.

***Bejo Sheetal Seeds*** : Lincoln.

***Nunhems Seeds*** : Carlton, Roxton, Parton.

Being highly cross-pollinated crop, the prospects of exploiting heterosis is very high. The scope of evolving hybrid is bright and some hybrids have also been evolved. 'Regins 930/57' (St. George x Special), $F_1$ hybrid is good cropper especially in late autumn and early winter.

**Detail characteristics of varieties and hybrids are given below:**

***London Flag*** : Tall with a thick stem, leaves broad, it is most generally cultivated.

***Musselburg*** : It is very long and hardy; leaves broad and tall; stems long and thick.

***Large Roven*** : Its leaves are very dark green, broad and thick. Stem rather short, but remarkably thick. It is said to grow as thick as man's arm. It is found to be the best kind of forcing as it acquires a sufficient thickness of stem sooner than any other varieties.

***Lincoln*** : It matures in 80 days after transplanting. It is early Bulgarian giant type leek, which gives uniform long and thick stems. It is suitable for *rabi* season planting.

***Carlton*** : This hybrid variety is very fast in growth, leaves are erect and green in colour. Long shoft length have excellent uniformity.

***Roxton*** : Long shoft length with excellent uniformity. The growth of this hybrid is very fast with green to dark green irect leaf colour.

***Parton*** : Medium growth of this hybrid produces medium shoft length of excellent uniformity. Leaves are erect and dark green in colour.

***France*** : It shows medium maturity. The leaves are dark-blue green. Shanks are 18-20 cm long and 2.8 to 3.3 cm in diameter.

***Large long summer*** : It is very early type and matures in four months. Foliage, medium green in colour. Plant long, medium sized shanks (20-25 cm long and 2.3 to 3.0 cm diameter), widely adapted and very fast growing.

***Palam Paushtik :*** It has long, broad, flat garlic like bluish green leaves and long white pseudostem thickening cylindrically to attain 3-4 cm diameter at harvesting. It takes about 140-150 days to reach marketable stage. It has mild digestive flavour due to organosulphur compounds. It can be consumed like green onions after cooking with potatoes or in any other form of Indian dishes. Besides, it can be enjoyed raw as salad and in soup. It has healthful digester nutritive substances, especially protein, fibre, carbohydrates, vitamins and minerals. Its average yield is 30-35 tonnes/ha.

## MULTIPLIER ONION

Multiplier onion syn Aggregatum onion (*Allium cepa* var. aggregatum); 2n = 2x = 16, is used mainly for seasoning of curries. Unlike common onion, it is propagated through bulb lets. It comes up well under tropical condition with sufficient soil moisture.

### Varieties and Hybrids

There are few varieties have been developed in this crop by different institution/organizations, which details are given below:

***TNAU*** : Co1, Co2, Co3 and Co4.

***NHRDF*** : Agrifound Red.

***Co1*** : Bulbs large and pink, 7-9 bulbs/clump. Yield 9-10 t/ha in 85 days.

***Co2 :*** Bulbs large and crimson, 7-9 bulbs/clump, yield 12 t/ha in 65-70 days.

***Co3*** : Bulbs pink with a weight of 75 g, 8-10 bulbs/plant, moderately resistant to thrips, good storage quality, yield 15.8 t/ha in 65 days.

***Co4*** : Bulbs attractive light pink, 9-13 bulbs/plant. Average weight 90 g/clump. Yield 19 t/ha in 65 days.

*Agrifound Red* : Each clump weighs 65 g with 5-6 light red bulblets, TSS 15-16 °Brix, yield 18-20 t/ha in 65 days.

## ONION

Onion (*Allium cepa* L.), 2n = 16; Syn. Piyaz, Piaj, Palandu, Payaz, Kanda, Dungli, Gauda, Ulli, Vengagam, Erangayam, Erulli; belongs to the family Alliaceae. The species is divided into 3 groups which are sometimes given the rank of botanical varieties, as follows :

***Allium cepa* L.** : Common onion group

***A. Cepa*** var. *aggregatum G. Don* : Aggregatum group – multiplier onion, potato onion, shallots.

***A. cepa* var. *proliferum*** (Moench), Aschers, syn. *A. Cepa* var. *Viviparum* (Metz) Alef : Proliferum group- tree onion.

### Varieties and Hybrids

Onion varieties are classified on the basis of bulb colour and response to photoperiodism.

***Bulb colour*** : Red Onions (Pusa Red), White onion (Patna white),

**Response to photoperiodism**

*The Bermuda* – Granex – Grano short day type, flat to conical shaped, mild and early maturing group, Texas Early, Grano No.502, Early yellow Cape Flat, Yellow Bermuda, Red creole, Tropicana, El Toro, Bombay Red; late-maturing types, long day type, globe onions (White Globe, Sweet Spinach) and the creole type, with very strong flavour, day neutral types (Ebenezer, White Portugal).

*Export types* : Red Creols, Granex, Bombay Red, Red Kano, Patna Red, Texas Grano, Yellow Bermuda and Early Cape Flat are suitable for export.

Poona Red, Nasik Red, Bellary Red, Patna Red and Patna White are common in onion trade.

Onion varieties differ in size, colour of skin, pungency, maturation etc of bulbs. Large sized bulbs are mild in pungency and are sweet in taste compared to small sized onion. Red coloured cultivars are more pungent than silver skinned varieties and keep better in storage. Yellow cultivars have less demand in the market. The local cultivars are known and marketed after the names of places where they are grown.

Almost all high yielding varieties are having medium to large sized bulb. The name of a variety given with the name of a place where it was found successfully growing viz., Patna Red, white patna, Poona Red, Nasik Red, Yadgiri or Bellary Red etc.

The list of varieties along with developing institutions is given below :

***IARI, New Delhi*** : Pusa Red, Pusa Ratnar, Pusa White Round, Pusa Madhvi, Pusa White Flat, Early Grano.

***IARI Regional Station, Katrain*** : Brown Spanish.

***IIHR, Bangalore*** : Arka Niketan (Sel-13), Arka Kalyan (Sel-14), Arka Bindu (Rose Onion), Arka Pragati.

***PAU, Ludhiana*** : Punjab Naroya, Punjab Selection, Punjab Red Round, S-148.

***MPKV, Rahuri*** : N-2-4-1, N-257-9-1, Baswant-780, N-53 (Nishad 53).

***HAU, Hisar*** : Hisar-2

***UPKAS, Almora, UP*** : VL 3, VL Piaz 67 (F1).

***CSAUA&T, Kanpur*** : Kalyanpur Red Round.

***RAU, Rajasthan*** : Udaipur 101, Udaipur 102, Udaipur 103.

***NHRDF, Nasik*** : Agrifound Dark Red, Agrifound Light Red, Agrifound Rose.

Onion varieties have been divided into two groups according to colour of skin.

***Red colour*** : Punjab Selection, Pusa Ratnar, Pusa Red, Pusa Madhavi, Kalyanpur Red Round, Hisar-2, Punjab Red Round, Agrifound light red, Agrifound Dark Red, N-53, Agrifound Rose, Baswant 780, Arka Bindu, Arka Kalyan, Arka Niketan, Arka Pragati, CO 1, CO 2, CO 3, CO 4.

***Yellow/White coloured varieties*** : Early grano, Pusa White Round, Pusa White Flat, Punjab, 48 (S-48), Udaipur 101, Udaipur 102, Udaipur 103, Punjab Selection, N 257-9-1, Arka Pragati, Hisar 2.

**Varieties resistant to purple blotch (*Alternaria porri*)** : Italian Red and Local Brazilian.

***Varieties resistant to Pink root*** : Granex and Excel 35.

***Varieties good for dehydration*** : Punjab selection and N2-4-1.

***Varieties best for storage*** : Pusa Red and Punjab Red Round.

***Promising hybrids*** : Abundance, Aristocrat, Bananza, Champion and Contender., F1 hybrids developed in India namely purple x red, Purple x white, and red x white have more weight of plant and increased bulb diameter.

***Varieties*** : There are commercially two groups of Onion they are:

***Big sized bulb varieties*** : Red skin, silver or white skin. yellow skin. Big sized have a mild flavour, sweet in taste, less pungent. Big sized are Red Globe, White Globe, Yellow Globe, White Patna, Large Red, Bombay Red, Patna Red, Poona Red, Nasik Red, Bellary Red, Early Grano, Yellow Scaled Onion and Pusa Red; N-53, VL-1, VL-67, VL-3, Udaipur 103, 101, 102, Pusa white flat, Arka Niketan, Arka Kalyan, Pusa Ratnar, N-2-4-1, Gujarat white, N-257-91.

***Medium size*** : Hisar-2, Punjab Red Round, Co-4.

***Varieties with small sized bulb*** : Prize Taber, Market gardner, Yellow Globe, Queen and Paris, Silverskinned; Co1, Co2, Co3, Co4, MDU 1, Pukekohe (for dehydration); White Imperial, Spanish and Riverina-Late Brown; Early Lockyar Brown; Multiple (With 3-4 bulbs of small size).

## Hybrids

***IIHR, Bangalore*** : Arka Kirthiman, Arka Lalima, Arka Pitamber.

***Tropica Seeds (Green CO)*** : Granex, Texas, Yellow Grano.

***MYHYCO Seeds*** : MOH-2

***Bejo Sheetal Seeds*** : Lucifer, Flare, Orient,BSS 262.

***Nunhems Seeds*** : Rojo 9539, 9730.

***Malav Seeds*** : Malav-21.

***Ankur Seeds*** : Ankur White.

**Detail characteristics of varieties and hybrids are given below:**

***Pusa red*** : It require short to intermediate day length, plant height 30 cm. Bulb medium sized, average bulb weight 70-90 g, bronze in colour, flat to globular shaped less pungent. Less bolting, keeping quality good, duration 125-140 days after transplanting. Average yield 250 q/ha.

***Pusa Ratnar*** : Bulb large round a deep red, keeping quality average with higher yield potential (300 q/ha), duration 125 days after transplanting.

***Pusa White Round*** : Bulbs are white and roundish flat in shape, TSS 12-13 per cent. Average yield 300 q/ha. Suitable for dehydration. Maturity 125-130 days and very good keeping quality in storage.

***Pusa White Flat*** : Bulbs medium to large in size, flatish round with attractive white colour, TSS 12-14 %. Maturity 120-130 days. Yield ranged from 335-350 q/ha. Good keeping quality.

***Pusa Madhivi (Sel.102-1)*** : It has been bred by selfing and massing method from the material collected from kairana, dist. Muzaffarnagar, in Uttar Pradesh. It produces light red bulb with a rounded flat shape, maturing in 130-135 days after transplanting. It yields 31.5 tonnes of bulbs per hectare. The TSS is 11 to 12 per cent. It is suitable for growing in the rabi season.

***Early Grano*** : Introduction from USA, large gobular bulbs, yellow coloured. Storage quality poor, mild pungent, good for salad purpose. Attain full size in 95 days after transplanting and matures in 115-120 days, heavy yielder (around 500 q/ha).

***Brown Spanish*** : Bulbs solid, reddish brown, oblong, mild pungent with high TSS and good storage ability. A long day type suitable for cultivation at an altitude of 1000 m and above.

***Arka Niketan*** : It is a common big onion, has been developed to meet the export requirement of medium large sized, lighter bulbs with long storage life; developed from local collection 'IHR 135' through mass selection; bulbs are light red, uniform medium to big sized, firm, high TSS, with a good storage quality, free from splits, bolters and internal doubles. It is suitable for both kahrif and rabi season providing a bulb yield of about 37-40 tonnes/ha. It has export market for Gulf countries, Asian countries.

***Arka Kalyan*** : Bulbs are dark red, globular, 4-6 cm in size and moderately pungent. Average Bulb weight 130 g and TSS 11-12 %. Moderately resistant to the purple blotch disease, can be grown in kharif and rabi. Mature in 100-110 days after planting and produce an average yield of 35 tonnes/ha.

***Arka Bindu*** : Developed from local rose onion collection 'IHR 402' through mass selection, exclusively meet our export requirement. Its bulbs are deep red, small sized, uniform in shape, size and maturity and flattish - globe shape, having high TSS, pungency and high dry-matter content from bulbs with good storage quality. This variety has export markets in Singapore, Malaysia, Sri Lanka and Bangladesh.

***Arka Pragati*** : Bulb globe shaped with thin neck and deep pink outer skin with high pungency. An early variety with good keeping quality. Yield 450 q/ha in 130 days. Can be successfully grown both in Rabi and Kharif.

***Arka Kirthiman (hybrid)*** : It has been developed at IIHR, Bangalore, Bulbs are attractive red with firm texture, light red flesh with 10 % TSS; takes 120-150 days to mature; shelf life is 5 months in ambient condition; good transport quality; yield potential of 40-45 tonnes/ha; can be grown in both kharif and rabi season; recommended for MP, AP, eastern part of MP Female parent (MS65): plants are 50-60 cm tall; bulbs are deep red, medium and flattish globe; takes about 120-125 days to mature. Male parent (Sel 13-1-1) : plants are tall, bulbs are deep red big and globe; takes 130-135 days to mature. Suitable for Orissa.

***Arka Lalima (hybrid)*** : Arka Lalima has been developed at 'IIHR', Bangalore. Bulbs are deep red, globe with firm texture; red flesh with 12 % TSS; shelf life of 5 months at ambient condition, good transport quality; resistant to biotic stresses under field condition; free from splits and bolters; takes about 130-135 days to mature; gives 28% higher yield than the check N-53, average yield is 45-50 tonnes/ha, suitable for both kharif and rabi cultivation, female parent (MS-48) : plants are 60-65 cm tall. Bulbs are red, medium oval globe and take 140-145 days to mature. Male parent (Sel-14-1-1) : plants are 40-50 cm tall. Bulbs are deep red, big, flattish globe and takes 135-140 days to mature.

***Arka Pitamber*** : Yellow onion 'Arka Pitamber' (short day) has been developed through pedigree selection by crossing between two short days local genotypes "UD 102 (White) and IHR 396 (red), its bulbs are uniformly yellow in colour with perfect globe shape, very firm single centered, thin neck, mild pungent and medium sized with good keeping quality. It yield 35 tonnes/ha, high TSS firm bulb, free from splits, bolters, internal doubles. It is export oriented onion for the European countries, Japan and the USA.

***Punjab Naroya*** **:** Bulbs red, medium to large, round with thin neck, tolerant to purple blotch, yield 37.5 t/ha in 123 days.

***Punjab Selection*** : Plant height 55-65 cm, 6-9 leaves. Bulb red, globular in shape and measure 5-6 cm across. Average weight 65-70 g/bulb, quite firm with good keeping quality, suitable for dehydration. Tolerant to purple blotch and thrips.

***Punjab Red Round*** **:** Bulbs shining red, globular, medium size with thin neck, yield 28-30 t/ha.

*S-48* : Bulb flattish round and white in colour, good texture and falvour. Suitable for dehydration, good keeping and storage quality.

*Gujarat White* : Flattish round, medium in size and good keeping quality.

*N-2-4-1* : Averge plant height 46 cm, bulb light red, globe shaped, average length and breadth 6.1 and 55 cm. Pungent in taste, TSS 11.3 %. High yield potentiality with good keeping quality. Tolerant to Alternaria blight and thrips.

*N-257-9-1* : Globe shaped bulbs with light red colour, high yield potential with good keeping quality. Suitable for cultivation in rabi season.

***Baswant 780*** **:** Bulb attractive red, globular, TSS 11-12 %, keeping quality good, average bulb yield 200-250 q/ha. Crop duration 100-120 days, suitable for kharif season in Maharashtra. Developed by Mahatma Phule, Krishi Viswavidiyalaya, Rahuri (Maharashtra).

***Phule Samarth*** : It is an early-maturing and suitable for kharif and late kharif season. It gives higher marketable bulb yield during regular and late kharif. Natural topfall, thin bulb neck, controlled vegetative growth. Dark red and globular bulbs with good keeping quality (2-3 month). It has lower premature bolting.

***Phule Suvarna*** **:** It is a yellow onion developed exclusively for export markets with the distinguish characteristics of mild pungency and excellent keeping quality. It is developed by crossing of yellow Texas X N-2-4-1 through pedigree method. The bulbs are medium (62 g with size 5.30 x 4.50 cm) type, T.S.S. 11.30 % and bulb can be harvested within 110 days. It can be stored for 3 to 6 months with an average storage losses of 20 % and 48 % respectively. The fresh bulb is yellow, but it develops a red-tinge during the successive storage period and can be used for dehydration.

*N-53* : Bulb bright scarlet red, attain full size in about 95 days after planting and average keeping quality. Yield 200-250 q/ha. Suitable for planting in kharif.

*Hissar 2* : Bulbs are bronze red in colour, top sided flattish-globular, sweet but pungent. Maturity in 165 days after sowing. TSS, 11.5-13.9%. Bulb yield 300 q/ha.

*VL Piaz 3* : Medium sized, closed neck, round shape and tight skinned, light red bulbs weighing 80-120 g and measuring 5 cm across, maturity 160-170 days after planting. Resistant to purple blotch. Low incidence of thrips. It is a cross from Line BYG 2207 and local collection.

*VL hybrid piaz 6* : Uniform colour and medium size, higher yielder hybrid. Recommended for cultivation in UP hills and plains and other places.

*Kalyanpur Red Round* : Bulbs are bronz-red coloured, globular in shape, moderately sweet, moderately pungent. Mature in 130-150 days from sowing. TSS 13-14.9 % and bulb yield 250-300 q/ha.

*Udaipur 101* : Bulbs are deep red, flattish, globular and sweet with less pungency. Good for salad. Yield 200-300 q/ha. Matures in 150-160 days after sowing. TSS 12-14 %.

*Udaipur 102* : Bulbs white suitable for dehydration, yield 25-30 t/ha.

*Udaipur 103* : Bulbs are less red than Udaipur 101. Oblate-globular in shape, sweet but slightly more pungent. Yield 250-300 q/ha. Maturity in 150-165 days after sowing. TSS 10.5-13 %.

*Agrifound Dark Red* : It is a selection of AADF, New Delhi. Bulbs are dark red, globular in shape with tight skin, moderately pungent. TSS 12-13 %. Mature in 150-160 days from date of Transplanting. Suitable for kharif and late kharif. Best storing capacity. Bulb yield ranged from 300-400 q/ha.

*Agrifound Light Red* : Bulbs are light red and globular in shape. Bolting is less. It has good keeping quality. Maturity 160-165 days after sowing. T.S.S. about 13%. Yield ranged from 300-385 q/ha.

*Agrifound Rose* : Suitable for export, bulbs dark red, yield 19-20 t/ha. It is pickling type. Bulbs are flattish round in shape, deep scarlet red in colour and 2.5-3.5 cm of size. It matures in 95-100 days from sowing. TSS is 15-16 %. Suitable for kharif season.

*Bangalore Rose* : Small onion, bulb uniform in colour, shape, size and neck thickness, developed by IIHR, Bangalore. Most suitable for export. Three crops may be harvested in irrigated condition. However, best quality bulbs for export are obtained from the rabi crop.

*Rajo* : This red hybrid produces large thick flat to flattened globe bulbs with excellent exterior colour. It has medium storage and good resistance to Pink root. Higher yield, superior colour and wide adaptability makes this hybrid an excellent choice for farmers and the customers.

*9539* : It has very good plant vigour and produce medium large oval round bulbs with pink colour, medium pungency, good uniformly firmness and medium neck.

*9730* : This hybrid is less pungent has green to dark green foliage colour. Bulb is globe in shape, medium in size, dark red exterior colour, very firm, good uniformity with narrow neck.

*MOH-2* : It has globe, dark red, medium bulb with average weight of 85-100 g each. Crop matures in 110-120 days. Suitable for fresh market and grown in both rabi and kharif seasons. Yield potential 25-30 t/ha during rabi and 20-25 t/ha during kharif. Developed by Maharashtra Hybrid Seed Company Limited.

*Lucifer* : Bulbs attractive dark red in colour, round; average weight 100 g of each bulb. It matures in 100-110 days after transplanting. Suitable for both the seasons i.e., kharif and rabi. Good for fresh market and very high yielding.

*Flare* : It is a uniform sized hybrid, good for storage, as well as for fresh market. Bulb round red, average weight 100 g, mature in 110-115 days after transplanting. Good for storage.

*Orient* : It is a very productive hybrid, which gives uniform shape, size and colour of bulbs. It has round, red bulbs with average weight 100 g of each. Mature in 100-120 days after transplanting and good for storage.

*BSS 262* : It gives uniform sized, round bulb with TSS 12 to 14 %. Good for dehydration. Bulbs round, milky white with average weight of 100 g. Mature in 105 days after transplanting. Suitable for growing both seasons i.e., kharif and rabi. Good for dehydration.

***Texas Yellow Grano*** : It is a short day onion. Medium early maturity (110 days). Top shape, large size, light yellow colour. The flesh is mildly sweet. Resistant to pink rot disease.

***Granex (F1)*** : Hybrid, short day onion. Early maturity (100 days). Round top shape, large size, brownish yellow colour. The flesh is slightly pungent and firm. Good storage ability.

***Rio Santiago*** : An early maturing hybrid (160-170 days) produces round bulbs thus makes this hybrid unique in the short day red onion market. Globe shape bulbs are medium large in size and medium in pungency.

***Ankur white*** : Erect growth habit and medium green leaf. Medium duration. Bulbs are globular in shape. Medium sized bulb. Bulb skin colour is white. Medium pungent. Average bulb weight ranges from 60-80 g. Good storage capacity.

***NHRDF Red*** : It is developed through mass selection from the germplasm collected from Patiala (Punjab), has been developed by NHRDF Regional Research Station, Karnal. Plant leaves are dark green with waxiness. Bulbs are attractive dark red, globular round in shape, and 5-6 cm in diameter. The average bulbs weight is 100-120 g, TSS being 13-14%. This is moderately resistant to major diseases. It matures MS-120 days after transplanting. Its average yield is 250-300q/ha. This variety is suitable for rabi season (sowing in October and harvesting in April) and recommended for growing in northern, central and Western parts of India.

## POTATO

Potato (*Solanum tuberosum* L.); Syn. Aaloo, Batata, Urulakki-zhangu, Urulagadda, Urulagadde, Urala kizangu; is one of the most important food crop of the world. The potato is a dicot plant belongs to family solanaceae and the genus *solanum*. The commercial cultivated potato belongs the species *solanum tuberosum*, which is tetraploid (2n=48) species. In addition to *S. Tuberosum*, seven other cultivated species and 154 wild species of potato are important. The commercial potato generally belongs to two species.

(1) *Solanum andigenum* : Plants have thin and long stem, small and narrow leaflets. Flowers are produced more profusely. It has long stolons and mostly coloured deep-eyed tubers. The tuberization takes place only under short-day conditions.

(2) *Solanum tuberosum* : Plants have shorter and thicker stem, larger and wider leaflets. The tubers are generally white and oval. This species is suitable for long as well as short day conditions for tuberization.

In addition to above two species, *S. demissum* and *S. stenotonum* are also of some importance as they are resistant to some forms of virus and diseases. Potato is perennial but as a crop it is grown as annual by means of vegetative propagation through tuber. The tubers is an enlarged underground modified stem produced at the end of a stolon. They posses "eyes" which contain multiple buds. The tubers also contain lenticels (respiratory structure) like stems of other plants. All cultivated species belong to series *Tuberosa* of subsection Potatoe. About 72% of the species are diploid (2n = 24) and nearly 12% Tetraploid (2n=48). The rest are triploid (2n=36), pentaploid (2n=60) and hexaploid (2n=72). The widely cultivated potato belongs to tetraploid species *Solanum tuberosum.*

## Varieties and Hybrids

There are many varieties of potatoes either introduced into or bred in India. They have grouped on the basis of their maturity period viz., early, medium and late maturing. Knowledge of maturity period helps in getting higher prices in the markets; it becomes easy to follow or accommodate varieties in intensive cropping systems. Further, varieties of different maturity periods can escape easily from incidence of insect-pests, diseases and disorders up to some extent.

Most of the improved varieties are developed at the Central Potato Research Institute (CPRI), Kufri, Shimla. Majority of early varieties are clonal selections derived from foreign varieties or from popular acclimatized varieties that survived after introduction. Varieties are distinguished based on their habit, pigmentation of stem, leaf, flower and shape, size, colour and depth of eyes and flesh colour of tubers.

A large number of earlier varieties either developed at CPCRI, Shimla or evolved as famous selections from improved varieties were replaced by new improved varieties. Potato varieties are released by CPRI, Shimla having tested on multilocational trials at its sub-centres, agricultural universities and/or farmers' field. There are three maturity group of varieties available in potato cultivation. They are :

***Early maturing varieties*** : Satha, Gola, Up-To-Date, Kufri Kuber, Kufri Kundan, Great Scot, Kufri Khasi-Garo, Kufri Naveen, Kufri Neelmani, Kufri Jyoti, Kufri Alankar, Kufri Lavkar, Kufri Ashoka, Kufri Chandramukhi, Kufri Navatal, Kufri Jawahar, Kufri Surya, Kufri Khyati.

***Medium maturing varieties*** : Darjeeling Red Round, Kufri Red, Craig's Defiance, President, Kufri Sheetman, Kufri Chamatkar, Kufri Moti, Kufri Sutleg, Kufri Pukhraj, Kufri Anand, Kufri Swarna, Kufri Muthu, Kufri Sherpa, Kufri Giriraj, Kufri Megha, Kufri Kanchan, Kufri Chipsona-1, Kufri Chipsona-2, Kufri Bahar, Kufri Arun, Kufri Pushkar, Kufri Sadabahar.

***Late maturing varieties*** : Phulwa, Kufri Kisan, Kufri Safed, Kufri Kumar, Kufri Neela, Kufri Sindhuri, Kufri Jeevan, Kufri Dewa, Kufri Thenmalai, Kufri Badshah.

**Detail characteristics of potato varieties of different maturing groups are given below :**

## Early Maturing Varieties

***Satha*** : It is an early variety and becomes ready in about 2 to 2½ months. This variety is grown in Bihar. The yields are low but can be remunerative owing to high prices of the early market.

***Gola*** : This is an old Indian potato, probably introduced a long time back in this country. Tubers are medium, smooth white, oval, irregular, slightly shouldered, tapering sharply towards the crown (pear shaped), occasionally slightly flattened, eyes deep flesh colour white. The commercial 'Gola' is a mixture of three varieties, of which 'Gola B' is dominant and has large, pear-shaped, white tubers with deep- set eyes. The flesh is cream-coloured. This is an early variety.

***Up-To-Date*** *(Victoria (Peterson's) x Blue Don) (Disease-free clone)*: It is an early maturing variety (90days) and is adapted to a wide range of environmental conditions and is particularly suited for Uttar Pradesh, Himachal Pradesh, the Punjab, Madhya Pradesh, Bihar and Maharashtra. Its tubers are of a large size, have a smooth surface, an oval and flattened shape, fleet eyes and white flesh. It is one of the most popular variety in cultivation in India.

***Kufri Kuber*** : It is developed from a cross of the $F_1$ seedling of (*S. curtilobum* x *S. tuberosum*) with *S. andigenum*. This variety is suited for both the plains and the hills of Maharashtra, the Punjab, and Bihar. It is also suitable as an early and main crop variety. It yields fairly well under long days, but tuberises better under short-day conditions. The tubers are of medium size, have smooth surface, oval, flattened shape, fleet eyes and white flesh and skin.

***Kufri Kundan*** *(Ekishirazu x Katahdin)* : It is an alternate variety to up-To-Date. This is particularly suitable for hill areas, heavy and wet soils. It performs well under long-day conditions in the hills. It has tubers of medium size, having a smooth surface, round shape, white skin, fleet eyes and yellow flesh. It has excellent keeping qualities.

***Great Scot*** *(Imperator x Champion)* : Introduced from England by the Madras Department of Agriculture. It is an early (3 to 3½ months) variety with oval, white tubers. It is very popular in the Nilgiri Hills because of its ability to give high yields and rapid bulking under conditions prevailing there. It is however, susceptible to late blight, mosaic and leaf-roll viruses and other diseases.

***Kufri Chandramukhi*** *(Sd 4485 x Kufri Kuber)* : Early maturing (matures 10 days earlier than Up-To-Date), with attractive oval, white tubers. It degenerates slowly and keeps well in storage. It gives high yields (higher than Up-To-Date) and cooks easily and does not degenerate rapidly in the plains.

***Kufri Khasi-Garo*** *(Taborky x Sd. 698-D)* : An early maturing variety, possessing a good degree of field resistance to late blight and moderate resistance to early blight and viruses. Fit for growing in the hilly regions of Assam.

***Kufri Naveen*** *(D 692 x 3070 d(4)* : A fertilizer-responsive variety for the Khasi Hills region of Assam; possesses a high degree of field resistance to late blight, resistance to wart and moderate resistance to *cercospora* leaf spot and early blight.

***Kufri Navatal*** *(G-2524)* :The plants are medium, tall, erect and medium compact. The tubers are large to medium sized and oval in shape. The skin is white and flesh yellow in colour. The eyes (buds) are fleet. The keeping quality of tubers is medium. This variety is suitable for hill and resistant to leaf blight disease and tolerant to frost.

***Kufri Neelmani*** **:** Suited for the Nilgiri Hills in the south. It has given 100 per cent more yields than the variety Great Scot, which has long been popular in the Nilgiri area. The new variety can be grown in all the three potato crop seasons in that area.

***Kufri Jyoti*** *(3069d (4) x 2814(a) (1)* : A widely adaptable, fertilizer-responsive variety, possessing a high degree of field resistance to late blight disease in the foliage and also a good degree of tuber resistance to the same disease. This variety is also resistant to the wart and moderately resistant to *Cercospora* leaf spot. It is recommended for Himachal Pradesh and Kumaon Hills of Uttar Pradesh.

***Kufri Alankar*** *(Kennebec x 0.N. 2090)* : A very early-tubering (75 days), photo- insensitive variety capable of giving high yields, suitable for cultivation in the plains of northern India. This variety, by virtue of its earliness, can be included in the multiple cropping patterns which are coming into vogue with the introduction of the short-duration, dwarf, high-yielding varieties of rice and wheat.

***Kufri Lavkar*** *(Serkov x Adina)* : It is an early maturing variety (Plateau-70; Plains 100 Days) with round white tubers and light cream and firm flesh. It is suitable for growing in the Deccan plateau in Maharashtra.

***Kufri Ashoka*** *(EM/C – 1021 x CP 1468)* : This is an early (Hybrid PJ-376) maturing hybrid. White skinned, matures in 80 to 90 days after planting. Tubers are oval and eyes are shallow with attractive appearance. The yield is 250-280 q/ha better than K. Chandramukhi. Suitable for Eastern U.P., Bihar, West Bengal.

***Kufri Jawahar*** *(Kufri neelmani x Kufri Jyoti)* : Kufri Jawahar is an early maturing variety (80 days). It is recommended for cultivation as a main season variety in Punjab, Haryana, Uttar Pradesh, Gujarat and Karnataka. Its plants are short, erect and compact with light green foliage. Its tubers are oval, medium sized with fleet eyes. This is moderately resistant to late blight. Its tubers have good keeping quality. Due to compact foliage and erect habit, this is suitable for intercropping and mechanized cultivation. It is suitable for growing in plateau region as it requires less number of irrigation due to its early maturity. It yields 25-31 tonnes /ha.

***Kufri Surya*** : An early maturing (70-80 days) and suitable for North-Western plains and rabi and kharif seasons. It can produce about 20-25 tonnes / ha tubers. The tubers are medium to large, white, oblong with shallow eyes and pale-yellow flesh. This is a heat tolerant variety and can be grown in areas having night temperature above 20$^0$C and is field resistant to hoppers. Due to large oblong tubers and high dry matter content it can be used for preparation of French fries.

***Kufri Khyati*** : This is early maturing (70-80 days) variety for cultivation in North Indian plains. It can produce 25-30 tonnes / ha tubers. The tubers are medium, white, round-oval, with medium eyes and white flesh.

## Medium Maturing Varieties

***Darjeeling Red Round :*** This is an old indigenous variety. It was probably introduced into this country from Europe a long time back. It is slightly earlier than phulwa. It has also a wide adaptability. The tubers are medium-sized, round with deep red-coloured eyes. It does not keep as well as phulwa.

***Kufri Red :*** This clonal selection from Darjeeling Red Round was first released in 1957. This variety gives excellent yields in the plains and is particularly suited for U.P., Bihar and West Bengal. Its tubers are of medium size, have a smooth and red skin, round shape, yellow flesh and medium deep eyes. It has a good keeping quality under ordinary conditions.

***Kufri Sherpa*** *(Utimus x Adina) :* The plants are tall and erect. The tubers are medium sized, round flattened in shape. The skin colour is white and flesh is pale yellow.The eyes are medium deep and medium maturing variety. It is moderately resistant to early and late blight and virus diseases; immune to wart disease. Suitable for hill region of Bengal.

***Craig's Defiance :*** This variety was introduced from the United Kingdom in the year 1953. It is suited for culture in the hill areas of Himachal Pradesh and the Punjab. The tubers of this variety are of a large size, have a smooth surface, long shape, white skin colour, fleet, eyes, and white flesh.

***President*** *(Richter Imperator x Wilhelm Korn) :* Introduced from England, it is a medium maturing (3½ to 4½ months) variety with round, white tubers. This variety is also grown in the Nilgiri Hills. It is field resistant to blight. A clone of this variety has been selected at the central Potato Research Institute. This clone has given a better performance than the parent variety and has been named as the president-Kufri clone.

***Kufri Muthu*** *(3046(1) x M 109-3) :* This is medium maturing variety. The plants are tall erect and compact. The size of tuber is large, round-oval in shape. Both the skin and flesh are white and the eyes are medium deep. The sprouts are blue, purple in colour. It is moderately resistant to late blight, suitable to grow in the Nilgiris hills of Tamil Nadu.

***Kufri Chamatkar** (Ekishirazu x Phulwa)* : An early bulking variety (110-120 days) with uniform-sized, shining and smooth tubers; suitable for the Punjab, Haryana, uttar Pradesh and Madhya Pradesh.

***Kufri Sheetman** (Craig's Defiance x Phulwa)* : A frost-resistant variety, suitable especially for the Punjab, Rajasthan, Haryana and Uttar Pradesh as also for Madhya Pradesh and West Bengal. It can be planted late in the northern plains owing to its high degree of resistance to frost.

***Kufri Moti*** : It is suitable for growing in the Nilgiri Hills. The tubers are oval white and round and takes 110-120 days to mature. It is field immune to race 1 of late blight.

***Kufri Bahar** (Kufri Red x Gineke)* : This is medium maturing (90-100 days) variety for cultivation in plains of Uttar Pradesh. Haryana and Punjab. Plants tall, medium size. Tubers white, medium in size, round to oval with medium deep eye. Flesh white, texture floury. The variety matures in 90-100 days and yield 200-250q/ha.

***Kufri Lalima** (Kufri Red x AG14 (Wis x 37)* : Plants tall, erect, compact, vigorous, foliage dark gray-green. Tuber size medium, shape round, skin red, smooth, eyes medium deep, flesh white. It matures in about 100-110 days. The variety is moderately resistant to early blight. The average yield is about 200-300 q/ha.

***Kufri Sutlej** (Kufri Bahar x Kufri Alankar)* : This variety, notified in 1996, has been recommended for cultivation in the western and central Indo-Gangetic plains. The variety gives high yields (30-35 tonnes/ha) in about 90-100 days and is moderately resistant to late blight. It produces very attractive, large white, oval and shallow-eyed tubers. The tubers are white fleshed and have good keeping quality. Because of the oval tuber and a good dry-matter content of about 19%, it has been found quite suitable for production of french fries. Its early tuberization and bulking character makes it more popular than other varieties and helps in the production of good yield in 60-70 days. In this regard, it is a stiff competitor for Kurfi Chandramukhi, kufri Bahar and Kufri Ashoka, the three early maturing /bulking varieties.

***Kufri Pukhraj** (Craig's Defiance x JEX B 687)* : This early-bulking, medium maturing (80-90 days) variety was released in 1998, and has andigena blood in it. It is very high yielder, producing 35-40 tonnes/ ha in about 90-100days. The variety is field resistant to late blight and produces attractive yellow, oval, shallow-eyed tuber with light yellow flesh. The tubers have a good keeping quality, with a dry-matter content of about 18 %. The variety has been found suitable for the northern plains, Gujarat and plateau region.

***Kufri Anand*** *(PJ-376 x 698-D)* : This is the latest of the releases (1998) and has been recommended for cultivation in Uttar Pradesh by the Uttar Pradesh State Variety Release Committee. It is a medium maturing (90-100 days) variety and has higher degree of resistance against late blight and some frost hardiness. The variety is a higher yielder (35-40 tonnes/ha) and produces very attractive, white, long oval, flattened, shallow-eyed tubers with white flesh. The tubers have about 19% dry matter and possess good keeping quality. Like Kufri Sutlej, this variety too has been found very sutiable for French fries. Though the variety is recommended for Uttar Pradesh, it is finding wide acceptance in surrounding states, viz, Punjab, Haryana, Rajasthan and Gujarat.

***Kufri Swarna*** *(Kufri Jyoti x UT $n^2$)* : It is medium maturing variety. The plants are tall and erect. Foliage dark green in colour. The nematode - resistant culture 'SON 76 /110' was released for general cultivation in the hills of Tamil Nadu as "Kufri Swarna". Its tuber yield is 30.9 tonnes / ha compared with 23.8 tonnes/ha of "Kufri Jyoti". Tubers are medium-sized, oval to round in shape. The colour of both skin and flesh is white. The eyes are fleet. The sprouts are blue purpole in colour. It is resistant to early and late blight and is highly resistant to cyst nematodes (multiple disease resistance).

***Kufri Arun*** : It is a medium-maturing (90-100 days) variety for cultivation in North Indian plans. The variety can produce 35-40 tonnes / ha tubers, possessing about 20% dry-matter. The tubers are medium, red, oval with shallow-medium eyes and creamy flesh. The variety is field resistant to late blight and tolerant to frost.

***Kufri Pushkar*** : It is a medium-maturing (90-100 days) variety for cultivation in Haryana, Punjab, Rajasthan, Uttar Pradesh, Bihar, Maharashtra and Karnataka. It produces around 30-35 tonnes / ha tubers, possessing good keeping quality. The tubers are medium, white, round-oval with fleet to medium eyes and light yellow flesh. The variety is field resistant to late blight and moderately resistant to phoma and early blight.

***Kufri Sadabahar*:** This is a medium-maturing (90-100 days) variety for cultivation in Uttar Pradesh. It can be grown in adjoining states as well and is capable of yielding 35-40 tonnes / ha tubers. The tubers are medium to large, white, oval-oblong with shallow eyes and white flesh. The variety possess field resistance to late blight and tolerance to frost. Due to large oblong tubers and high dry matters, it can be used for preparation of Fresh fries.

***Kufri Chipsona-1*** *(MP/90-83)* : This is medium in maturity. Tubers are medium, smooth and round-oval with shallow eyes. Particularly suitable for processing.The content of dry matter is high (23 %). Keeping quality is good. The variety is resistant to late blight. The average yield is 250 q/ha. It is released in 1997.

***Kufri Chipson-2*** *(MP/91-G)* : This variety is also excellent for processing and keeping quality is good. White skinned variety having dry matter content of 21-24% and suitable for chips making. Crop duration is 90-110 days. Suitable for sowing in North Indian plains. This variety is resistant to late blight and suitable for growing in Uttar Pradesh and Bihar.

***Kufri Giriraj*** *(SM/85-45)* : Kufri Giriraj released for hills of Himachal Pradesh, however, it performs well even in south and north-eastern hills. It is a medium-maturing (130-135 days) variety producing medium-tall, semi-erect, medium compact and vigorous plants with light purple flowers. Tubers are medium to large, white, oval, with fleet eyes and unlike Kufri Jyoti do not show cracking. It is field resistant to late blight and immune to wart, average yields being 25-30 tonnes /ha.

***Kufri Megha*** : Kufri Megha released for north-eastern hills, however, it could not become so popular among farmers because of its longer maturity period (140-145 days). It produces medium-tall, semi-erect, medium compact plants with white flowers. Tubers are medium, round-oval, white with fleet eyes. It is highly resistant to late blight and yields on an average 22-28 tonnes /ha.

***Kufri Kanchan*:** Kufri Kanchan has been released for north Bengal hills and Sikkim. It is a medium-maturing (110-135 days in spring and 90-100 days in autumn) variety, producing medium-tall, semi-erect, medium compact, vigorous plants with red purple flowers. Tubers are medium-large, oval-oblong and red with fleet eyes. The variety is field resistance to late blight, resistant to wart and slow rate of degeneration with an average yield of about 25-28 tonnes /ha.

*Kufri Sindhuri (Kufri Red x Kufri Kundan)* : Medium maturing (3½ to 4½ months) with round, light red medium-sized tubers. It is suitable for cultivation as a main crop variety in the plains, as a replacement to Kufri Red and other late varieties.

***BSS296*** **:** Tuber, round to oval; Skin yellow; Flesh yellowish; eye fleet; mature in 90 days; field tolerance to late blight and dry matter-21.9% (approx.). It is good for processing.

## Late Maturing Varieties

***Phulwa*** **:** It is probably an old European potato variety which has lost its identity. It has been under cultivation in this country for more than a century. It produces a good yield under a wide range throughout North India. The tubers are small to medium, skin smooth, colour white, shape round, eyes moderate in number, deep flesh, colour yellow, it keeps well.

***Kufri Kisan*** *(Ekishirazu x Katahdin) → (Seedling $F_1$ - 17 x Phulwa) → (Up-to-date x Seedling 16)* : This is a late maturing variety and gives heavy yields in the plains. It is particularly suitable for Uttar Pradesh, Madhya Pradesh, Delhi and the Punjab. The tubers are of medium size, with a smooth surface, deep eyes, white skin and round shape. It has a very good keeping quality.

***Kufri safed*** **:** A late (4½ to 5½ months) variety. It produces small, round, white tubers with deep eyes and a light, creamy-yellow flesh. A clonal selection of phulwa, it is an excellent substitute for the latter. It is a hardy variety, suitable for a poor soil, with meagre facilities of irrigation. It has an excellent keeping quality. It is recommended for the plains of Uttar Pradesh.

***Kufri Kumar*** **:** A late maturing (4½ to 5½ months) variety, with large, white, oblong tubers. It is a good yielder both in the hills and on the plains and possesses a high degree of field resistance to late blight. It has, therefore, been recommended for the hills, particularly where late blight is a problem.

***Kufri Neela*** **:** A late maturing (4½ to 5½ months) variety with round, white tubers. As a substitute for Great Scot, it has several virtues. It answers the trade needs and possesses Great Scot type of tubers. Besides, it yields about 70 per cent higher than Great Scot. The greatest value of this variety is that it is resistant to late blight.

***Kufri Jeevan*** *(M 109-3 x 698-D)* : A late-matuing, high-yielding variety, possessing a high degree of field resistance to late blight, resistance to the wart and *Cercospora* leaf spot disease. It is recommended for cultivation in Himachal Pradesh.

***Kufri Thenmalai*** *(D/79-56)* : This hybrid is released by Tamil Nadu Variety Release Committee. Suitable for cultivation in the part of Tamil Nadu hills. The colour of sprout is red. It is late maturing and high yielding variety. Moderately resistant to early and late blight

***Kufri Dewa*** *(Craigs' Defiance x Phulwa)* : It is recommended for growing in the Tarai area of Western Uttar Pradesh. The tubers are oval round and white with yellow flesh and takes about 130 days for harvest in the plains and 150-160 days in the hills.

***Kufri Badshah*** *(Kufri Jyoti x Kufri Alankar)* : It is a medium maturing (100-110 days) variety for cultivation in parts of Gujarat, Haryana, Jammu and Kashmir, Madhya Pradesh, Punjab and Uttar Pradesh. Plant tall, erect, medium compact. Tubers medium to large, oval, skin white, eye shallow, flesh dull white, floury in texture. The variety matures in 100-110 days in the plains and yield 250-500 q/ha. The variety possesses moderate resistance to late blight, early blight and potato virus.

## RHUBARB

Rhubarb (*Rheum rhaponticum* L.); 2n = 44 (4x); belongs to the family polygonaceae, also known as pie plant or garden rhubarb,is a long lived herbaceous perennial with large fleshy underground rhizome having a fibrous root system. The economic part of rhubarb is the large thick leaf stalk or petiole. The stem contain oxalic acid like spinach. Rhubarb is protandrous, self pollination does not take place. Lower flowers are fertilized by pollen from upper flowers. The large, thick leaf stalks are treated as a vegetable. They can be stewed or baked in the oven, made into pies, sauces, pudding or crumbles and served as a desert. They can be preserved by bottling, freezing or making into jam or chutney and also used to prepare home made white wine.

Rhubarb is a cool season perennial vegetable and tolerate cold and dry conditions. Production of rhubarb stalks during the winter is called "Forcing of rhubarb in regions where the climate is suitable for production of vigorous crowns.

## Varieties and Hybrids

Chipman's Canada Red, Mac Donald, Ruby Valentine, Sutton Pink, Victoria, Gient Cherry, Sun rise, Strawberry and Cherry Red are some promising varieties.

# SWEET POTATO

Sweet potato (*Ipomoea batata* L.); 2n = 90, syn. Sakarkand, Shakaria, Mitha aloo, Misthi alu, Kandamula, Seenikilangu, Cheenikizangu, Genasu, Ratalee, Ratalu; is the second largest cultivated tuber crop in India. It is grown as root tuber vegetable and also for manufacturing starch used in industry. It is mainly used for human food after boiling or steaming, baking or frying and also as animal feed.

It belongs to the family Convolvulaceae, genus Ipomoea, *Ipomoea batatas* (L.) Lam. There are 45 genera and about 1,000 species in the family convolvulaceae, only *Ipomoea batatas* is of major economic importance. Another species producing vegetable only of minor importance is clove bean (*I. muricata* (L.) Jacq., non Cav. The edible part of sweet potato is the modified root. Unlike other crops, the roots of sweet potato produce adventitious shoots. The shoot and root come from each node along the main stem. Development of root depends considerably on temperature. The crop is highly cross-pollinated due to self-incompatibility and male sterility. Cross-incompatibility has also been reported. Vines are prostrate and spreading, Porto Rico type; and vines are of bunching type. Texture of tuber flesh with firm, dry and mealy flesh after cooking, with soft, moist and gelatinous flesh after cooking and with very course tuber suitable only for cattle feed or for industrial uses.

Sweet potato is an autohexaploid with 2n = 90 (x -15). High sterility is observed in the crop due to autopolyploidy.

## Varieties and Hybrids

Sweet potato varieties differ in shape, size and colour of leaves, tubers and nature of tuber flesh. Most of varieties have been developed from clonal selection and are self incompatible. A number of cultivars from mutation as chance seedlings or deliberate hybridization. Varieties are two types (i) white types (ii) red types. These are Pusa Safed, Pusa Lal, Pusa Sunheri, Kalmegh, Samrat, Kanghangad Local, Rajendra Sakarkand-5, Konkan Ashwini, CO1, CO2, CO3, Sree Nandini, Sree

Vardhini, Kiran, Varsha, Sree Bhadra, Gouri, Shankar, RS-5, S-30, H-41, H-42, H-268,H-620, H-635, Cross-4, V-35, RS-35, RS 43, S-30, OP-1.

**Drought resistant strains** are DR-17, DR-56, DR-62, DR-64 and other high yielding strains are 82-2, 85-16, 85/70 and 90/704.

***Other selection*** : SP-3, SP-9, SP-Lalka, V 12, Hosur Red, Hosur Green, Chuvala, Anakomban, Bhadrakali, Velmon, Ranger, B 4306, Gold Rush and Centennial, Scarlet.

*Varieties Porto* - Rico,Triumph, Hall and Nancy are introductions to India.

**Detail characteristics of varieties and hybrids are given below :**

***Pusa Safed*** : This is a selection of a white variety originally introduced from Taiwan (FA 17). It is high yielding and produces medium sized tubers of excellent quality.

***Pusa Lal*** : This is a selection from a Japanese variety (Norin).The tubers are medium-sized thick in the middle.Theskin colour is red but the cooked flesh is white. It stores well.

***Pusa Sunheri*** : The flesh colour is light orange, indicating higher carotene than other white-fleshed varieties.

***Sree Nandini*** : This variety was identified from an open pollinated seed progeny. The plants are semi-spreading, leaves unifoliate, margin entire with petiole of medium length. Emerging leaves are light green. The plants are shy-flowering. Tuber skin and flesh are white. Each plant produces 3-4 medium sized tubers. The tuber neck is medium long. It is early maturing and can be harvested 100-105 days after planting. The tuber yield is 20-25 t/ha. It cooks well and tastes sweet.

***Sree Vardhini*** : This variety is isolated from an open pollinated seed progeny. The plants are semi-spreading, the leaves are simple with 3-5 shallow lobes. The emerging leaves are light brown in colour. The plants are shy flowering. The tubers are medium in size with characteristic pink colour. The tuber flesh is light orange with carotene content of 1200 IU and hence more nutritious. The crop period is 100-105 days and it is considered as an early variety. Each plant produces 3-4 tubers and the yield recorded with the recommended package of practices is 20-25 t/ha. The tuber cooks well with attractive colour and has good consumer acceptability.

***H-41( Hybrid 41)* :** This is a hybrid , developed from a single cross involving a Japanese cultivar. 'Norin' and an indigenous clone. The plants are semi-spreading and the leaves are unifoliate, with entire margin. The well developed tubers are fusiform and measure 16-20 cm in length and 4-5 cm in diameter at maturity. The tuber skin is red with white rind and flesh. The tuber cooks easily, taste sweet and is low in fibre. The crop matures by 120 days after planting and yields 20-25 t/hectare under good management.

***H-42* :** This is a hybrid derived from a cross between an indigenous cultivar Vella damph and an American variety 'Triumph'. The plants are semi-spreading with unifoliate leaf having toothed margin. H-42 possesses medium long (14-20 cm), medium thick (4-6cm diameter) and fusiform tubers. The colour of tuber rind is pink and that of flesh is white. The crop matures by 120 days after planting. The tuber cooks well, tastes sweet and is almost free from fibre. The crop is relatively tolerant to weevil and tuber yield is 20-25 t/ha.

***H-268* :** This is a hybrid developed following a double cross involving two exotic and two indigenous cultivars. The plants are semi-spreading with thick vine is dark green in colour. Leaves are simple, 3-5 lobed with emerging leaf light sepia in colour. Tubers are fusiform, 14-20 cm long and 6-8 cm in diameter. The tuber skin is pink and flesh is light yellow. H 268 can withstand certain amount of drought. The crop period is 120 days and yield is 20-25 t/ha under good management.

***CO-1* :** It is a clonal selection. It yields on an average 28 t/ha. It exhibits tolerance to root weevil. The crop duration is 135 days. The tubers contain 24 per cent starch.

***H-620* :** It has long red tuber, less affected by weevil. Crop matures in 120-125 days after planting. Tuber yield varies from 220-250 q/ha.

***S-30* :** Tubers are round, light brown and susceptible to weevil. Crop maturity is 90-110 days after planting of vine. Tuber yield ranged from 225-300 q/ha.

***CO-3* :** Tuebrs has medium white skin and fusiform shape. Crop duration is 110-120 days after planting of vine. Tuber yield ranged from 22-28 t/ha.

***RS 35* :** Tubers fusiform and white skinned. Crop matures in 105-120 days after planting. Tubers yield varies from 20-25 t/ha.

*RS 43* : Tubers shape is fusiform and white in colour. Crop mature in 110-120 days after planting of vine. Tuber yield ranged from 20-25 t/ha.

*Samrat* : Tubers round with light pink skin. Crop mature in 90-110 days after planting of vine.Tubers yield varies from 20-28 t/ha.

*Kiran* : Tuber long and white in colour. Crop duration is 110-120 days. Tubers yield varies from 20-29 t/ha.

*Varsha* : A semi-spreading hybrid with reddish purple skin. Tuber fusiform and white in colour, light yellow flesh and excellent cooking quality. Crop duration is 115-120 days. Tubers yield ranged from 17-22 t/ha.

*Kanghangad local* : It has long, reddish purple; light yellow flesh colour of tubers. Crop mature in 110-120 days after planting of vine. Tubers yield varies from 12-20 t/ha.

*Gouri* : It is rich in vitamin 'A' and tubers can be used as salad. Tuber has orange flesh and crop duration is 105-120 days after planting of vine. Tuber yield ranged from 8-10 t/ha.

*Shankar* : It is a hybrid type, has high Vit. 'A'.Crop mature in 110-120 days after planting of vine. Tuber yield ranged from 6-8 t/ha.

*Sree Bhadra* : It is a spreading variety with pink skin early and high yielding crop mature in 90 days after planting of tubers. Tuber yield ranged from 20-27 t/ha.

*Ayamurasaki* : It is a high pigment Japanese cultivar, popularized in Japan for extraction of anthocyanins for food industries.

*Sree Ratna* : A spreading variety with purple skin, orange flesh and excellent cooking quality; yield 20-26 t/ha in 90-105 days.

*Sree Arun* : A spreading variety with pink skin, cream flesh and good cooking quality. Yield; 20-28 t/ha in 90-100 days.

*Sree Varun* : A spreading variety with cream skin, cream flesh and good coking quality; yield 20-28 t/ha in 90-100 days.

*V2* : The leaves of this variety are broad with darkgreen colour, stem is long, thick and hairy. White skinned, white fleshed and large sized tubers are produced. The average yield is approximately 150 q/ha.

*V6* : The base of the leaves is purple. The tubers of this variety are large in size, thick, red in colour and non-fibrous. The average yield is around 100-125 q/ha. These above two (V2 and V6) are recommended by Punjab Agril. University, Ludhiana.

*No.4004* : This variety is recommended by Department of Agriculture, Maharashtra state, Government. It produces white skinned tubers at a shallow depth. It is a high yielding variety. The average yield is around 200 q/ha.

*V8* : This variety was introduced in India from U.S.A. It is an early and take only 120 days from transplanting to harvest. The vines are long, trailing, with a purple shade at the internodes. The stem is round, smooth and hairy. The tubers are small in size and white skinned. It is recommended for cultivation in Punjab.

*V 12* : This is an early variety and takes from 105 to 120 days for complete maturity. The vines are semi-erect in nature. It produces big sized, elongated white skinned tubers. Tubers are very sweet in taste and non-fibrous. It is recommended for cultivation in Punjab.

*Jawahar Sakarkand 145* : It is recommended from Jawaharlal Krishi Vishvavidhyalaya. Jabalpur. It is early variety and ready to harvest 130-145 days after planting. Roots are copper red in colour.

*Yellow Jersey* : Leaves are green but small in size. The stem is green and slender. The tubers are yellow, smooth with yellow flesh.

*VL-Shakarkand-6* : It was developed from an exotic collection, B-219 from USA through clonal selection. It is suitable to grow in the hill areas of Uttar Pradesh. The tubers are elongated (15-18 cm long with 10-14 cm girth). The skin colour of tuber is pink and flesh is light yellow in colour. The tuber yield potential is 18-20 t/ha. Cooking quality of tuber is good.

*Co-2 :* It is a seeding clone obtained from open pollinated seeds of a type in germplasm bank. It yields on an average 32 tonnes of tuber per ha; the crop duration is only 110 to 115 days. Plants are dwarf and erect. The tubers are of medium sized with light pink skin and white flesh containing a high starch of 29.5% and carotenoids of 3.2 mg / 100g. The root weevil incidence is less i.e 10 to 15 percent. It is suitable for Sept.-Oct. and June-July planting.

*Cross 4 :* It is a high yielding variety which was popular earlier. It is susceptible to weevil attack and is not preferred either by landless labours or small and marginal farmers.

***Rajendra Shakarakand*** 5 : It is a cross between 'Cross 4' x 'H5', is high yielding and tolerant to weevil infestation. It has attractive white cylindrical tubers. It is very much suitable to northern condition for planting in September.

***V35*** : It is widely adaptable and is suited for Chhotanagapur Plateau region for south Bihar.

***Kalmegh*** : A quick bulking, high yielding variety for South Bihar conditions.It is suitable for multiple cropping. The tubers are roundish, pink and bold.

***'Konkan Ashwini'*** : It plants are spreading type with 240 cm vine growth. Light green stem with purple tinge at nodes. Four branches / plant, stem hairiness, no spines on stem having round shape. Leaves are heart shaped, alternate non shiny and non browsing. It is non - flowering type having elliptical, long tubers with mean length of 15cm and diameter of 4cm. The tuber weight is 128g having no root lets with dark purple colour, 3-4 tubers / plant, weighing 433g each. Tuber surface is smooth with deep dark purple colour, tubers are easy to cut having cream coloured flesh. Shelf life is two months. Tubers have 10 minutes cooking time, protein content is 3.6 %, reducing sugar 3.7 %, duration 105 days, can be grown both in kharif and rabi seasons. Yield ranged form 16–24 tonnes/ha under rainfed condition whereas 20-27 tonnes/ha under irrigated condition.

## TANNIA

Tannia (*Xanthosoma sagittifolium* (L.) Schott; 2n = 26; belongs to the family Araceae, is a tall herbaceous plant growing to 1.5 to 2.7 m and has about 6 large outspreading hastate leaves in contrast to peltate leaves in taro. In tannia, mother corm is more or less cylindrical and surrounded by cormels which develop from lateral buds adjacent to leaf scars on the corm. Just like taro, mother corms and cormels are used for planting. Since plant grow vigorously, tannia is widely spaced at 1 x 1m. It looks like colocasia but it is much bigger and attains a height of 2m. It matches colocasia in nutritive and medicinal values.

### Varieties and Hybrids

There are no specific released varieties of tannia in India. Only local selections with low acridity, good texture and suitability of leaves for making rolls are available. Where corms and cormels yields are being preferred, local collections with high yield and less acridity are desired. For making rolls, cultivars with high leaf yield and less acridity are preferred.

## TAPIOCA

Tapioca (*Manihot esculenta* (Crantz); 2n = 2x=36; syn. Cassava, Yuca, Manioc, Mandioc and Maravali; is the most important starchy root tuber crop grown in the tropics and is mainly cultivated in southern peninsular India. The tubers form a part of diet of the people in Asia and Africa. With the prospect of production of industrial alcohol from tapioca, its cultivation has been extended to many countries. The tapioca leaves are a rich source of vegetable protein (7 per cent on fresh-weight basis). It is also useful as a feed in sericulture. Underground tuber is easily digestible and mainly consumed after cooking. Processed products like chips, sago and vermicelli made of tapioca are also popular in the country.

It belongs to the family Euphorbiaceae, genus *Manihot M. esculanta Crantz* (syn. *M. utilissima* Pohl).The family contains mainly laticifers and produces latex.

Cultivars have been distinguished on the basis of morphological characteristics like leaf shape, size, plant height and tuber shape. Earliness in the maturity of the crop, root yield and cyanogenic glucoside content of roots have also been used to distinguish varieties. Colour of the young leaf bearing part of the stem is generally utilized for botanical description of cultivars. Colour and size of roots are useful for identification.

Tapioca clones are often classified by taste as sweet or bitter, the former being called *M. esculanta* and the latter *M. palmata* or *M. dulcis*. The toxicity of cultivar varies according to environmental growth conditions. The basis for classification of cultivars is (1) Hairiness or otherwise on the upper side of the unfolding leaves and petioles, colour of the tender stem, petiole, leaf veins and mid ribs, skin of the main stem,colour of leaf scars and the proportion of width to length of the main lobe of leaf. (2) Stamen colour (red or not), petiole length (sessile or not), colour of the tuber peel, the branches and perianth. (3) Bitter or sweet types based on the degree of bitterness of tubers due to HCN content. The sweet types are commonly grown for home consumption while the bitter ones are cultivated for the industrial market. Presence of winged-seed capsule is the morphological characteristics of bitter types.

## Varieties and Hybrids

A good number of cultivars have been introduced to India from Brazil, Madagascar, Nigeria, Uganda, Kenya, Ghana, Thailand, Malayasia, Srilanka and Fiji. M4 is an introduction from Malayasia.

The tapioca varieties are classified as sweet and bitter based on the content of cyanogenic glucoside. The bitter type produce high yield and contain more HCN content. Sweet types have low NCN content and used as raw. Varieties differ in colour of rind and flesh, size of tubers, colour of stem, leaf and petiole, branching pattern, duration of crops and resistance tomosaic disease. High amount of cross-pollination results in heterozygous nature. Vegetative method of propagation resulted in development of a number of polyploid varieties and hybrids.

In tapioca large number of cultivars are in existence in the major tapioca growing regions of the world. These cultivars vary widely in their yield potential as well as in other economic characters.

Sree Harsha, Sree Vijaya, Sree Prakash, Sree Visakham, Sree Sahya, Sree Jaya, M-4, Co.1, Co.2, Sel. 265, H-578, H-119, H-97, H-226, H-2304, H-2502, H-312.

**The detail characteristics of varieties and hybrids are given below**

***Sree Prakash*** *(S-856)* : It is a selection from the germplasm collection from indigenous origin. The plants are erect, relatively short (1.0-1.5 m), generally non-branching but rarely branching at the top. The leaves are dark green, lobes 7-9, with dark brown petiole. Leaf retention is high and is very shy flowering. Tubers are medium (25-35 cm) and necked. Tuber is shallow bulking and radial in arrangement. 'Sree Prakash' has been identified as an early variety and can be harvested by 7-8 months after planting. An yield of 35-40 t/ha has been recorded with good management. The plants are highly tolerant to CMD. Starch content is 29-31 per cent in the fresh tubers. The cooking quality of the tuber is fairly good and is non bitter.

***Sree Visakham*** *(H-1687)* : This is a hybrid resulted from crossing an indigenous collection 'Acc No.1501' and an exotic cultivar 'S-2312' introduced from Madagascar Republic.The former is characterized by unbranched habit with light yellow tuber flesh while the latter is a heavy yielder with relatively good culinary qualities. "Sree Visakham" is predominantly non-branching and tall (2-2.5 m). Emerging leaf is light brown in colour. Tubers are medium in length (25-35 cm) conical in shape and compact in arrangement. Tuber skin is brown and rind is cream in colour. Flesh is light yellow and recorded carotene content of about 466 IU and hence more nutritious. The starch content in the fresh tuber is 25-27 per cent. The crop period is about 10 months and the tuber yield is 35-38 t/ha with reasonably fair culinary qualities. It shows high tolerance to CMD.

***Sree Sahya*** *(H-2304)* : It was developed following multiple crossing in succession involving five parents of which two were from exotic and three from indigenous sources. Plants are tall (2-2.5 m), predominantly non-branching, rarely top branching with relatively compact plant type. Petiole is dark brown and stipular mark is spiny and prominent. Tubers are medium long (30-40 cm), long necked, occassionally with the characteristic beading. Tuber skin is light brown with cream coloured rind. Tuber flesh is white and starch content is 29-31 per cent. The yield range is 35-40 t/ha.

***H-97 (Hybrid-97)*** : It is a hybrid developed by crossing an indigenous cultivar 'Manjavella' and an exotic selection "Acc. No.300", which is Brazillian in origin. The plants are medium tall, (1.5 to 2.0 m) and branched at middle level. The hybrid is characterized by light sepia colour of the emerging leaf. The tubers are conical, medium sized (25-35 cm) and with light brown skin. The rind is cream coloured and the tuber flesh is white with 27-29 per cent starch content. It gives a yield of 25-35 t/ha under good management and will be ready for harvest 10 months after planting. The hybrid is field tolerant to cassava mosaic disease (CMD).

***H-165*** : This is a hybrid which resulted from a single cross between two indigenous cultivars 'Chadayamanagalam Vella' and a type similar to 'Kalikalan' known for its tubers with good culinary qualities. The plants are medium tall (1.5-2.5 m) generally non-branching are rarely top branching with mature lower leaves and the emerging leaves are light brown. The plant has divergent growth habit and is found to be suitable for raising after paddy crop in rotation cropping. The tuber is relatively short, without neck and conical in shape. Tuber skin is golden brown and the rind cream coloured. The tuber flesh is creamy with starch content of 23-25 per cent on fresh weight basis. The crop attains maturity in 8-9 months and it is recommended for early harvesting. Yield ranges from 33-38 t/ha. It is noticed to be field tolerant to CMD.

***H-226*** : This hybrid was developed from a single cross involving a popular cultivar M-4 introduced from Malaya and a local cultivar

'Ethakkakaruppan'. Plants are medium tall (2-2.5m), occasionally branching and the leaves have characteristic green colour. Tubers are medium (25-35 cm) with creamy skin having purplish patches. Tubers rind is light purple and tuber flesh is white with 28-30 per cent starch on fresh weight basis. It is preferred in starch industry. The crop period is about 10 months and gives an yield of 30-35 t/ha. It has been found to be highly susceptible to CMD in recent years. Cooking quality of tuber is fair.

***Co.1*** **:** It yields 35 t/ha. The crop duration is eight and a half to nine months. It exhibits a high amount of field tolerant to mosaic virus and contains 35 per cent starch with a low HCN content of 18.5 µg/g.

***Co.2*** **:** It yields 37.55 t/ha with a crop duration of eight and a half to nine months. The tubers possess a high starch content, having low HCN content (10 µg/g). The tubers are suitable both for consumption and industry.

***Sree Harsha*** **:** This is a triploid clone developed by crossing a diploid with an induced tetraploid clone of 'Sree Sahya', plants are stout, erect and non-branching with tubers of good cooking quality and high starch content (38-41 %). Yield 35-40 t/ha in10 months.

***Sree Jaya*** **:** This is a short duration (6 months) clonal selection suitable for low land cultivation as a rotation crop in a paddy-based intercropping system; tubers are with brown skin and purple rind and have good cooking quality; yield 26-30 t/ha; susceptible to CMD.

***Sree Vijaya*** **:** This is also a short duration (6-7 months) clonal selection suitable for low land cultivation as a rotation crop in a paddy based intercropping system.; tubers are with cream coloured rind and light yellow flesh due to high carotene; susceptible to mite and scale insect; yield 25-28 t/ha in 7 months.

***M-4*** **:** It is introduced from Malaya. This is a non branching variety with excellent cooking quality; susceptible to mites; yield 18-23 t/ha in 10 months.

***Sree Rekha (TCH-1)*** **:** This variety has been developed through pedigree method of breeding and released in 2000. Plants are erect and with light green of mature stem. Leaves are medium size, lanceolate and acuminate tip. Tubers are conical in shape with light brown in colour. Erect branching variety with good cooking quality; yield 35-40 t/ha in 10 months.

***Sree Prabha (TCH-2) :*** Plants are medium in height. Semi-spreading variety with good taste and quality; yield 35-40 t/ha in 10 months. It has been developed through pedigree method of breeding and released in 2000. Tubers are conical in shape with brown in colour. This variety is susceptible to cassava moraic disease, field tolerant to cercospora leaf spot and field resistant to scale insect. This variety is adoptable for Kerala, Tamil Nadu, Karnataka and Andhra Pradesh.

***AC-2380*** : It is a new selection. Its tubers are white in colour and ready to harvest after 300 days of planting.

***H-2304*** : It is a hybrid released from Central Tuebr Crop Research Institute, Thiruvananthapuram (Kerala).The average yield is 325 to 450 q/ha.

***H 1687*** : It is also a hybrid from CTCRI, Trivandrum, Kerala. The average yield is 325 to 450 q/ha.

***Nidhi*** : Yield 25.1 t/ha, tolerant to mosaic, short duration (5-6 months) grayish white stem, petiole white with red shade skin light pink.

***Kalpaka*** *(KMC-1) :* Yield 28.4 t/ha, short duration (6 months), non-branching stem, pink tuber rind.

***Vellayani Harswa*** : High yield (44.01 t/ha), short duration (5-6 months), pink tuber rind, excellent cooking quality.

***Co 3*** : A branching variety having tubers with brown skin and 35.6 % starch; yield 42.6 t/ha in a months. Developed by T.N.A.U., Coimbatore.

***Travancore Red*** : Tubers are slightly red in colour and ready to harvest after 275 days of planting. Stem is slightly pink.

***Travancore white*** : Tubers are white, medium in size, sweet and ready to harvest after 290 days of planting. Stem is dark green.

***ADIRA-1 :*** It is a sweet cultivar with only 50 ppm HCN in the tuber and yield of 250-300 q/ha. It is recommended for small scale farmers, where cassava is intercropped with other food crops. It can be harvested in 7-8 months.

***ADIRA-2*** : It is high yielding cultivar. Tuber is medium in starch content and high in HCN. It is tolerant to cassava bacterial blight (CBB).

***ADIRA-4 :*** It is a bitter cultivar with approximately 90 ppm HCN, but capabla of producing high yield and widely adopted. It is suitable for large planting of industrial use. The crop can be harvested at 9-10 months with fresh tuber yield of 350-450 q/ha.

***Colombia (PR-C62) :*** It is an edible purpose cultivar with high yields potential of 460 q/ha in 10-12 months duration.

***Faroka :*** This cultivar yielded 190 q/ha and is adopted for dry areas. The tubers are low in HCN with good table quality.

***Golden Yellow (PR-C24) :*** This cultivar gives yield both under open and partially shaded conditions prevent in coconut garden. Tubers are low in HCN and suitable for table purpose. It is an early maturing (8-10 months) cultivars with a yield potential of 430 q/ha under good management.

***Kadabao (PRC 13) :*** This cultivar is suitable for open and partical shade. It has high starch content and fairly low HCN. It matures in 10-12 months with a yield of 420 q/ha.

***Kalikalan :*** It is early maturing cultivar with good table quality of tubers. It is highly susceptible to CMD. It is well adapted to low fertility conditions.

***Karuthakaliyan :*** This cultivar is spreading type with 10 months duration. It produces non-bitter tuber with an average yield of 320 q/ha. It is susceptible to CMD, bacterial leaf-spot (BLS) and mites.

***Rayong-1 :*** It is high yielding (250 q/ha) widely adopted bitter cultivar. Since tuber has high dry matter and starch content, it is well accepted as an industrial cultivar.

***Rayong-2 :*** It was selected from hybrid seeds of CM 305 introduced from CIAT, colombia. In tuber dry matter and starch content are superior to Rayong-1. These qualities fetch higher price for market. This cultivar is table type, good for making fried chips and other convenient foods. It has yellow flesh content, high carotene and very low HCN.

***Rayong-3 :*** It is a selection from CM 407 cross which was introduced from CIAT, Columbia. In tuber dry matter and starch content are superior to Rayong-1. These qualities fetch higher price for Rayong-3. The tuber yield of this cultivar is an at par with Rayong-1.

***Rayong-60 :*** It is a derivative of CIAT and Thai clones (M Col-1684 x Rayong-1). This cultivar is superior to Rayong-1 in fresh root yield and root dry matter content. It is early maturing and adapted to fertility.

***TMS-30001 :*** It is early maturing and sweet cultivar. Tubers contain low HCN and well accepted by consumers. Tuber yield is around 220 q/ha. It is resistant to CMD but susceptible to root rot and insect pest.

***TMS-30555 :*** It is a high yielding and widely adopted cultivar. It is resistant to CmD and CBB, with a tuber yield of 400 q/ha.

***TMS-30572 :*** It is widely adopted cultivar, because of high yield (400 q/ha) and good quality of tubers. It is resistant to CMD and CBB. It is susceptible to cassava mealy bug and cassava green mites.

***TMS-50395 :*** This cultivar is resistant to CMD, CBB and tolerant to cassava green mites. It is widely adopted though the tubers are high in HCN. Tuber yield is 330 q/ha.

## WELSH ONION

Welsh onion (*Allium fistulosum* L.); 2n = 16; belongs to the family Liliaceae; is a perennial, herbaceous plant, usually grown as an annual for its edible tops, long leaf bases or young shoots, it does not form area! bulb but only a small enlargement at the base; the Welsh onion is thought to have been in cultivation for many centuries; references, can be found in Chinese lilerature dating from the third century BC. It was probably introduced into Europe in the middle Ages through Germany where it was given the, name 'Welsche' which translates as foreign. The common name 'Welsh Onion' is considered to be a corruption of this. The other commonly used name is 'Japanese bunching onion', but there are numerous other local names for the species, which may be expected given its wide distribution. Production of Welsh Onion is centred on Japan, Korea, China and Taiwan, where it ranks in importance among the top ten vegetable crops.

Welsh onion contains alliin, a precursor of allicin which is known to play a role in uptake of vitamin $B_1$, and also has an antimicrobial effect. Ancient Chinese literature indicates that this plant does have important properties, such as improving eyesight, enhancing metabolism and prolonging life. The eating qualities improve under low temperature, because both sugar and protein content increase.

*Allium fistulosum* is self-compatible and normally insect pollinated, resulting in a high level of outcrossing. Flowering is initiated after vernalization , which varies between cultivars. Temperature below 13ºC are required when seedlings have more than 12 leaves or are more than 7 mm in pseudostem diameter. Together with low temperature short day condition are also necessary.

## Varieties and Hybrids

The many local and commercial cultivars in existence are adapted to a variety of climatic conditions. Different systems of cultivars classification exist, but one is widely accepted in Japan and divides the cultivars in four groups, namely Kaga, Senju, Kujyo, and Yagura Negi. This division is based partly upon ecological performance and reflects the degree of winter growth.

Cultivars of the Kaga group are bred or selected primarily for their thick pseudostems. The Senju group cultivars are somewhat intermediate in terms of the characteristics for which they have been selected or bred. On the other hand,in the Kujyo group, cultivars have tender, green leaf blades with excellent eating quality.

In Japan there are over 120 registered cultivars of Welsh Onion, only some of which have been developed as $F_1$ hybrids using male sterility. The Nebuka (He-Shi-Ko or Japanese Bunching) variety is grown commercially as a green onion. It is hardy and resistant to pink root and onion smut.

# YAMS

Yams (*Dioscorea* spp.); 2n = 30-80, belongs to the family Dioscoreaceae, are usually herbaceous climber with rhizomes or tubers and form an important part of diet in Asia and Africa. Plants are perennial and deciduous in nature. Vine may be winged or round and with or without prickles or hairs.

The genus Dioscorea is sub-divided into different sections in which different species fall.

*Section-I :* *Enantiophyllum* (*D.rotundata* Poir, *D.alata,* L. *D.cayenensis* Lam., *D.opposita,* Thumb and *D.japonica* Thumb.)

*Section II :* *Lasyophyton* ( *D. dumetorum* (Kunth) Pax., *D. hispida* Dennst).

*Section III* : *Oxophyton*. (*D. bulbifera* L.)

*Section IV* : *Combilium* (*D. esculenta* (Lour). Burk.

*Section V* : *Macrogynodium* (*D. trifida* L.)

The section Enantiophyllum is characterized by the twining of vines to the right, i.e., in a clock-wise direction when viewed from ground upwards. The other sections are characterized by twining of the vines to the left, i.e. in an anti-clockwise direction when viewed from ground upwards. Among the species listed above only 4 are mainly used as vegetable.They are *D. esculenta* (Lesser yam), *D. alata* (Larger yam). *D. bulbifera* (Potato yam) and *D. rotundata* (white yam).

Yams are food crops of considerable importance in many countries. They are widely consumed in the form of tubers (by boiling or frying or baked or mashed), flour 'Fufu' Kokonte and yam pottage. Starch is extracted on a commercial scale.

The morphological characters of the four important species , which are mainly used as vegetable and edibles are given below :

## Greater yam or water yam or larger yam (*Dioscorea alata*)

It has winged, spineless stems twining to the right, leaves are large and arranged in opposite manner. The tubers of cultivated types are normally round, oval or irregular in shape. The skin of the tuber is black or brown while flesh is white, yellowish or purplish. Plant produces usually 1-3 tubers at the end of a stolon.

***Lesser yam or Chinese yam (Dioscorea esculanta)*** : This species has thin, pubscent and spiny stems which twine to the left, leaves are small cordate and alternate, in arrangement. It produce tubers in bunches (10-15 tubers) at the base of the stem. The tuber is small and rounded, thin yellowish, brown skin and white flesh. They bruise easily and do not store well, sprout again in a few months.

***White yam (Dioscorea rotundata)*** : It is characterized by sturdy, spiny and roughly circular stem which twines to the right. Leaves are small, simple, cordate and opposite or alternate in arrangement. Tubers are normally cylindrical having brown skin and white flesh. This is the principal yam cultivated in west Africa. This species was introduced into India in the form of seeds.

***Areal yam, potato yam (Dioscorea bulbifera)*** : Stem has no hairs or spines, leaf simple, large, devoid of hairs or spines. Under ground tuber is small, bulbils (aerial tubers) are originated in the axil of leaves grown for bulbils.

## Varieties and Hybrids

There are two edible species of yams namely *D. alata* and *D. esculanta* popularly cultivated in India for under ground tubers. These tubers are primarily used as a vegetable. The other species *D. rotundata* is an introduction from Africa. The crops was introduced in the form of true seeds.

**Detail Characteristics of Varieties and Species are given below:**

### Greater yam *(D. alata) (2n = 20, 30, 40-80) (Hindi-Ratula)*

***Sree Keerthi*** **:** It was released in 1987 for cultivation. It is a selection from the germplasm collection. Up to 30 tonnes/ha of tubers are ready for harvest in 9-10 months. Tubers are conical in shape with white flesh and good cooking quality. Starch content ranged from 20-22 percent. The plants are climbing in habit, 4-5m tall, vines twining in the clockwise directions on the support and emerging leaf is light brown.

***Sree Roopa*** **:** This is a selection from the germplasm collection. The plant is a climber, 4-5m long and vine twine in the clockwise direction. The leaves are cordate. Emerging leaf is green and stem is purple. Tubers are digitate, with black skin and white flesh and favourable culinary qualities. The tubers contain 17-19 percent starch and 1-2 percent protein. The yield potential is 25-30 tonnes/ha.

***Sree Shilpa*** **:** Tubers are swollen, oval and smooth with black skin and white flesh, yield is 28 t/ha in 8 months

Some more high yielding varieties are Da 42, Da 48, Da 60, Da 80, Da122. Da 60 has large conical shaped tubers, Da 80 medium size irregular tubers and Da 122 medium sized round to oval tubers.

### White yam *(D. rotundata)*

***Sree Subhra*** **:** A selection from the open pollinated progeny of the exotic cultivar and an introduction form Africa. It produces 1-2 cylindrical tubers with brown skin , white flesh possess good cooking quality and 21-23 % starch. The duration is 9-10 months and average yield 35-40 tonnes of tubers per hectare.

*Sree Priya (U-195-2)* : It has been identified from a population derived from open pollinated seeds of variety "Umidika" received from Africa. The plant is a climber, twining to the right on the support. The vine is spiny. It has dark glossy leaves, unifoliate with wavy margin. The plant produces 2 or 3 long tubers with good cooking quality. The crop period is 9-10 months. An average yield is 35-40 tonnes of tubers per hectare, the starch content is 20-21 % and protein 2-3 per cent.

*Sree Dhanya* : It is a dwarf bushy selection from seedlings progeny. The main advantage of this variety is that it does not require staking as the plant do not grow beyond 50 cm high. The tubers yield is 20-25 tonnes/ha.

### Lesser yam (*Dioscorea esculenta*) (2n = 40, 60, 90, 100) (Hindi – Suthni)

*Sree Latha (De-II)* : It is a selection from *D. esculenta* germplasm. The plants are climbing in nature with a height of 2.5-3.0 m. Vine is green with prickles. Leaves are simple, cordate, in shape, with entire margin, petiole is brownish green and swollen at both ends. The crop duration is 7½ to 8 months. The variety yields more than 25 tonnes/ha. Tuber length is 13.5 cm with girth 12.2 cm. Tuber shape is oblong to fusiform, skin greyish brown and flesh creamy white and a starch content of 18.4 % and shelf life of tuber is 2½ to 3 months.

*Sree Kala* : It is a selection from an introduction from puerto rico. "Sree kala" has been released for commercial cultivation. The tubers ready for harvest in 8 months are smooth and possess excellent cooking quality. On an average 20 tonnes of tubers are produced in a hectare. Some more, De23, Lathani, Chaparia and De 17 have been identified as high yielding varieties.

Some other cultivars are *D. cayensis,* which is a white flesh cultivar very popular in Puerto Rico. "Congo yellow" or Guniea yellow, potato and tango *( D. asculenta)* and agua ( *D. alata*) are white flesh cultivars. Purple ceylon is a good purple cultivar. Florida, Forastero, Gamelos, Leone Globe, Veveen *etc.* are *of D. alata.*

## YAM BEAN

Yam bean (*Pachyrrhizus erosus* L.); 2n = 22; known as Sakalu; belongs to the family Fabaceae is grown for its tender tubers which are used as vegetable. Young tubers are crisp, succulent and sweet and are preferred for salad purpose. In Philippines, China, Indonesia,

Nepal, Bhutan and Myanmar tubers are processed, canned and many sweet preparations are made. In Asia, specially in Indian regions like West Bengal, Bihar and Orissa grown during hot humid climate. Yambean also need long days for vegetative growth and short days for better tuberisation.

## Varieties and Hybrids

***Rajendra Mishrikand*** : It is an improved variety yielding 36-40 t/ha and is being propagated through seeds. Seeds @ 3-5 per hill or sown at a spacing of 0.75-1.0 m during June-July. It flowers in 75 days after sowing. It is desirable to remove flowers for encouraging tuber yield. Provide trellis or props for plants to trail. It will be ready for harvest in 150 days. If harvesting is delayed, it results in cracking of tubers. Harvested tubers can be stored for 2-3 days without any deterioration.

□□□

# 5

# Solanaceous Vegetables

## BRINJAL

Brinjal (*Solanum melongena* L.); 2n=24; syn. Egg Plant, Baingan, Bengena, Begun, Balgan, Vange, Brinjal, Badane Kayi, Vangi, Vazhuthana, Bataun, Mulukutakali, Vayiugana, Guinea squash, Aubergine, ringna, Waangum; belong to the family solanaceae, is one of the most common tropical vegetable grown in India. India is the second largest producer of brinjal in the world next to China. Plant is erect, semi-erect or prostrate, herbaceous and branched. Flower bisexual and self pollinated crop, cross pollination to varying degree has been reported in brinjal due to its heteromorphic flower structure. Fruit is berry with wide variation in shape, colour and size.

Genus *solanum* comprises approximately 2000 species which include both tuber bearing and non-tuber bearing forms. Important edible species under non tuber bearing forms are *S.melongena, S.torvum, S.nigrum, S.macrocarpon, S.ferox* and *S.aethiopicum.*

*S. torvum* used for its small clustered fruits for curry purpose and for drying. It is grown as a wild plant and due to its resistance to

Fusarium wilt and bacterial wilt, it can be a resistant root stock for grafting cultivated *S. melongena.*

*S.macrocarpon* and *S.aethiopicum* – grown for edible fruits and leaves.

*Solanum nigrum* – Small clustered acidic fruits are also edible and are harvested at ripening stage when it turns purple in colour. In Tamil Nadu this species is cultivated for use as leaf vegetable. Ripe fruits are used in curries and for frying. Leaves and flowers of *S. nigrum* have more resemblence to that of chilli.

Several species of *Solanum* are used in native medicines throughout tropics.

*S.melongena* var. *insanum, S.incanum, S.indicum, S.viarum* are few among them. Based on growth habit and fruit shape, four botanical varieties are reported under *S. melongena.*

1. *S.melongena* var. *melongena* (Syn. *S.melongena* var. *esculenta* Nees)- includes cultivars with round egg shaped fruits
2. *S.melongena* var. *serpentinum* Desf. Includes long and slender fruited cultivars.
3. *S.melongena* var. *depressum* includes early and dwarf cultivars.
4. *S.melongena* var. *insanum* includes wild and prickly plants with small fruits.

## Varieties and Hybrids

According to Bailey (1949) brinjal is having three botanical varieties, var. *esculentum* – fruits are round or egg-shaped; var. *Serpentinum* – fruits are long slender; var. *depressum* – early and dwarf types. Brinjal varieties are grouped on the basis of their colour, fruit shape and size. The selection of varieties depends on the climatic condition and consumers preference. In West Bengal, North UP, Orissa regions green fruited brinjal is more preferred. The preference is vary according to the season also as the white fruited brinjal is mostly used in summer months as there is very less bitterness and fibre at that time. Besides improved varieties, there are promising native eggplants, which are grown in different regions of the country.

**Table 1** : The promising native eggplants grown in different regions of the country

|  | |  |
|---|---|---|
| Uttar Pradesh and North Central Region | : | Muktakeshi, Ramnagar Giant, Batia, Bholanath, Sheonath, Bundelkhand Desi, Dindhya Bangan, Ramnagar Baigan, Kucjilachia Baigan, Jafrabadi, Balfahwas Jetliwa, Bhanta. |
| Bihar | : | Hyderpur, Improved Muktakeshi. |
| West Bengal | : | Makra, Elokeshi, Krishnanagar Purple Round, Krishnanagar Green Long, Muktakeshi, Baromashi, Kalyani Green, Nurki. |
| Maharashtra | : | Manjari Gota, Bourad Local, Borgona, Local Khandesh, KhedShivapur, Muleti, Malkapuri, Shirur Khandala, Khanpur, Rajangona, Songona Kagal, Tambalwadi Local, Chikhalgaon Local, Dorli. |
| Gujarat | : | Junagadh Bhattu, Surati Gota, Gulabi Dorla. |
| Karnataka | : | Coorg Local, Irangeri, West Coast Green Round. |
| Madhya Pradesh | : | Nurki Round, Nurki Long, Banjari |
| Tamil Nadu | : | Wynad Giant, Pattabiram, Annamalai |
| Punjab | : | Ludhiana Local, Batia |
| Other cultivars (Mainly from Bihar and Orissa) | : | Changulia, Okhla, Very long, Lamba, Round Bangalore, Satputia Hara, Satputia Kathvatia, Grasevindy, Local Kalimpong, Gola Hara, Long Saharapur, Local oval, Kusumsila, Bheda Tupri, Bada Tupri, Rattan Mani, Burupada, Borigumma, Kutiguda, Dhenkanal, Batparsi, Balugaon Market, Banmalipur, Narayanpur, Nisan Local, Jaganathpur (Adivasi), Malipondi, Anchla-B, Bheda Ramchandrapur, Chekkaguda, Narikiri Tholi, Randha, Pathuria, Chitrakonda, Rayagada Baigan, Kharagpuri, Bogma, Bhola, Bara Mahanian, Surajmukhi, Bhanta, Bor Baingna, Pol Baigan. |

Source : Swarup (1998).

## Varieties/Hybrids grouped on the basis of shape and colour of fruits

***Long purple :*** Pusa Purple Long, Pusa Purple Cluster, Pusa Kranti, KT-4, DBSR-31, Arka Sheel, Azad Kranti, ARU-1, S-16, Punjab Barsati, Punjab Sadabahar, Punjab 129-5, Pant Samrat, H-4, ARU-2C, H-7, NDB-25, BB-26, Sel-4, BB-13, KS-331, JB-15, NDB-28-2, Pusa Hybrid-

5, NDBH-6, ARBH-201, PBH-6, Pant Hybrid-1, MHB-1, Long 13, Junagadh Long, Pusa Bhairav,

***Long Green*** : Arka Kusumakar, Arka Shirish, Krishnanagar Green Long.

***Round purple*** : PB-91-2, T-3, K-202-9, H-8, KS-224, DBR-8, AB-1, CHBR-1, D-2-88-6, Pusa Hybrid-6, Vijay Hybrid, NDBH-1, Pusa Hybrid-9, JVH-1, MHB-2, MHB-10, Round 14, Azad Hybrid, Pusa Purple Round, Black Beauty, Jamuni Gola, Krishnanagar Purple Round, Pant Rituraj, Shurti Gota, $T_3$.

***Round green*** : Banaras Giant, Ramnagar Giant.

***Small round*** : Aruna, DBSR-44, PLR-1, DBSR-91, JB-64-1-2, ABH-1, ABH-2, Phule Hybrid-2.

***Oval or oblong purple*** : Arka Navneet, Junagadh long, H-4, Pusa Anmol, Suphal ($F_1$ hybrid).

***Round with strips*** *(Purple white)* : Manjari, Manjari Gota.

***Round white*** : Swarna Shree.

**Brinjal varieties grouped on the basis of their morphological distinctions :**

i) ***Fruits borne in cluster*** : Arka Kusumakar, Arka Neelkanth, Pusa Purple cluster, Pant Samrat, ARU-2C (Almora).
ii) ***Spines Present*** : H-4, Manjari, Manjari Gota.
iii) ***Plant growth erect*** : Pant samrat, H-4.
iv) ***Plant growth spreading*** : $T_3$-Pant Rituraj, Surti Gota.

**Table 2** : Morphological Characters of Brinjal Cultivars / Hybrids

| Cultivars /hybrids | Developing Institute/ Centres | Yield (q/ha) | Characteristics |
|---|---|---|---|
| Pusa Anmol | IARI | 385.00 | It is a hybrid in the attractive dark purple oblong fruits, early and high yielder. Yield increased to 80% over PPL. |
| Pusa Kranti | IARI | 400.00 | Oblong, deep purple glossy fruits, high yielding variety |

Contd...

| | | | |
|---|---|---|---|
| | | | with shining, green calyx and less seeded. It is a across from three; PPL, Hyderpur and Wynard Giant. |
| Pusa Purple Long | IARI | 300.00 | Long purple glossy and tender fruits, suitable for summer and autumn crops. It is a selection from the mixed Baria variety grown in the Punjab, Delhi and Western Uttar Pradesh. |
| Pusa Purple Cluster | IARI | 250.00 | Fruits, long deep purple born in cluster, moderately resistant to bacterial wilt, suitable for southern and northern hills. |
| Pusa Bahairav | IARI | 275.00 | Fruits, deep purple, glossy long, early variety, resistant to phomopsis fruit rot. It is a across of (11a x PPL) x2-4-8-2. |
| Pusa Purple Round | IARI | 325.00 | Fruits, attractive round purple, wt. 450 g / fruit, tolerant to little leaf. |
| Pusa Anupam | IARI | 350.00 | Fruits, cylindrical, medium long, purple, borne in cluster, resistant to bacterial wilt, duration 160 days. |
| Pusa Upkar | IARI | 400.00 | Fruits, pendent, solitary round, medium size, glossy, dark purple, average weight 200 g. |
| Pusa Uttam | IARI | 410.00 | Fruits, pendent, oval medium large sized, glossy dark purple, average weight 250 g. |
| Pusa Bindu | IARI | 300.00 | Fruits, pendent, small, oval round, glossy, violet purple, solitary to partial cluster. About 14 fruits/kg wt. |

Contd...

| | | | |
|---|---|---|---|
| Pusa Ankur | IARI | 337.00 | Fruits, small, oval round, dark purple, glossy, 60-80 g av. wt. Early bearer, first picking 45 days after transplanting. |
| Pusa Hybrid-5 | IARI | 500.00 | Fruits, long, glossy, dark purple, wt. 100g , 80-85 days from sowing to fruit picking. |
| Pusa Hybrid-6 | IARI | 650.00 | Fruits, round, glossy, violet purple, weight 200 g/fruit. 85-90 days sowing to first picking, early bearing. |
| Pusa Hybrid-9 | IARI | 670.00 | Fruits, oval-round, glossy, dark purple, weight 300 g/fruit, 90 days from sowing to first picking. |
| DBR-8 | IARI | 410.00 | Fruits, round, glossy, dark purple , medium size, 5 fruits/kg wt., medium duration (95 days for first picking). |
| DBR-31 | IARI | 430.00 | Fruits, oval-round, glossy purple, medium size, 4-5 fruits/kg wt. Medium duration (85 days for first harvest). |
| DBR-44 | IARI | 300.00 | Fruits, small oval round, glossy dark violet purple, 14 fruits/kg wt. Medium duration (85 days for first harvest). |
| Swarnashree | CHES, Ranchi | 650.00 | Fruits, white round, less seeded, flesh spongy, early variety, moderately tolerant to bacterial wilt and shoot and fruit borer (Line CHES-157). |
| Azad B3 | CSAUAT | 350.00 | Fruits, medium thick, long, shining purple, glossy attractive, high yielding, good for fresh market. |

Contd...

| | | | |
|---|---|---|---|
| Azad hybrid | CSAUA T | 450.00 | Fruits, bold, bright purple, early bearer. Cross from Azad $B_1$ (round ) x Kalyanpur $T_3$ (Round). |
| Azad Kranti | CSAUA T | 425.00 | Fruits, uniform, thick, oblong (15.20 cm long), dark purple, high yielder, suitable for Zone I and IV. |
| K-202-9 | CSAUAT | 380.00 | Fruits, bright purple, mature in 70-75 DTP. Tolerant to fruit rot, wilt and blight. |
| $T_3$ | CSAUAT | 340.00 | Fruits, round, light purple with whitish green colour at stigmatic end. Resistant to little leaf and bacterial blight. |
| KT-4 | IARI, Katrain | 370.00 | Fruits, cylindrical, tender, firm, purple colour, good keeping quality. Variety resistant to wilt. |
| Arka Sirish | IIHR | 384.00 | Fruits, thick (2-3 cm), long dark green, borne in cluster (6-8 fruits). |
| Arka Kusumakar | IIHR | 490.00 | Fruits, small, light green, borne in cluster (5-7) good quality. Maturity 110 days after transplanting. |
| Arka sheel | IIHR | 480.00 | Fruits, medium-long, tender, deep purple, Maturity 120 days after transplanting. |
| Arka Nidhi | IIHR | 485.00 | Fruits, blue-black, glossy, green purple, medium-long in cluster. Excellent keeping and transport quality (25 days at 25-30°C). Resistant to bacterial wilt. Maturity in 120 days after transplanting. |
| Arka Neelkanth | IIHR | 530.00 | Fruits, short in cluster, glossy, violet-blue, good keeping |

Contd...

| | | | |
|---|---|---|---|
| | | | quality, free from bitter principle. Resistant to bacterial wilt, duration 150 days. |
| Arka Keshav | IIHR | 540.00 | Fruits, long in cluster, glossy, red purple, free from bitter principle, good cooking and keeping quality. Resistant to bacterial wilt. Duration 150 days. |
| Arka Navneet | IIHR | 700.00 | Fruits, deep purple black, glossy, wt. 350g/fruit, white flesh, less seeded, good cooking and keeping quality, mature in 120 days. |
| Neelachal Kranti (CHBL-54) | CHES | 1694.0 / plant | High yielding, good quality, green, oblong fruit. |
| Neelachal Shyama (CHBL-30) | CHES | 1352.00 / plant | High yield, round light purple fruit with soft pulp suitable for frying. |
| Punjab Chamkila | PAU | 500.00 | High yielder, developed from local material. |
| Punjab Barsati | PAU | 350.00 | Fruits, dark-purple, 18-20 cm long with 4-6 cm dia, tender, high yielder. |
| Punjab Sadabahar | PAU | 600.00 | Fruits, 20-25 cm long, 3-4 cm dia, dark purplish to blackish in colour. Heavy yielder. |
| Punjab-8 | Punjab (Agril. Dept.) | 300.00 | Fruits, round, medium sized light purple, early bearer, first fruiting 65 days after transplanting. |
| Jamuni Gola | Ludhiana | 275.00 | Fruits, round , plumpy, purple and early bearer (65 days after transplanting). |
| PH-4 | PAU | 450.00 | Fruits, medium-long and thin, dark purple, flesh light-green, cross of Hyderpur x PPL. |

Contd...

| | | | |
|---|---|---|---|
| Hisar Jamuni | HAU | 350.00 | Fruits, oblong, bulging at the blossom end, 10-15 cm long, deep bright, purple- blackish, dual season Variety, most suitable as ratoon crop in early spring. Product of Aushey round brinjal. |
| Surya (SM 6-7) | KAU | - | Resistant to bacterial wilt, small purple oval fruits. |
| Swetha (Sm6-6) | KAU | - | Resistant to bacterial wilt, small white elongated fruits. |
| Haritha | KAU | - | Resistant to bacterial wilt, long light green elongated fruits, long duration. |
| CO1 | TNAU | - | Oblong, pale green fruits. |
| CO2 | TNAU | - | Oblong fruits having dark purple streaks under pale back ground. |
| PKM1 | TNAU | - | Small ovate fruits with green stripes, developed through mutation breeding. |
| PLR1 | TNAU | - | Small to medium sized egg shaped fruits with purple colour. |
| MDU1 | TNAU | - | Large, round and purple fruits. |
| KKM1 | TNAU | - | Small, white coloured and egg shaped fruits borne in cluster of 2-4. |
| Annamalai | Annamalai | - | Aphid resistant. Fruits oblong, purple, few thorns on the calyx. |
| Gulabi (Sel 4) | APAU, Hyderabad | - | Light purple, medium long fruits borne in clusters of 3-5, suitable for long distance transport. |
| Shyamala | APAU, Hyderabad | - | Very small purple around fruits. |

Contd...

| | | | |
|---|---|---|---|
| Bhagyamathi | APAU, Hyderabad | - | Oblong and deep purple fruits. |
| Hisar shyamal | HAU | 325.00 | Fruits, round, fleshy, bright purple, tolerant to little leaf, lesser prone to shoot and fruit borer. |
| Ks 224 | - | 400.00 | Fruits, round, attractive, purple, wt. 300 g/fruit, tall variety. |
| Doli 5 | RAU | 300.00 | Fruits, oval, stocky, dark purple, wt. 150 g/fruit, dual season, early variety, duration 120 days. |
| Rajendra Baigan | RAU | 325.00 | Fruits long, light green, slight curve, heavy bearer, dual season, tolerant to wilt and phomopsis blight. |
| Aruna | - | 275.00 | Fruits, round, purple, attractive. |
| Pant Samrat | GBPUAT | 450.00 | Fruits, medium long, dark purple, borne in cluster. First picking in 70 days after transplanting. Tolerant to phomopsis blight, bacterial wilt, shoot and fruit borers and tassids (PBR-129-5). |
| Pant Rituraj | GBPUAT | 400.00 | Fruits, round, dark purple, shiny and medium sized, field resistant to bacterial wilt. Early variety, first picking 60 days after transplanting (T-3 x PPC). |
| Ramnagar Giant | Ramanagar Varanasi | 300.00 | Fruits, round, shoft, light green, very big size, less seeded, suitable for Bharta preparation. |
| Junagadh long | GAU | 270.00 | Fruits, medium, size, pinkish coloured, long shaped fruits |

Contd...

| | | | |
|---|---|---|---|
| | | | with tapering ends. Early variety (first picking 50 days after transplanting) cross of Doli 5 x Panchmahal. |
| Surti Gota | MS (Agril. Dept.) | 200.00 | Fruits, medium to big , round with purple colour, soft and watery in taste. |
| Manjari Gota | MS (Agril.Dept.) | 250.00 | Fruits, medium, round, rosy colour, white strips and very tasty. Resistant to bacterial wilt. |
| MHB-39 ($F_1$) | MAHYCO | 350.00 | Fruits, oval shining, reddish purple, clustered (4-6 fruits), early bearing hybrid, (60-65 days for first picking), duration 130 days cross of MBA 149 x MBR 1285. |
| MHB-56 ($F_1$) | MAHYCO | 300.00 | Fruits, oval, greenish to yellow, matures in 130 days. |
| MHB-63 ($F_1$) | MAHYCO | 550.00 | Fruits, round, violet with white streaks, high yielder. |
| MHB-70 ($F_1$) | MAHYCO | 575.00 | Fruits, very long, medium duration, high yielder. |
| MHB-80 ($F_1$) | MAHYCO | 585.00 | Fruits, round, reddish-black colour, rough skin, good keeping quality. |
| MHB-82 ($F_1$) | MAHYCO | 580.00 | Fruits, big and heavy, 150-200 g/fruit, mature in 60-65 days, early hybrid. |
| MHB-111(3) ($F_1$) | MAHYCO | 565.00 | Fruits, long, blackish violet, less seeded. |
| Narendra Hybrid Brinjal-1 (NDBH-1) $F_1$ | NDUAT | 570.00 | Fruits, round, purple, shining, soft fleshy, less seeded, wt. 150-200 g/fruit, harvested 70-75 DATP, resistant to little leaf, cross of PBR91-1 x K 202-14-90-110. |

Contd...

| | | | |
|---|---|---|---|
| Brinjal-6 (NDBH-6) | NDUAT | 600.00 | Fruits, long-oval, purple soft, fleshy, wt. 150-200g, bear in bunch, moderately resistant to shoot and fruit borer, phomopsis blight and little leaf. Cross of $T_3$ x Pant Samrat. |
| Kashi Komal ($F_1$) | IIVR | 720.00 | An-excellent long-fruited and prolific bearing hybrid. Fruits are light purple, long, soft textured, picking starts 65-70 days after transplanting. Suitable for growing in all brinjal growing parts of the country. |
| Kashi Sandesh ($F_1$) | IIVR | 780.00 | Fruits, purple, medium-sized, round (12.4 cm long x 10.2 cm diameter) with average weight of 225 g, picking starts 76 days after transplanting. |
| Kashi Prakash | IIVR | 700.00 | An open pollinated variety, plants semi-upright, with green stems and leaves, fruits oblong, attractive with light green spots, calyx spiny and average wt. 190 g. The picking starts 80-82 days after transplanting. |
| Kashi Taru ($F_1$) | IIVR | 750.00 | An open pollinated variety, tall and erect, Fruits long, dark purple, glossy and tender, weigh 75-85 g. Picking starts 75-80 days after transplanting. Suitable for summer and autumn seasons. |
| VRBHR 1 ($F_1$) | PDVR | 850.00 | Fruits, ideal for exports purpose. |
| PLR-1 | | 300.00 | Fruits, small round, purple and clustered (1-2 fruits). |
| 47 PPL ($F_1$) | Golden Seeds | 360.00 | Fruits, long, cylindrical, deep purple, wt. 90-100 g /fruit. |

Contd...

| | | | |
|---|---|---|---|
| | | | First harvest commences 40-50 days after planting. |
| 321 ($F_1$) | Golden Seeds | 410.00 | Fruits, medium long, light purple, wt. 80-100 g/fruit, first harvest 55 to 60 DAP. |
| Nishant ($F_1$) | Golden Seeds | 450.00 | Fruits, elongated oval, dark violet, glossy, clustered (3-4/cluster), wt. 100-120 g/fruit. |
| Green Long ($F_1$) | Golden Seeds | 420.00 | Fruits, green, long, slender, clustered (3-4/cl), wt. 100-120 g/fruit, resistant to bacterial wilt. |
| Preeti ($F_1$) | Golden Seeds | 400.00 | Fruits, small, round to oval, lustrous purple, clustered (3-5/cl), 40-60 g each first harvest after 40 DAP. |
| Krishna ($F_1$) | Golden Seeds | 450.00 | Fruits, oval-round, dark purple (blackish), clustered (2-3/cl) wt 80-100 g/fruit, first picking 40-50 DAP, potential to set fruit in adverse climatic condition. |
| New Kiran ($F_1$) | Golden Seeds | 400.00 | Fruits, round, bright purple, solitary, weigh 180-200 g each, first picking 62 DAP. |
| 310 ($F_1$) | Golden Seeds | 420.00 | Fruits, large sized, attractive, round, deep black, solitary, wt. 200-250 g each, first harvest 60-65 DAP. |
| Green Moti ($F_1$) | Golden Seeds | 450.00 | Fruits, small, oval-round, green with white streaks, weigh 30-40 g each , borne in cluster (3-4 cl), first harvest 40-42 DAP. |
| Rangeela ($F_1$) | Golden Seeds | 450.00 | Fruits, oval-round, light purple with narrow white streaks, borne in cluster (1-3/cl), weigh 100-125 g each, early (55 days first plucking), A prolific yielder. |

Contd...

| | | | |
|---|---|---|---|
| Rangoli ($F_1$) | Golden Seeds | 350.00 | Fruits, oval round, lustrous purple with white streaks, weigh 100-125 g each, borne in cluster (1-3/cluster). |
| Vijay ($F_1$) | Golden Seeds | 400.00 | Fruits attractive, oval-round, light purple with white stripes, weigh 80-100 g each, first harvest 45-50 DAP. |
| BE 701 ($F_1$) | Nunhems Seeds | 320.00 | Fruits, round, purple black, attractive shining, large (300 g), excellent flavoured pulp quality, partial resistant to mites. |
| Sandhya (257) ($F_1$) | Nunhems Seeds | 330.00 | Fruits, round, attractive, purplish black, weigh 250-300 g each, less seeded, good shelf life, early maturing (45-50 DAP). |
| Black Boy (175) | Nunhems Seeds | 310.00 | Fruits, oval-round, purple black, shining, weigh 200-250 g each, good shelf life, partial resistant to sucking pests. |
| Jaya (1001) | Nunhems Seeds | 315.00 | Fruits, firm round, deep purple, weight 40-50 g each, borne in clusters, suitable for stuffing, frying and pickles. |
| Kareena(1002) ($F_1$) | Nunhems Seeds | 320.00 | Fruits, glossy, oval- round, blackish, purple, weigh 80-100g each, early maturing (50-55 days), partial resistant to sucking pest and high temperature. |
| Shilpa (743) ($F_1$) | Nunhems Seeds | 345.00 | Fruits, glossy blackish, long, tender flesh, weigh 100-125g each, partial resistant to sucking pests and bacterial wilt. |
| Yamini (177) ($F_1$) | Nunhems Seeds | 335.00 | Fruits, oval, blackish, weigh 100-150 g, early maturing |

Contd...

| | | | |
|---|---|---|---|
| | | | (45-50 days), partial resistant to sucking pests. |
| BE 708 ($F_1$) | Nunhems Seeds | 325.00 | Fruits, elongated-oval, purplish black, weight 200-250 g each, high yielder. |
| Abhishek (256) ($F_1$) | Nunhems Seeds | 345.00 | Fruits, deep purple, round, weigh 300g each. |
| BE 925 ($F_1$) | Nunhems Seeds | 330.00 | Fruits, glossy green, long cylindrical straight, weigh 80-100 g each, partial resistant to sucking pests. |
| Chaya (3715) ($F_1$) | Nunhems Seeds | 325.00 | Fruits, half long, blackish purple, glossy, borne in clusters, resistant to bacterial wilt and partly resistant to sucking pests. |
| BE 706 ($F_1$) | Nunhems seeds | 335.00 | Fruits, long (15-20 cm), glossy purple, early maturing (50-55 days), wide adaptability. |
| Kanha (091) ($F_1$) | Nunhems Seeds | 315.00 | Fruits, small (50 g), oval round, deep purple, borne in cluster, partly resistant to sucking pests. |
| Mohini ($F_1$) | Nunhems Seeds | 328.00 | Fruits, round to oval, violet with white stripes, borne in clusters and less seeded. |
| Super Mohini ($F_1$) | Nunhems Seeds | 370.00 | Fruits, round, pink with white stripes , glossy, firm and attractive, weighing 75-80 g each, resistant to sucking pests. |
| Kamini ($F_1$) | Nunhems Seeds | 365.00 | Fruits, round, pink with white stripes, borne in clusters and spines on calyx, partial resistant to sucking pests. |
| Harshita (1346) ($F_1$) | Nunhems Seeds | 355.00 | Fruits, oval-round with white stripes, glossy, green in colour, borne in cluster |

Contd...

| | | | |
|---|---|---|---|
| | | | (2-3 fruits), medium maturity (50-55 days), high resistant to bacterial wilt. |
| Pinki (1364) ($F_1$) | Nunhems Seeds | 360.00 | Fruits, medium large (15-16 cm), attractive, pink, borne in cluster. Field resistant to bacterial wilt, medium maturing hybrid. |
| Mahakranti (PK 123) ($F_1$) | Sungro Seeds | 345.00 | Fruit, long club shaped, bright purple, wt. 75 g each, first picking 60 days after transplanting, tolerant to fruit borer. |
| Pragati (Chu-Chu ) ($F_1$) | Sungro Seeds | 675.00 | Fruit, oval-round, purple, weight 30 g each, borne in cluster, first picking 45 DAP , high yielder hybrid, tolerant to fruit borers. |
| Unnati (Chu-Chu) ($F_1$) | Sungro Seeds | 610.00 | Fruits, oval-round, black, weigh 40 g each, high yielding hybrid, tolerant to fruit borer, first picking 45 DAP, suitable for stuffing. |
| Green long ($F_1$) | Sungo Seeds | 520.00 | Fruits, long green, weigh 100 g each , borne in clusters, high yielder hybrid, first picking 60 DAP. |
| Kanhaiya ($F_1$) | Sungro Seeds | 560.00 | Fruits, oval-round, uniform, dark -black, weight, 120 g/ fruit, tolerant to jassids, first picking 60 DAP high yielder. |
| Navkiran ($F_1$) | Sungro Seeds | 465.00 | Fruits, round, bright purple, weigh 160 g each, first picking 65 DAP, tolerant to fruit borer and jassids, recommended for rainy season. |
| Mukta ($F_1$) | Sungro Seeds | 475.00 | Fruits, oval, dark purple, weigh 100 g each, first picking |

Contd...

| | | | |
|---|---|---|---|
| | | | 65 DAP, do well in rainy season also, continuous fruiting. |
| PPL-74 ($F_1$) | Sungro Seeds | 485.00 | Fruits, cylindrical, dark purple, weight 90 g each, less seeded, first picking 55 DAP. |
| Ankur 888 ($F_1$) | Ankur Seeds | 440.00 | Fruits, round and blackish-purple, solitary, wt. 150-200 g each, first picking 105-115 DAS. Good for bharata purpose. |
| Ankur-786 ($F_1$) | Ankur Seeds | 430.00 | Fruit, long slender with purple colour, cluster fruiting, wt. 60-70 g each, first picking in 100-110 DAS. |
| Hirak ($F_1$) | Ankur Seeds | 450.00 | Fruits, oval and green with light green strips, clusterous fruiting, wt. 50-60 g each, first picking starts 105 –115 DAS. |
| Abhimannyu ($F_1$) | Ankur Seeds | 475.00 | Fruits, oval-round and purple with white strips, clusterous fruiting, wt. 50-70 g each, first picking 110-115 DAS. |
| Ajay ($F_1$) | Ankur Seeds | 415.00 | Fruits, oval, medium size, clusterous fruiting, purple with green streaks, good shelf life. |
| Kuldeep ($F_1$) | Ankur Seeds | 435.00 | Fruits, oval-round, green colour, medium size, good fruit quality. |
| Kaustubh ($F_1$) | Ankur Seeds | 385.00 | Fruits oval-round, purple with white strips, single to pair fruits. |
| Utkarsha ($F_1$) | Ankur Seeds | 465.00 | Fruits, oval, purple, medium sized, bearing in clusters. |
| Kushal ($F_1$) | Ankur Seeds | 375.00 | Fruits, long, slender, medium, light green colour, fruiting clusterous. |

Contd...

| | | | |
|---|---|---|---|
| Kirti ($F_1$) | Ankur Seeds | 455.00 | Fruits, oval, medium to small size, green with light green strips, good quality. |
| Shymal ($F_1$) | Ankur Seeds | 565.00 | Fruits, oblong, purple, clusterous fruiting, good cooking quality, excellent yield potential. |
| Abhishek ($F_1$) | Ankur Seeds | 438.00 | Fruits, oval round, purple with white stripes, clusterous fruiting, wt. 55-65 g each. |
| Ajinky ($F_1$) | Ankur Seeds | 415.00 | Fruits, round, green with purple strips, solitary, wt. 50-70 g each, first picking 100-110 DAS. |
| Nakul ($F_1$) | Ankur Seeds | 385.00 | Fruits, oval, green with light green strips, wt. 45-60 g. each, first picking 90-105 DAS. |
| NS 509 ($F_1$) | Namdhari Seeds | 425.00 | Fruits, dark purple with white stripes, solitary bearing, glossy, attractive, medium sized (70 g), good performer. |
| Apsara ($F_1$) | Namdhari Seeds | 450.00 | Fruits, round to oval, bright purple with white stripes, glossy, medium sized (80 g), clusterous, suitable for growing throughout the year. |
| NS 1720 ($F_1$) | Namdhari Seeds | 465.00 | Fruits, oval, uniform purple, size 60-70 g each, cluster bearing, early fruiting, profuse yielder. |
| Nisha ($F_1$) | Namdhari Seeds | 520.00 | Fruits, bell shaped, deep purple glossy, uniform, wt. 350 g each, bear solitary. |
| Sunder ($F_1$) | Namdhari Seeds | 540.00 | Fruits, bell-shaped, deep purple colour, shiny, wt. 350-400 g each, solitary, good adaptability. |

Contd...

| | | | |
|---|---|---|---|
| BSS 471 ($F_1$) | Bejo Sheetal Seed | 410.00 | Fruits, oval to slightly high oval, purple with green as well as white stripes. |
| BSS 472 ($F_1$) | Bejo Sheetal Seed | 370.00 | Fruits, round to oval, purple with white as well as green stripes, suitable for all three seasons. |
| Kalyan ($F_1$) | Bejo Sheetal Seed | 425.00 | Fruits, reddish purple with white stripes, round to oblong. first harvest 65-70 DAP. Suitable for all seasons. |
| Arpan ($F_1$) | Bejo Sheetal Seed | 415.00 | Fruits, round to oblong, dark purple with white stripes. first harvest 55-60 DAP. Suitable for all seasons. |
| BSS 331 ($F_1$) | Bejo Sheetal Seed | 421.00 | Fruits, oval to oblong, purple, early maturing, very productive, maturity in 50 DAP. |
| BSS 332 ($F_1$) | Bejo Sheetal Seed | 410.00 | Fruits, round to oval, dark, purple, thorn less, suitable for growing in all season, mature in 55-65 DAP. |
| BSS 351 ($F_1$) | Bejo Sheetal Seed | 395.00 | Fruits, oblong, black, thorn less, first harvest 55-60 DAP. Suitable for growing in all seasons. |
| BSS 426 ($F_1$) | Bejo Sheetal Seed | 400.00 | Fruits, oval, dark purple, small, cluster- bearing, suitable for all seasons, fruits chu-chu type. |
| BSS 427 ($F_1$) | Bejo Sheetal Seed | 380.00 | Fruits, oval-round, dark purple, first harvest 55-60 DAP, suitable for round the year cultivation. Fruits chu-chu type. |
| BSS 388 ($F_1$) | Bejo Sheetal Seed | 390.00 | Fruits, medium long and dark purple, thick, clusterous, |

Contd...

| | | | |
|---|---|---|---|
| | | | profuse fruiting. Duel season hybrid. |
| BSS 431 ($F_1$) | Bejo Sheetal Seed | 415.00 | Fruits, long to medium long, purple thorn less, first harvest 55-60 DAP, suitable for all seasons. |
| Kokila (BSS 387) ($F_1$) | Bejo Sheetal Seed | 425.00 | Fruits, oblong, black, first harvest 55-60 DAP, Thorn less chu-chu type, suitable for all seasons. |
| Sumo ($F_1$) (BSS 425) | Bejo Sheetal Seed | 410.00 | Fruits, oblong, black, jumbo sized, as big as 1.0 kg, less seeded. First harvest 70-75 DAP. Suitable for all seasons. |
| Kajol ($F_1$) | Bejo Sheetal Seed | 418.00 | Fruit, round, black, medium sized, profuse bearing, first harvest 55-60 DAP. |
| Kajri (BSS382) ($F_1$) | Bejo Sheetal Seed | 390.00 | Fruits, oval, black, medium-biggest size, slightly oblong, less seeded. First harvest 70-75 DAP. Suitable for all round the years. |
| Nisha (BSS 383) ($F_1$) | Bejo Sheetal Seed | 410.00 | Fruits, medium long, dark purple, prolong fruiting, first harvest 55-60 DAP, all season growing hybrid. |
| BSS 330 ($F_1$) | Bejo Sheetal Seed | 465.00 | Fruits, medium long, purple, slightly thick first harvest 70 DAP, Duel season hybrid. |
| BSS 434 ($F_1$) | Bejo Sheetal Seed | 460.00 | Fruits, medium, long to oblong, reddish black to black, first harvest 70-75 DAP. Suitable for round the year cultivation, mid early hybrid. |
| BSS 465 ($F_1$) | Bejo Sheetal Seed | 470.00 | Fruits long, shining, dark black, first harvest 60 DAP, profuse fruiting, duel season hybrid. |

Contd...

| | | | |
|---|---|---|---|
| Preeti ($F_1$) | Bejo Sheetal Seed | 410.00 | Fruits, round to oval, green with white stripe, slightly purple shade. First picking 60-65 DAP, duel season hybrid. |
| BSS 428($F_1$) | Bejo Sheetal Seed | 405.00 | Fruits, oval, green with white stripes, small sized, first harvest 60-65 DAP, suitable for dual season. |
| Raveena ($F_1$) | Bejo Sheetal Seed | 435.00 | Fruits long green thorn less, first harvest 65-70 DAP, duel season hybrid. |
| Greenlong ($F_1$) | Century Seed | 445.00 | Fruits, long green, firm glossy, large sized, borne in bunches (2-3), early and prolong bearing, excellent flavour of pulp. |
| Kokila ($F_1$) | Century Seed | 425.00 | Fruits, long, deep purple-black, contrasted, less seeded. |
| Vardhan ($F_1$) | Century Seed | 385.00 | Fruits, attractive, purplish, clusterous, wt. 40-50 g, high yielding. |
| Mahadeva ($F_1$) | Century Seed | 415.00 | Fruits, globular, purplish-black, less seeded wt. 300 g each, good cropper. |
| Shiva ($F_1$) | Century Seed | 395.00 | Fruits, round, medium size, purplish black, weight 220-260 g, excellent flavour less seeded. |
| Hybrid-1 ($F_1$) | Century Seed | 425.00 | Fruits, cylindrical uniform, attractive, shiny purple black, good and continuous fruiting for longer period. |
| Nisha ($F_1$) | Century Seed | 405.00 | Fruits, attractive, bluish purple, medium sized, clusterous, very early maturing hybrid. |
| Choo-choo ($F_1$) | Century Seed | 370.00 | Fruits, elongated, small, glossy, purple black, |

Contd...

| | | | |
|---|---|---|---|
| | | | clusterous (3-5), profuse bearing continuous over a long period. |
| Nisha improved ($F_1$) | Century Seed | 395.00 | Fruits, medium size , slow maturing, adapted to monsoon cultivation. |
| Kamdhenu ($F_1$) | Century Seed | 415.00 | Fruits, small , purple with white stripes, A prolific cropper. |
| Sanyog ($F_1$) | Century seed | 428.00 | Fruits, dark blackish crimson, wt. 45-55 g each, good flavoured, less seeded, clusterous. |
| Angora ($F_1$) | Sandoz Seed | 895.00 | Fruits, oval-round, firm, fleshy, deep, purple black, wt 200-400 g, good shelf life, ideal for long distance transport and export. |
| Harit ($F_1$) | Sandoz Seed | 550.00 | Fruits, oblong-ovate, brilliant green with creamy streaks, wt. 80-100 g, early bearer (45-50 DAP). |
| Green Bunch ($F_1$) | Sandoz Seed | 800.00 | Fruits, oval, round, green with creamy streaks, wt. 75-90 g, clusterous, early maturing. |
| AUB -132 ($F_1$) | Sandoz Seed | 1000.00 | Fruits, oval-round, brilliant black, less seeded, thin skinned and solid flesh. |
| Kanha ($F_1$) | Sandoz Seed | 1100.00 | Fruits, long glossy, violet skin, wt. 90-110 g each, vigorous plant. |
| Manju ($F_1$) | Sandoz Seed | 1050.00 | Fruits, small, oval round, light purple with white stripes, wt. 80-100 g each. |
| Mew-Mew($F_1$) | Novartis Seed | 800.00 | Fruits, small, round, violet, wt. 60-80 g each, lustrous violet, clusterous, medium maturing. |

Contd...

| | | | |
|---|---|---|---|
| Sanju ($F_1$) | Novartis Seed | 750.00 | Fruits, oval round (8-10 cm long), clusterous, wt 90-100 g each, medium maturing. |
| Green Star | Sreemaa Seed | 350.00 | Fruits, long cylindrical, dark purple, wt. 150-200 g each, tolerant to long distance transport and bacterial and root knot wilt, clusterous. |
| Green Star Long | Sreemaa Seed | 410.00 | Fruits, oval-long green with white stripes, wt. 150-200 g each, resistant to bacterial wilt. |
| Cross-99 | Sreemaa Seed | 380.00 | Fruits, oval, whitish green, clusterous (4-5), wt. 200-250 g each, suitable for long distance transportation, resistant to bacterial and nematode wilt. |
| Uphar | Sreemaa Seed | 375.00 | Fruits, long (15-20 cm), cylindrical, dark purple, wt 15-20 g each resistant to long distant transportation, bacterial and nematode root knot wilt. |
| Sudha | Sreemaa Seed | 700.00 | Fruits oblong, dark purple, smaller (100 g) size, clusterous and profuse bearer. |
| Saptarath | Sreemaa Seed | 500.00 | Fruits, long, small, light purple, clusterous (8-10 fruits), resistant to bacterial and root-knot nematode wilting. Suitable for summer cultivation. |
| Vijaya | Sreemaa Seed | 385.00 | Fruits oblong, green-white, striped, wt. 100-150 g each, clusterous, tolerant to shoot and fruit borer, resistant to wilt. |
| Blue Star | Sreemaa Seed | 395.00 | Fruits, oval, glossy, purple colour, wt. 200-250g each, |

Contd...

| | | | |
|---|---|---|---|
| | | | wide adaptability, resistant to wilt, cultivated throughout the year, heavy cropper. |
| Chandrika | Sreemaa Seed | 390.00 | Fruits, round, dark green, white striped, wt. 200 g each, resistant to bacterial and nematode wilt. |
| Panchasakha | Sreemaa Seed | 375.00 | Fruits oval, blue, clusterous (8-10), suitable for summer cultivation. |
| Mukta | Sreemaa Seed | 345.00 | Fruits, oval, purple, glossy, wt. 100 g each, resistant to bacterial and nematode wilt. |
| Purnima | Sreemaa Seed | 350.00 | Fruits, cylindrical, whitish green, Clusterous (6-7) wt. 100-200 g, suitable to long distance transportation, resistant to bacterial and nematode wilt. |
| Komal ($F_1$) | Unicorn seeds | 318.00 | Fruits, long, dark purple, wt. 80-150 g each, medium size, solitary. |
| Kajal ($F_1$) | Unicorn Seeds | 327.00 | Fruits, oval-round, dark purple, wt. 300-400 g, first picking 60-65 days from transplanting. |
| Vijay ($F_1$) | Unicorn seeds | 328.00 | Fruits, oval, firm glossy, black purple attractive, wt. 300-450 g, uniform size. |
| Ghanshyam ($F_1$) | Unicorn seed | 325.00 | Fruits, oval, medium to big, dark purple skin, wt. 300 g each, tolerant to fruit borer and early bearer. |
| Shyam ($F_1$) | Unicorn seed | 320.00 | Fruits, round-oval, purple, small (30-45) g, uniform and early bearing, clusterous, suited for stuffing, first picking 45 days after planting. |

Contd...

| | | | |
|---|---|---|---|
| Prabha ($F_1$) | Unicorn seed | 342.00 | Fruits, round-oval, violet with white streaks. Wt. 80-100 g, early prolific bearer. |
| Gowri ($F_1$) | Unicorn seed | 335.00 | Fruits, long, shiny green, medium size, wt. 40-50 g each, clusterous (6 fruit). |
| Pratibha ($F_1$) | Unicorn seed | 340.00 | Fruits, round, violet with white streaks, size 100-125 g each. |
| Divya | Unicorn seed | 302.00 | Fruits, round, shiny purple, wt. 250-300 g, first picking in 65 days after transplanting. |
| Kiran | Unicorn seed | 325.00 | Fruits, round, shiny green with white spots, borne in clusters, wt. 40-50 g each. |
| BJT 103 | Unicorn seeds | 360.00 | Fruits, long, black purple, wt. 100-130 g each. |
| BJT 146 | Unicorn seeds | 345.00 | Fruits, round-oval, purple, wt. 500-600 g each. |
| Long-13 ($F_1$) | IAHS | 450.00 | Fruits, long, dark purple, early maturing hybrid. |
| Round-14 ($F_1$) | IAHS | 500.00 | Fruits, round, light purple, medium in size. |

## CAPSICUM

Capsicum (*Capsicum annuum* var. grossun); 2n = 24, syn. Bell Pepper, Sweet Pepper, Simla Mirch, as a vegetable is one of the recent introductions in to India from USA, Italy, Brazil, Hungary, Bulgaria, Czechoslovakia and USSR. Plant is an annual or short lived perennial herbs up to 1.50 m in height. It has well developed top root system with many laterals. Stems are branched, erect, often woody at the base, growth determinate or indeterminate. The fruit is a many seeded, berry, hollow with 2 or more locules. The varieties introduced had been bred to suit milder climates and therefore, lack in wider adaptability.

### Varieties and Hybrids

There are several cultivars/varieties and hybrids of bell shaped, non-pungent, mild, thick fleshed sweet peppers/Simla Mirch.

The important varieties and hybrids are California Wonder, Yolo Wonder , King of North, Early Giant, World Beater, Neapolitan , Early Calwonder, Windsor A, Harris Early Giant, Vinedale, Bullnose, Russian 449, Chinese Giant, Ruby King, Tara, California, Arka Basant, Arka Mohini, Arka Gaurav, Bharat, Oshkosh,Hungarian Wax, Elephant's Trunk, R 449, Golden Bell, Domino.

**Detail characteristics of some important improved varieties and hybrids are given below :**

| Varieties/ Hybrids | Developing Institute/ centre/ Agencies | Yield (q/ha) | Remarks (Special Features) |
|---|---|---|---|
| Arka Mohini | IIHR | 200.00 | It is a selection from variety Titan from USA, determinate, with medium to large fruits. blocky, 3-4 lobed, dark green, thick flesh, non pungent, good aroma, fruit 80-100 g each, suitable for both season of cultivation. |
| Arka Gaurav | IIHR | 230.00 | It is a pure line selection from Golden Calwonder from USA. Fruits, dark green, blocky, erect, 3-4 lobed, medium sized (70-80 g), good for both season crop. Tolerant to bacterial wilt diseases. |
| Arka Basant | IIHR | 210.00 | It is an improvement over an Hungarian variety, Soroksari. Fruits creamish white, conical, erect, thick, flesh and prolific continuous bearing, excellent keeping and cooking qualities. Good aroma and crisp texture, good for stuffing. Suitable for both season, good export value. |

Contd...

| | | | |
|---|---|---|---|
| California | IARI, RS, Katrain | 285.00 | Fruits, smooth blocky, 4 lobed, dark green, wt. 90-150 g, maturity 70-80 DAP, 3-4 picking, thick fleshy layer, most popular variety in India. It takes 170-195 days for maturity from seed to seed stage. |
| Kt-1 ($F_1$) | IARI | 320.00 | Very early and heavy yielder with yellowish green fruits pointed at the tip end. Suitable for growing in humid western Himalayan region. |
| Chinese Giant | Foreign Introduction | 145.00 | Plants vigorous and prolific bearer, fruits dark green, 3-4 lobed, medium flesh thickness and vigorous. |
| World Beater | Foreign Introduction | 125.00 | Plant upright, fruit 4 lobed with medium thick flesh, mild and sweet. |
| California Wonder | Foreign Introduction | 150.00 | Plants are vigorous, upright, prolific bearer; fruits 3-4 lobed, smooth with medium thick flesh, yielding about 120-150 q/ha in 125-140 days after transplanting. |
| Yolow Wonder | IARI, RS, | 120.00 | Fruits dark green , drooping, flesh medium thick 3-4 lobed fruit, fleshy sweet and fine flavour turn, bright crimson on maturity. |
| Bharat ($F_1$) | IAHS | 350.00 | Fruits are dark green, 4-lobed with thick flesh, good taste, large sized (150 g), plants are vigorous. Harvest 120-140 days after transplanting. |
| IAHS-P-2 | IAHS | 300.00 | It is relatively heat tolerant, early, compact hybrid. Fruits thick walled, tri lobed, dark |

Contd...

| | | | |
|---|---|---|---|
| | | | green, slightly elongated; light green in colour. Fruit matured in 65 days after transplanting. Average fruit weight, 140 g. |
| Pusa Deepti ($F_1$) | IARI, RS, Katrain | 380.00 | Fruits smooth, erect, conical, light green, thick flesh, long (10 x 4 cm), early to medium maturing (60-70 DAP), suitable for commercial purpose, tolerant to bacterial leaf spot and anthracnose under field condition. |
| NS 436 ($F_1$) | Namdhari Seeds | 300.00 | Fruits dark green, good size, very firm, attractive, 4-lobed and blocky, early bearer. |
| Sindhoori ($F_1$) | Namdhari Seeds | 320.00 | Fruits lustrous green, turn yellow to maturity, blocky, 8 x 8 cm size, firm and very smooth. |
| Ratan ($F_1$) | Namdhari Seeds | 310.00 | Fruits attractive, shiny dark green, blocky, 8 x 8 cm size, firm, smooth red at maturity, very good transport quality. |
| Arun ($F_1$) | Namdhari Seeds | 315.00 | Fruits shiny, dark green, blocky, 8 x 8 cm size and firm, red at maturity. Prolific bearer. |
| Gagan ($F_1$) | Namdhari Seeds | 370.00 | Fruits beautiful light green, blocky, 8 x 6 cm size., medium firm and thin walled, turn red at maturity. |
| Vikas ($F_1$) | Golden Seeds | 327.00 | Fruits large smooth, uniform, blocky, medium green and four lobed, wall thick and firm, av. fruit wt. 150-200 g each. Suitable for outdoor cultivation. |

Contd...

| | | | |
|---|---|---|---|
| Anupam ($F_1$) | Nunhem Seeds | 255.00 | Fruits, bright green, smooth, 3-4 lobed, thick walled, wt, 75-80 g, early (55 days) maturing. |
| Alankar ($F_1$) | Nunhem Seeds | 270.00 | Fruits bell shaped, blocky, excellent firmness, partial resistance to PVYs TMV. |
| Harit ($F_1$) | Nunhem Seeds | 265.00 | Fruits firm, green blocky, smooth and thick wall, suitable for processing. |
| Nutan($F_1$) | Nunhems Seeds | 278.00 | Fruits, lobed , firm, green and smooth surface, turn yellow at maturity. Recommended for green house cultivation. |
| Tanvi ($F_1$) | Unicorn Seeds | 210.00 | Fruits, 4 lobed, blocky, thick walled, firm, wt. 150-200 g each, good keeping quality, tolerant to tobacco mosaic virus. |
| Tanvi Plus($F_1$) | Unicorn Seeds | 215.00 | Fruits, dark green, elongated blocky, thick walled, wt. 200-250 g each, long pendency. |
| BSS 89 ($F_1$) | Bejo Sheetal Seeds | 225.00 | Fruits green, blocky, 4 lobed, wt. 100 g each, first picking 65 days after transplanting, suitable for all seasons growing. |
| BSS 399($F_1$) | Bejo Sheetal Seeds | 245.00 | Fruits green, blocky, 4 lobed, wt. 100 g, each, thick, skin first picking 55 DAP, good for long transportation, very early type and can grown in all seasons. |
| BSS 401($F_1$) | Bejo Sheetal Seeds | 240.00 | Fruits green, 4 lobed , wt 120 g each, mature in 85-90 DAP, vigorous. |
| Green gold ($F_1$) | MAHYCO | 250.00 | Fruits dark green, blocky, on maturity dark red, medium thick skin, 3-4 lobed, pendent, first picking 60-65 DAP. |

Contd...

| | | | |
|---|---|---|---|
| Sungold ($F_1$) | MAHYCO | 240.00 | Fruits long (10-12 cm), shiny dark green, blocky, medium thick skinned with 3-4 lobes first picking 65-70 DAP. |
| Bombay ($F_1$) | Novartis | 700.00 | Fruits dark green thick walled glossy wt. 140 g, 3-4 lobes, longer shelf life, ideal for long distance transportations, tolerant to TMV, PVY and early bearer. |
| Indra ($F_1$) | Novartis | 550.00 | Fruits dark green, thick walled and glossy, wt. 170 g, 3-4 lobes, fruiting starts 55 DAP, longer shelf life. Ideal for long distance transport. High export potential, tolerant to TMVS and PVY. |
| Lario ($F_1$) | Novartis | 800.00 | Fruit medium sized, dark green, thick walled, fleshy and blocky type, wt. 150 g, 3-4 lobes, first picking starts 55 DAP, long shelf life. Ideal for long distance-transport, tolerant to TMV. |
| Portos ($F_1$) | Novartis | 600.00 | Fruits blocky, 4 lobed, excellent, long distance transportability, tolerant to TMV and PMVY. |
| California | Foreign Introduction | 450.00 | Fruits dark green, blocky, thick and juicy flesh, 3-4 lobes, dark, red at ripening, long, shelf life, tolerate long distance shipping, tolerant to tobacco mosaic virus. |
| Tara | Novartis | 385.00 | Fruits medium large, 4-lobed, thick wall, wt. 250 g, dark green fruits turn to golden yellow on maturity. |
| Zenith | Tropica (Green Co.) | 375.00 | Fruits dark green to dark red at full maturity, size (10 x 8 cm), 3-4 lobes, thick flesh, ideal for export. |

Contd...

| | | | |
|---|---|---|---|
| DARL 202 (F1) | IARI | 422.00 | Its plants are with pendent fruits. Fruits are green, average fruit 18-20 / plant. Maturity days are 90-100 DAT. Resistant to fusarium wilt. Suitable for cultivation in open and protected conditions with normal package of practices, especially adapted for hilly regions. |

## CHILLI

Chilli (*Capsicum annum* var. *acuminatum* Linn.); 2n=24, syn. Lalmirchi, Lalmirch, Mirach, Mirapakaya, Galakonda, Kempu, Mensau, Malagay, Oerangimuluk,Marichiphalam, Katuvira, Spanish pepper, Red pepper, Cayenne pepper, Filfil , Surkh, Gasmiris, Mirch is a quintessential spice in every Indian couisine and is grown throughout length and breadth of country. Green chilli, Chile pepper, Cayenne pepper, Tabasco, Paprika, Sweet/Bell pepper, Pimentos, Serrano pepper are all derived from fruits of different species of *Capsicum*. Pungent forms are used as green chilli. Dried fruits a well as dried fruits powders and extensively used as spice. Where as some low pungent dry chilli varieties are used for colouring also.

### Varieties and Hybrids

There are many varieties of chilli according to shape size, colour and pungency which are grown in the country. Important among are: Pusa Jwala, Pusa Sadabahar, Jwahar Mirch-283, Resham Patti, Bharni, Japani Laungi, Surya Rekha, J-218, Musalwadi, X-235, K-1, K-2, Pant C-1, Pant C-2, WP 46 –A, BR-red, Sindhur (CA-960), CA-1068 (Aparna), X 200 (Kiran), CO-1, CO-2, MDU-1, Andhra Jyoti (G-5), Bhagyalakshmi (G-4), G-3, LCA-306, HC-44, HC-28, Utkal Rashmi, Utkal Ava, Utkal Ragini, Arka Lohit Arka Abir, Punjab Lal, Punjab, Surkh, Punjab Guchhedar, Ujwala, AAUM-1 and AAUM-2, Kt-pl 19, Kashi Anmol, Neelachal Agni, Neelachal Prabha.

The important varieties of hot chilli are : Sanauri, Patna Red, Sirhindi, NP-34, NP-41, NP-46 and NP-61, G-1, G-2, CA-451-1 with Indian Origin. Varieties such as Tobasco, Maxican chillies, Anaheim chillies, Hungarian wax etc. have been introduced to India from foreign countries. Besides, some local varieties are also grown in different regions of the country (Swarup, 1998). They are :

***Andhra Pradesh*** : Oosimilagi, Lavang Mirche, Golgonda Micle, S4, Mundi, Sadasivpet, Shivali, Nallapadu, India Dundicut, Warangal, Lanka Chilli, Tomato Chilli, Seema Mirapa, Gundu Type, India Long, Bagyalakshmi, Samba.

***Tamil Nadu*** : Salem, Ramanand, Samba Kodakal, Sattur Samba, Kovilpatti.

***Karnataka*** : Javari, Javagoal, Coilbil, Danur, Gowridumr, Bijadagi, Dubba Bijadagi, Kaddi, Sankeshwar Type, Chincholi, Kollegal Type, Arsikera, Madhugiri.

***Kerala*** : Kandhari, Achar, Vella Khandhari, Aana Mulagu, (Elephant chilli).

***Maharashtra*** : Musalvadi, Shankeshwar, Sindhur, Seema Mirapa, Dandicha, Achalpuri, Malkapuri, DPLC, Bombay-742.

***Gujarat*** : Gholar, Patta.

***Rajasthan*** : Mathania, Mandoria, Haripur, Lalpur.

***Madhya Pradesh*** : Elichip, Kalipeeth.

*Punjab* : Sirhindi, Sanauri Red, Sanauri Yellow, Patna Red, Punjab Lal, Rajpura Long Red, Jullundur, Shokati, Longi.

***Uttar Pradesh*** : Kalyanpur Red, Kandhari, Suryamukhi, Kashi Anmol.

***Bihar*** : Rahri, Sity, Patna Red, Sabour, Angon, Sabour Anal.

***Orissa*** : Puri Red, Puri Orange.

***Assam*** : Latabh, Tiger, Baghi.

***Sikkim*** : Bird's eye.

The important Characteristics of some of the leading varieties and hybrids of Chilli have been given in Table 1.

**Table 1:** Key characteristics of different varieties and hybrids of Chilli.

| Varieties/ hybrids | Source | Yield potential (q/ha) | Characters |
|---|---|---|---|
| Pusa Jwala (NP46-A x Puri Red) | IARI | 250.00 | Fruits long thing green, shrunken skin, usually curved, medium maturity, high capsaicin and red at maturity. Tolerant to virus and root knot nematodes, suitable for green chilli and powder also, highly pungent. |
| Pusa Sadabahar | IARI | 280.00 | Fruits medium , long, green, clusterous (6-14), capsaicin 7 %, oleoresin 12 %, good for green and powder. Tolerant to mosaic and leaf curl disease. |
| NP 46-A | IARI | | 120.00 Fruits long (10.5 cm), thin, pendent, wrinkled, green samba type, red on ripening, good capsaicin (0.53 mg/ g), suitable for both green and dry use. |
| Pant C-1 (NP-46-A x Local) | GBPUA&T | 210.00 | Fruits medium long (6-7 cm), erect, green, smooth, blunt apex, highly pungent, Ist picking 60 DAP, tolerant to leaf curl and mosaic disease. |
| Pant C-2 (NP-46-A x Local) | GBPUAT | 205.00 | Fruits medium long, pendent, green tolerant to leaf curl virus good for fresh market. |
| Sindhur (CA-960) | APAU | 200.00 | Fruits long (8.2 cm), thick, green, deep red at maturity, capsaicin 0.81 mg / g of fruits, duration 180 days. Suitable for green purpose. |
| JCA-154 | JNKVV | 195.00 | Fruits long, dark green, turn to bright red when ripe, duration 115-120 days. Most suitable for pickle making. |

Contd...

| | | | |
|---|---|---|---|
| Kiran (X.200) | APAU (Lam) | 335.00 | Fruits long, light green, thin, dry red on ripe, tolerant to thrips, mite and aphids. |
| Aparna (CA-1068) | APAU (Lam) | 330.00 | Fruits medium long blunt tip, green, turn to yellow on ripening, good for capsaicin extraction, tolerant to salinity. |
| Arka Lohit | IIHR | 280.00 | Fruits long, straight, dark green, smooth, highly pungent, deep red on maturity, duration 180-210. days. Tolerant to moisture stress. |
| Arka Harita (MSH-96) | IIHR | 310.00 | High yielding $F_1$ hybrid, developed by MS line. Fruits medium long (10 cm) with width 1.0cm.Fresh yield 31 t/ha and dry yield 6 t/ha in 150-160 days. Fruits dark green. Resistant to powdery mildew and viruses. |
| Arka Sweta (MSH-149) | IIHR | 384.00 | High yielding $F_1$ hybrid, developed by using MS line. Fruits long (13.2 cm) with 1.3 cm width. Fresh yield 38.4 t/ha and dry yield 6 t/ha in 140-150 days. Fruits are light green and turns red. |
| Arka Meghna (MSH-172) | IIHR | 335.00 | High yielding $F_1$ hybrid developed by using MS line. Fruits long (10.6 cm) with width 1.2 cm. Very early variety taking 24 days for 50 % flowering. |
| Bharni | JNKVV | 270.00 | Fruits long (10 x 7 cm), deep green, straight, bright red on ripe, suitable for all purposes. |
| Japani Laungi | JNKVV | 210.00 | Fruits long, green, clusterous (10-20), upright "Jhumka", flowering 60-65 DAP. Tolerant to pest and diseases. |

Contd...

| | | | |
|---|---|---|---|
| Surya Rekha | JNKVV | 150.00 | Fruits long, light green, single, erect, upward, flowering 45-50 DAP, maroon on ripe, red on drying, strong pungent, suitable for duel season. |
| Jawahar Mirch -218 | JNKVV | 220.00 | Fruits long, thin, light green, erect, purplish tinge, bright, red and glossy on ripe, highly pungent, early maturing. |
| CO-1 | TNAU | 225.00 | Fruits medium long (7.3 x 1.5 cm), green, samba type, high capsaicin, bright red at ripening, duration 210 days. |
| CO-2 | TNAU | 280.00 | Fruits oblong, green, bright red on ripening, capsaicin 0.56 mg/g duel purpose i.e. green and dry use. Suitable for all season cultivation. |
| K1 | TNAU | 290.00 | Fruits medium long (6.6 x 3.8 cm), green, good capsaicin, shiny red at ripening, duration 215 days, suitable for rainfed area. |
| K2 | TNAU | 235.00 | Fruits , medium long (7.3 x 5.5 cm), green, good in capsaicin (0.49 mg/g), bright red at ripening, duration 210 days. Tolerant to thrips and fruit rot. |
| MDU1 | TNAU | 210.00 | Fruits long (8 cm), green clusterous (4-9), good capsaicin. (0.70 mg/g), shiny red at ripening, developed through induced mutation (K1 by gamma irradiation). |
| Agni (F$^1$1) | Sandoz | 305.00 | Fruits long (10 x 1.5 cm), green attractive glossy, pungent, good for long transportation, tolerant to TMV, PVY. |
| Angare ($F_1$) | Sandoz | 310.00 | Fruits long (12 x 1.5 cm), cylindrical, pale green, moderate pungency. |

Contd...

| | | | |
|---|---|---|---|
| Bijlee ($F_1$) | Sandoz | 312.00 | Fruits long (13 x 2.0 cm), Conical, thick walled, green medium pungent, red at maturity early bearing, longer shelf life and good for long distance transportation tolerant to TMV. |
| Roshni ($F_1$) | Novartis | 314.00 | Fruits long (14 x 2.0 cm), cylindrical, dark green glossy, mature to dark red, medium pungent, thin skin, suitable for both fresh & dehydration. |
| Sholay ($F_1$) | Novartis | 317.00 | Fruits long (13 x 1.5 cm), dark green smooth, thin walled and uniform cylindrical, moderate pungency. |
| Agni Rekha ($F_1$) | Novartis | 315.00 | Fruits medium long (11 x 2.0 cm), faint green, lustrous, smooth, thick walled, uniform, solitary, medium pungency, deep red at maturity, good for long distance transport and fresh market, bearing at 55 DAP. |
| Flame Hot ($F_1$) | Novartis | 318.00 | Fruits small ( 8 x 1.5 cm) long, smooth, lustrous dark green, strong pungent, deep red on maturity, start bearing at 55 DAP, good for long transportation and dry use. |
| Hot Line ($F_1$) | Novartis | 315.00 | Fruits long (14 x 1.5 cm), medium thick walled, smooth, lustrous faint green, deep red on maturity. Strong pungent good for long transportation and dry use and powder. |
| Picador ($F_1$) | Novartis | 312.00 | Fruits long (20 x 8 cm), glossy, thick walled, green taste and flavour of capsicum, bearing start 45-50 DAP, good keeping quality, tolerant to TMV, suitable for pickling, salad and fry. |

Contd...

| | | | |
|---|---|---|---|
| Parbhani Tejas | MAU, Parbhani | 260.00 | Fruits long, green, smooth, red at maturity, good quality and use both for green and dry purpose high yielder. |
| Bhagya Lakshmi | APAU (RARS, Guntur) | 240.00 | Fruits medium long (8.8 cm), green, firm attachment, bright red on ripening, good capsaicin, suitable for both green and dry use. |
| Andhra Jyoti (G-5) (G2 x 1331) | APAU | 275.00 | Fruits short (5.1 x 6.3 cm), "gundu type" green smooth, bright, red on ripening, good capsaicin (0.65 mg/g). suitable for both green and dry purpose. |
| Bhaskar (LCA-235) | APAU | 550.00 | Early variety with thin, 5.6 cm long, highly pungent fruits with pointed tip.Tolerant to thrips and mites. |
| Arka Abir | IIHR | 220.00 | Long fruits, thin and green, red at maturity, paprika variety for oleoresin industry, red chilli yield 20 q/ha. |
| Neelachal Agni (CHCL-127) | CHES | 345.00 | Medium spreading plant with light green immature and light red mature fruits. The plant bears 250-300 fruits with 8.5 cm length and 0.6 cm width of each fruit. It is a duel purpose high yielding variety, best suited for rabi crop under Eastern Indian conditions. Resistant to mosaic virus. The green fruit yield of the variety is around 26 t/ha and dry fruit yield is around 6 t/ha. |
| Neelachal Prabha (CHCL-105) | CHES | 256.00 | Upright plant with dark green robust fruits of high pungency. Dark green fruits |

Contd...

| | | | |
|---|---|---|---|
| | | | at immature stage and crimson red at ripening stage. Bears 110 fruits/plant with 9.5 cm length and 0.7 cm width of each fruit. Best suited for kabi crop under western Indian conditions. The green fruit yield of the variety is 13.2 t/ha and red dry fruit yield is 3.6 t/ha. |
| Prakash (ICA 206) | APAU | 475.00 | Thin, long, glossy red pods with medium pungency. |
| LCA 304 | APAU | 180.00 | Fruits (7.8 cm) long, medium, stout, green mature fruit red, tolerant to virus, suitable for rainfed crop. |
| LCA 324 | APAU | 175.00 | Fruits long thin, tolerant to virus diseases, suitable for irrigated condition. |
| LCA 334 | APAU | 185.00 | Fruits medium long and stout, early maturing, ideal for green and red chillies. |
| Hisar Vijay (HC 28) | HAU | 209.00 | Fruits smooth, clustered, upright, straight, medium long (8-9 cm), thin, dark green, highly pungent, bright red on ripening, slight blackish tinge on fruit surface under low temperature, a marker character. First picking in 63 DAP for green and 82 DAP for ripe fruits, suitable for dry powder. Highly resistant to TMV,CMV, PVY and leaf curl, free from fruit rot, powdery mildew and die- back disease and suitable for dry powder, good for rainy, spring season. |
| Hisar Shakthi (HC 44) | HAU | 327.00 | Fruits, erect, clustered, medium long (8-9 cm), light green, red glossy at maturity, |

Contd...

| | | | |
|---|---|---|---|
| | | | high pungency, rich in capsicin, oleoresin, dry matter, pigment, suitable for dry powder and duel season variety. |
| J 218 | JNKVV | 215.00 | Fruits long (10 x 3 cm), thin, light green with purplish tinge, ripe fruits light bright red, glossy, suitable for green and dry use. |
| Musalwadi | MPKVP, Rahuri | 178.00 | Fruits, medium long (6.86 x 0.95 cm), dark green, black patches on ripening blood red, resistant to leaf curl. |
| X-235 | APAU (Lam) | 185.00 | Fruits, medium long green, highly pungent, suitable for pickling and export, early maturity. Tolerant to thrips. |
| Jayanti | MKV Akola | 186.00 | Fruits, long, green, bigger size, good ascorbic and capsaicin content, duration 215 DAP. |
| Punjab Surkh | PAU | 242.00 | Fruits, medium long, dark green, red at maturity rich in capsaicin, oleoresin and dry matter, red at ripening. Resistant to mosaic wilt and die back. |
| Punjab Guchhedar | PAU | 245.00 | Suited for processing. |
| Punjab Lal (Perennial x long red) | PAU | 245.00 | Fruits medium long (4.25 x 0.79 cm), erect, dark green, turning dark red at maturity, rich in capsaicin (0.7 %), oleoresin, dry matter and red pigment, resistant to TMV and CMV. |
| Akash, | IAHS | 369.00 | Long, bright green, glossy, good size, long bright textured and shape, tolerant to foliar diseases. |

Contd...

| | | | |
|---|---|---|---|
| No. 6 | IAHS | 518.00 | Long medium, green glossy, good shape and size, tolerant to leaf curl , die back diseases. |
| Phule Suryamukhi | MPKV | 280.00 | Fruits medium, green, clusterous, erect, red at maturity, tolerant to leaf curl and powdery mildew, duel type, suitable for kharif and summer planting. |
| Phule Jyoti | MPKV | 285.00 | Fruits medium long green, clusterous, pendent, red at ripening, duel purpose, suitable for kharif and summer planting, tolerant to leaf curl and powdery mildew, resistant to lodging. |
| Utkal Rashmi | OUAT | 295.00 | Fruits long, medium girth, straight, green, good pungency deep red on ripening, pendent, suitable for dry chilli, mature in 130-135 days. Resistant to bacterial wilt, fruit borer. |
| Utkal Ava | OUAT | 220.00 | Fruits medium size, green solitary, highly pungent, mature in 125-130 days, deep red on ripening good for both green and dry use. Resistance to bacterial wilt. |
| Utkal Ragini | OUAT | 230.00 | Fruits medium, small, culusterous, green, highly pungent, upright fruiting, mature in 140-145 days, suitable for green chilli. |
| BC-24 | OUAT | 250.00 | Fruits long, bold, green, attractive, pendent, deep red on ripening,. Mature in 125-135 days , good transport quality, suitable for dry chilli purpose. |

Contd...

| | | | |
|---|---|---|---|
| Resham Patty | JNKVV | 120.00 | Fruits long, thick and green and glossy, deep red on ripening. Mild pungent, less seeded, suitable for dry purpose. |
| Kashi Anmol | IIVR | 250.00 | Plants determinate, bushy (60-75 cm), umbrella type. Fruits pendent, attractive, compact, dark green at maturity, pungent, first picking at 50-55 DAT, early bearer, duration 150-160 days and suitable for green chilli and long distance transportation. |
| Kashi Vishwanath ($F_1$) | IIVR | 200.00 | Plants tall (120 cm), dull green stems and bear pendent, long (10-12 cm x 1.0 –1.2 cm) fruits, highly attractive, yellowish green and turn bright red at maturity, pungent with wrinkled surface. First picking of ripe red fruits at 90 DAT, duration 220 days. Suitable for dry fruit production. |
| Kashi Early ($F_1$) | IIVR | 300.00 | Plants tall (110 cm), dull green stem, fruits pendent, long (8-9 cm x 1-1.2 cm), highly attractive, dark green and turn bright red at maturity, pungent with smooth surface. First picking 45 DAT, duration 220 days. Suitable for green as well as dry fruit production. |
| Azad Mirch 1 | CSAUAT | 70.00 | Plants dwarf bushy (35-45 cm), 8-10 branches, bear erect and clusterd green fruits and red at maturity. 12-14 clusters per plant having green 8-12 |

Contd...

| | | | |
|---|---|---|---|
| | | | fruits in a cluster. First picking of green fruits 50 DAT. Takes 135-140 days from transplanting to full maturity. |
| Jwalamukhi | KAU | 235.00 | Fruit dark green with low pungency, suitable for green chilli purpose. |
| Jwalasakhi | KAU | 200.00 | Light green to cream coloured fruits with thick pericarp and low pungency. Suitable for green chilli and vegetable purpose. |
| Ujwala | KAU | 220.00 | Plants determinate, bushy and low branching, medium long (5-6 cm), highly pungent, fruits borne in erect clusters. Resistant to bacterial wilt. Tolerant to mosaic and leaf curl virus. |
| Anugraha | KAU | 270.00 | Hybrid derivative of Jwala x Ujwala. Resistant to bacterial wilt. Pods light green, long, wrinkled and medium pungent. |
| Shankeshwar-32 | MPKV | 220.00 | Plant tall, grown for dry chilli purpose. Fruits 20-30 cm long, thin, wiry, and highly wrinkled. |
| Agnirekha | MPKV | 240.00 | Long, bold, light green and medium pungent fruits. |
| Phule Sai | MPKV | 230.00 | Moderately tolerant to thrips and anthracnose. Fruits retain colour in storage for 6-7 months. |
| Surakta (AKC-86-89) | PKV, Akola | 680.00 | Pure line selection. Yield very high of red fruits. |
| KDC-1 | UAS, Dharwad | 720.00 | Hybrid derivative of Byadagi (*C. annum*) x *C. frutescens*). Deep red, semi wrinkled fruits with low pungency. Resistant to leaf curl. |

Contd...

| | | | |
|---|---|---|---|
| Sabour Arun | RAU, Bihar | 450.00 | Selection from a local genotype, fruits long (9.5 cm). High yielder. |
| CH1 ($F_1$) (MS12 x LLS) | PAU | 250.00 | Plants medium tall (92.6 cm) with medium sized (6.62 cm length and 1.6 cm width), light green fruits turning deep red on ripening, fruits are highly pungent (0.83 %). Colour value 88.75.ASTA units. Tolerant to fruit rot and wet rot. High yield of Red Chilli. |
| CH-3 ($F_1$) (MS-12xS-2530) | PAU | 250.00 | Fruit length more than CH-1(8.2 cm). Capsaicin content (0.52 %), colour value 145 ASTA units. |
| HCH-9646 ($F_1$) | UAS, Dharwad | 250.00 | Fruits medium long (10.11 cm), thick walled, pendulous and medium pungent. Tolerant to leaf curl complex disease. |
| Divya Jyoti ($F_1$) | Unicorn Seeds | 375.00 | Fruits medium, green, long , slender, highly pungent solitary downward. |
| Shama ($F_1$) | Unicorn Seed | 380.00 | Fruits long (20 x 2.0 cm), waxy light green, thick skin, suitable for fresh market, pickle and bhajia. |
| Asha Jyoti ($F_1$) | Unicorn Seeds | 387.00 | Fruits medium thick, pericarp are glossy, long (14 x 1.3 cm), medium thick, slender. |
| Usha Jyoti ($F_1$) | Unicorn Seeds | 370.00 | Fruits long (16 x 1.3 cm), slender medium green, pungent, shiny red at maturity, very smooth, good for transport, suitable for both fresh market and delay duration. |

Contd...

| | | | |
|---|---|---|---|
| Nav Jyoti ($F_1$) | Unicorn Seeds | 367.00 | Fruits long (15 x 1.2 cm), slim, dark green, very pungent, turn bright red at maturity, good for fresh market and drying. |
| Red flame ($F_1$) | Unicorn Seeds | 389.00 | Fruits long (16 x 1.8 cm), medium thick, light green, medium pungent, excellent fruit setting ability. |
| Maha Jwala ($F_1$) | Unicorn Seeds | 372.00 | Fruits long (15 x 0.8cm), slender, light, green and highly pungent. |
| Wonder Jyoti | Unicorn Seeds | 375.0 | Fruits long (16 x 1.5 cm), slender, medium thick skin, shining, excellent for both fresh and dry market. |
| CHT-367 ($F_1$) | Unicorn Seeds | 372.00 | Fruits medium long, green, upright, clustered highly pungent, excellent, upright hybrid. |
| CHT 298 ($F_1$) | Unicorn Seeds | 376.00 | Fruits long (13 x 1.3 cm) light green to green, high pungency, suitable for fresh market. |
| CHT 347 ($F_1$) | Unicorn Seeds | 379.00 | Fruits long (14 x 1.3 cm), light green to green, high pungency, suitable for processing. |
| CHT 380 ($F_1$) | Unicorn Seeds | 377.00 | Fruits long (13.5 x 0.75 cm), green, high pungency good for fresh market. |
| CHT 365 ($F_1$) | Unicorn Seeds | 365.00 | Fruits (14.5 x 1.5 cm), green, high pungency, good for fresh market. |
| CHT 399($F_1$) | Unicorn Seeds | 372.00 | Fruits long (15 x 2.0 cm), light green, medium pungency, suitable for fresh market. |
| CHT 248 ($F_1$) | Unicorn seeds | 370.00 | Fruits (10 x 1.0 cm), green, high pungency, good for fresh market and processing. |
| MHP 58 ($F_1$) | MAHYCO | 310.00 | Fruits long, deep green, pungent, first picking 75 DAP, suitable for both season. |

Contd...

| | | | |
|---|---|---|---|
| MHP 59 ($F_1$) | MAHYCO | 305.00 | Fruits long, deep green, highly pungent, first picking 75 DAP, duel type. |
| Agnishikha ($F_1$) | MAHYCO | 315.00 | Fruits medium long (10 x 2 cm), thick skin, green, solitary, first picking 55 DAP, suitable for green and dry purpose. |
| MHP-62 ($F_1$) | MAHYCO | 332.00 | Fruits medium long (12.5cm), light green, wrinkles, pendent, moderate pungency, first picking 72 DAP, red at maturity, ideal for green chilli summer cultivation. |
| MHP-63 ($F_1$) | MAHYCO | 330.00 | Fruits long (13 cm), light green shiny, wrinkles, pendent, moderate pungency, red at maturity, first picking 70 DAP, suitable for green chilli and for summer cultivation. |
| MHCP 155 ($F_1$) | MAHYCO | 315.00 | Fruits long (13 cm), shiny dark green, smooth, pendent, highly pungent, first picking 72 DAP, deep red at maturity, duel use i.e., green and red purpose. |
| MHCP-157 ($F_1$) | MAHYCO | 320.00 | Fruits medium long (10.5 cm), shiny, dark green, smooth, pendent orientation, low pungency, first picking 70 DAP, deep red at maturity. |
| MHCP-158 ($F_1$) | MAHYCO | 330.00 | Fruits medium long (10-11 cm), light green, medium pungency, first picking 65 DAP, light red at maturity. |
| Tejaswini ($F_1$) | MAHYCO | 375.00 | Fruits long, green to deep green, first picking 75 DAP, highly pungent. |
| Sujata ($F_1$) | MAHYCO | 364.00 | Fruits long, light green, medium pungent, first picking, 70 DAP, duel type. |

Contd...

| | | | |
|---|---|---|---|
| Surya ($F_1$) | MAHYCO | 385.00 | Fruits long, green, upright, highly pungent, first picking 75 DAP. |
| MHCP-159 ($F_1$) | MAHYCO | 337.00 | Fruits medium long (10-11 cm), green, pendent, highly pungent, first picking, 78 DAP, red on maturity. |
| CCH-1 ($F_1$) | Century | 340.00 | Fruits, medium size (6.6 cm), green, pungent, deep red on ripening, good VHC, capsaicin, moderate resistant to mosaic, leaf curls wet rot, resistant to die-back. |
| CCH-2 ($F_1$) | Century | 335.00 | Fruits long (12 x 1 cm), deep green, upright clusters, extremely pungent, mature 70 DAP, scarlet red at maturity, early-medium type. |
| Agnivaan ($F_1$) | Century | 345.00 | Fruits long (12 x 1 cm), medium-green colour, early maturing (65-70 DAP), excellent quality crimson on ripening. |
| Agnidev ($F_1$) | Century | 342.00 | Fruits medium long (7 x 1 cm), excellent deep green colour and very high pungency, solitary pendent fruits, good for dehydration, maturity (68-72 DAP). |
| Agnipath ($F_1$) | Century | 338.00 | Fruits long (10 x 1.6 cm), green, smooth, tight skin, medium-thick wall, high pungency, medium maturing (68-70 DAP), red on maturity, suitable for fresh and dehydration purpose. |
| CCH-791 ($F_1$) | Century | 358.00 | Fruits long (8.0 cm), thin, light green, medium maturing (70-75 days). Red on ripening, good for fresh use and dehydration. |

Contd...

| | | | |
|---|---|---|---|
| CCH-792 ($F_1$) | Century | 348.00 | Fruits long (8-9 cm), medium thickness, light green colour, maturity at 75-80 DAP, deep red at maturity. |
| Anaheim TMR 23 ($F_1$) | IAES | 325.00 | Fruits long (19 x 5 cm) bright green, medium pungent, turning red at maturity, smooth pods, resistant to T.M.V. Mature in 74-76 days. |
| Cayenne large red thick ($F_1$) | IAES | 335.00 | Fruits medium, dark green, thin wall wrinkled (15 x 2 cm) two celled, dark red at maturity, medium thin wall, pendent, mature in 73-76 DAP, excellent for processing and fresh market. |
| Hungarian yellow wax hot ($F_1$) | IAES | 321.00 | Fruits long (14 x 4 cm), light yellow to orange red , smooth conical, upright, pendent, medium thick wall, early type, good for processing and home garden. |
| Cayenne long Slim ($F_1$) | IAES | 330.00 | Fruits long (13 x 1 cm), dark green to red, thin walled, 2-celled, slim, wrinkled, pendent, early-medium maturing (65-70 days) good for home garden. |
| BSS 269 ($F_1$) | Bejo sheetal Seeds | 359.00 | Fruits long (16 x 2.0 cm), less wrinkled, light green, pungent, first picking 75 DAP, dark red at ripening, Hungarian wax type, medium thick, skin, good for fresh market. |
| BSS 356 ($F_1$) | Bejo Sheetal Seeds | 348.00 | Fruits long (15 x 30 cm), rough, green, mild pungent red at ripening, first picking 80 DAP, Hungarian wax type, thick skin, good for all seasons crop. |

Contd...

| | | | |
|---|---|---|---|
| BSS 377 ($F_1$) | Bejo Sheetal Seeds | 385.00 | Fruits long (14 x 1.4 cm), green, smooth, pungent, first picking 85-90 DAP, dark red at ripening, tolerant to virus complex, good for fresh and dry use. |
| BSS 345 ($F_1$) | Bejo Sheetal Seeds | 415.00 | Fruits long (13x1.1 cm), smooth, green, medium pungent, first picking 65 DAP good for dry purpose. |
| BSS 219 ($F_1$) | Bejo sheetal Seeds | 379.00 | Fruits long (14x1.5 cm), green to light green, thick and smooth surface, medium pungent, first picking 70-75 DAP, dark red at ripening, suitable for both green and dry use. |
| BSS 308 ($F_1$) | Bejo Sheetal Seeds | 435.00 | Fruits long (14 x 1.8 cm), smooth, green to light green, pungent, thick skin, first picking 75 –80 DAP, used for green. |
| Laxmi ($F_1$) | Bejo Sheetal Seeds | 395.00 | Fruits long (14 x 1.5 cm), smooth blackish green, mild pungent, medium, thick skin, dark red at ripening first picking 75 DAP, good for fresh and processing use, tolerant to virus. |
| BSS 411 ($F_1$) | Bejo Sheetal Seeds | 379.00 | Fruits long (15 x 1.5 cm) smooth, green, medium pungent, dark red at ripening first picking 75 DAP, good for both fresh and dry purpose. |
| BSS 413 ($F_1$) | Bejo Sheetal Seeds | 359.00 | Fruits long (18 x 1.4 cm) smooth dark green, medium pungent, dark red at ripening first picking 70 DAP, excellent for dry chilli purpose. |

Contd.

| | | | |
|---|---|---|---|
| BSS 346 ($F_1$) | Bejo sheetal Seeds | 430.00 | Fruits long (14 x 1.2 cm), smooth, dark green, mild pungent, first picking 75 DAP, dark red at ripening, good for dry purpose. |
| Arjun ($F_1$) | Bejo Sheetal Seeds | 438.00 | Fruits long (16 x 1.6 cm), less wrinkled, green medium pungency, thick skin, red at ripening, first Picking, 60 DAP, good for fresh use. |
| BSS 213($F_1$) | Bejo Sheetal Seeds | 387.00 | Fruits long fresh use (16 x 3 cm), rough, green, mild pungent, first picking 75 DAP. thick skin, suitable for all season crop. dark red at ripening, Hungarian wax type. |
| BSS 414 ($F_1$) | Bejo Sheetal Seeds | 365.00 | Fruits long (10 x 1.2 cm), light green to green, medium smooth, highly pungent, first picking 75-80 DAP, suitable for all seasons. |
| BSS 140 ($F_1$) | Bejo Sheetal Seeds | 368.00 | Fruits long (10 x 1.2 cm), green to light green, very pungent, medium smooth surface, first picking 60-70 DAP, red at maturity, early maturing, suitable for rainy season for fresh market. |
| BSS 202 ($F_1$) | Bejo Sheetal Seeds | 355.00 | Fruits long (10 x 1.5 cm), green to dark green, medium group, highly pungent, first picking 70-75 DAP, dark red at maturity, good for rainy and summer season. |
| BSS 344 ($F_1$) | Bejo Sheetal Seeds | 360.00 | Fruits long (10 x 1.0 cm), yellowish green wrinkled, highly pungent, first picking 70 DAP, thin skin, red at maturity, good for fresh as well as dry purpose. |

Contd...

| | | | |
|---|---|---|---|
| BSS 375 ($F_1$) | Bejo sheetal Seeds | 367.00 | Fruits long (13 x 1.2 cm), yellowish green smooth, highly pungent, first picking 75-80 DAP, dark red at ripening, tolerant to virus and other diseases good for all season crop. |
| Garima ($F_1$) | Bejo Sheetal Seeds | 378.00 | Fruits long (15 x 1.2 cm), yellow green, smooth, highly pungent, first picking 70-75 DAP, tolerant to virus complex, good for all season cultivation. |
| BSS 141 ($F_1$) | Bejo Sheetal Seeds | 367.00 | Fruits long (132 x 1.2 cm), dark green, less wrinkled, pungent, thin skin, first picking 70-75 DAP, dark red at ripening good for green as well as dry chilli. |
| BSS 376 ($F_1$) | Bejo Sheetal Seeds | 359.00 | Fruits long (12-14 cm), green, smooth, medium pungent, dark red at ripening, suitable for dry chilli. |
| Anmol | Bejo Sheetal Seeds | 362.00 | Fruits long (16 x 1.3 cm), dark green, smooth, pungent. first picking 75 DAP, dark red at ripening, thin skin, excellent for dry chilli. |
| BSS 305 ($F_1$) | Bejo Sheetal Seeds | 364.00 | Fruits long (14 x1.4 cm), medium wrinkled, dark green, medium pungent, first picking 80-85 DAP, thin, skin, good for dry purpose, tolerant to virus. |
| BSS 306 ($F_1$) | Bejo Sheetal Seeds | 360.00 | Fruits long (18 x 1.2 cm), attractive dark green, tough surface, medium pungent, good for dry purpose and good tolerance to virus complex. |

Contd...

| | | | |
|---|---|---|---|
| BSS 355 ($F_1$) | Bejo Sheetal Seeds | 425.00 | Fruits long (16 x 1.2 cm), rough, dark green, pungent. first picking 70 DAP, dark red at ripening, good for dry purpose. |
| Suraj ($F_1$) | Bejo Sheetal Seeds | 395.00 | Fruits long (14 x 1.2 cm), wrinkled , light green to green pungent, first picking 75 DAP, bright red at ripening thin skin, can also used for dry purpose. |
| Savitri ($F_1$) | Bejo Sheetal Seeds | 380.00 | Fruits long (10 x 1 cm), wrinkled, yellowish green, high pungent, thin skin, red at ripening, first picking 70 DAP, high pungent, good for green and dry use. |
| BSS 267 ($F_1$) | Bejo Sheetal Seeds | 375.00 | Fruits long (10 x 1 cm), smooth to slightly wrinkled, light green to green, highly pungent, first picking 75 to 80 DAP, good for dry chilli. |
| BSS 268 ($F_1$) | Bejo Sheetal Seeds | 364.00 | Fruits long (10 x 1 cm), smooth, green, highly pungent, first picking 75 DAP, good for green as well as dry purposes. |
| Gagan ($F_1$) | Golden Seeds | 280 | Fruits small, green upright, very high pungency, first picking 65-70 DAP, suitable for both fresh and dry purposes. Tolerant to mosaic and leaf curl virus diseases. |
| Alankar ($F_1$) | Golden Seeds | 300 | Fruits light green, pendent, high pungency, Ist picking 70-75 DAP, prolific bearer, suitable for both fresh and dry use. |
| Byadagi ($F_1$) | Golden Seeds | 290 | Fruits long light green, pendent, prominently wrinkled, medium pungent, |

Contd...

| | | | |
|---|---|---|---|
| | | | medium early maturing. Suitable for both green and dry fruit. Tolerant to virus diseases. |
| A.K. 47 ($F_1$) | Golden Seeds | 320 | Fruits long, green, and pendent. Pungency medium to high, suitable for fresh and dry market, first picking 70 DAP, adaptation to dry and wet conditions, do well in hot regions and in long rainy period. Tolerant to mosaic and leaf curl diseases. |
| Farmers No.1 ($F_1$) | Golden Seeds | 320 | Fruits long, dark green, pendent, 12-15 cm x 1.0 – 1.2 cm size, high pungency, first picking 68 DAP, suitable for fresh and dry purpose. Tolerant to viruses and anthracnose diseases. |
| Farmers Choice ($F_1$) | Golden Seeds | 300 | Fruits long green shiny, 12-14 cm x 1.3 x 1.3-1.5 cm size, pendent, medium pungency, mid early maturity, suitable for both fresh and dry use. Tolerant to virus diseases. |
| GC-39 ($F_1$) | Golden Seeds | 290 | Fruits large, 12-14 cm x 1.8 – 2.0 cm size, light green, pendent, light to medium pungent, mid early prolific bearer, suitable for dressings, stuffing, and pickles. Tolerant to phytophthora root rot and bacterial diseases. |
| S-16 ($F_1$) | Sungro Seeds | 270.00 | Fruits pendent, dark green, bright red at maturity, high pungency, skin thick, length 8-9 cm. Tolerant to thrips and mites. |

Contd...

| | | | |
|---|---|---|---|
| S-86235 ($F_1$) | Sungro Seeds | 275.00 | Fruits long (10-12 cm), dark green, pendent, pungency medium, thick skin. Compact umbrella type. |
| New Vardan (693) | Nunhems Seeds | 245. 00 | Fruits long (12.5 x 1.1 cm), pendent, light green, thin skin, very pungent, early maturing, strong, suitable for fresh and dry use. Cayenne type, partial resistant to PVY, mites and stress. |
| Sona ($F_1$) | Nunhems Seeds | 325.00 | Fruits long (8 x 0.8 cm) , light green, pendent, thin skin, very pungent, suitable for fresh and dry use, Cayenne type, resistant to PVY. |
| Mashal ($F_1$) | Nunhems Seeds | 345.00 | Fruits long (15 x 1.2 cm), light green, mild pungent, early maturing, pendent, good for fresh use, Cayenne type, Strong and stall. |
| Red star ($F_1$) | Nunhems Seeds | 435.00 | Fruit long (14 x 1.0 cm ), dark green, mild pungent, good for fresh and dry use, early maturing, pendently cayenne type, resistant to CVMV. |
| Sankranti (695) | Nunhems Seeds | 430.00 | Fruits long (13 x 1.2 cm), pendent, dark green, mild pungent, good for fresh and dry use, early maturing, strong framework, cayenne type, resistant to CVMV. |
| Dushhera ($F_1$) | Nunhems Seeds | 415.00 | Fruit long (11.5 x 1.2cm) green, pungent, medium, early maturing, pendent, good for fresh and dry use, cayenne type, good for processing and storage. |
| Devanur Deluxe ($F_1$) | Nunhems Seeds | 427.00 | Fruits long (12 x 1.3 cm), green pendent, medium skin |

Contd...

| | | | |
|---|---|---|---|
| | | | thickness, very pungent, good for fresh and dry use, cayenne type, resistant to CVMV. |
| Soldier ($F_1$) | Nunhems Seeds | 435.00 | Fruits long (10 x 1 cm), green, thin skin, pendent, very pungent, good fresh and dry use, cayenne, resistant to CVMV. |
| Deepali ($F_1$) | Nunhems Seeds | 417.00 | Fruits long ( 8 x 0.9 cm), green, thin skinned, very pungent, medium early, pendent, good for fresh and dry use cayenne type. |
| Green Wonder ($F_1$) | Nunhems Seeds | 422.00 | Fruits long (15 x 1.7 cm), green, thick skinned, mild pungent, good for fresh and dry use, medium-early maturity, pendent, anaheim type, partial resistant to high temperature. |
| Rudra ($F_1$) | Nunhems Seeds | 427.00 | Fruits long (12 x 1.4 cm), light green, thick skinned, mild pungent, suitable for fresh and dry use, early maturity, pendent, anaheim type. |
| Kranti ($F_1$) | Nunhems Seeds | 445.00 | Fruits long (13.5 x 1.4 cm) light green, thick skinned, mild pungent, early maturing, pendent, good for fresh and dry use, anaheim type. |
| Veeru ($F_1$) | Nunhems Seeds | 425.00 | Fruits long (12 x 2 cm), green thick skinned, pendent, mild pungent, good for fresh and dry use, early maturing, anaheim type. |
| Shikha ($F_1$) | Nunhems Seeds | 435.00 | Fruits long (8 x 0.8 cm), green, thin skinned, very pungent, early maturing, upright, pendent, cluster bearing, good for fresh and dry use cayenne type. |

Contd...

| | | | |
|---|---|---|---|
| Anjula ($F_1$) | Nunhems Seeds | 427.00 | Fruits long (6 x 1.2 cm) , dark green, medium skin, very pungent, early maturing, pendent, good for processing, Jalopeno type. |
| Sindoor ($F_1$) | Nunhems Seeds | 437.00 | Fruits long (13 x 1.2 cm), green medium skin, pendent, early maturing, mild pungent, good for fresh and dry use, cayenne type, partial resistant to mites. |
| ARCH -236 ($F_1$) | Ankur Seeds | 430.00 | Fruit, elongated , undulated, pointed and green, turn red at maturity, good shelf life and good for fresh market, high capsaicin content, duration 110-140 days. |
| ARCH-226 ($F_1$) | Ankur Seeds | 415.00 | Fruits long, medium, green shining, turn deep red at maturity. Stable red colour, suitable for summer cultivation. |
| ARCH-228 ($F_1$) | Ankur Seeds | 415.00 | Fruits medium long, attractive dark green colour, later deep red, high capsaicin content in the good colour value. |
| ARCH-001 ($F_1$) | Ankur Seeds | 422.00 | Fruits medium long, dark green smooth, deep red at maturity, suitable for green and red dry fruits, good colour, high capsaicin content. |
| ARCH-112 ($F_1$) | Ankur Seeds | 416.00 | Fruits attractive, lustrous light green, medium long, bright red colour at maturity and good pungency. |
| ARCH-162 ($F_1$) | Ankur Seeds | 410.00 | Fruits, medium, long attractive, light green wrinkled, dark red at maturity, suitable for fresh market green chillies. |

Contd...

| | | | |
|---|---|---|---|
| ARCH-9 ($F_1$) | Ankur Seeds | 405.00 | Fruit medium size, green, compact, deep red at maturity, highly pungent, excellent flowering and fruiting, good transportation. |
| ARCH 55 ($F_1$) | Ankur Seeds | 435.00 | Fruits, lustrous green, medium long, bright red at maturity, medium capsaicin content and high oleoresin, suitable for green and red chilli. |
| ARCH -313 ($F_1$) | Ankur Seeds | 425.00 | Fruits bold triangular shaped, blunt end, attractive green, dark red at maturity, medium pungent, good for green chillies. |
| ARCH-83 ($F_1$) | Ankur Seeds | 405.00 | Fruits long, dark green, lustrous, deep red at maturity, medium pungent, good for red chilli. |
| CHILLI-82 ($F_1$) | Ankur Seeds | 422.00 | Fruits medium long, light green, lustrous, smooth, medium pungency, good for fresh green chilli. |
| NS 101 ($F_1$) | Namdhari Seeds | 401.00 | Fruits long (12 x 15 cm), dark green, thick skinned, medium pungent, maturity 70 days (DAP), suitable for green and dry chilli purpose. |
| NS 1101 ($F_1$) | Namdhari Seeds | 415.00 | Fruits medium long ( 8 x 1 cm), light green, thin skinned, very high pungent, dark red at maturity, mature in 70 DAP, tolerant to virus and thrips. |
| NS 170 ($F_1$) | Namdhari Seeds | 407.00 | Fruits medium long ( 8 x 0.8 cm) light green, thin skinned, very high pungency, dark red at maturity, fruits mature in 75 DAP. tolerant to viruses |

Contd...

| | | | |
|---|---|---|---|
| | | | and thrips, suitable for all seasons. |
| NS 555 ($F_1$) | Namdhari Seeds | 375.00 | Fruits long (13 x 2.25 cm), light green, skin medium thick, medium pungent, dark red at maturity, early bearer. |
| NS 999 ($F_1$) | Namdhari Seeds | 390.00 | Fruits long (14. X 2 cm), light green, medium thick skinned, pungent, dark red at maturity, early bearing. |
| Balwant ($F_1$) | Namdhari Seeds | 385.00 | Fruits long (13 x 1.1 cm), light green, thick skin, dark red at maturity, highly pungent, early bearer, tolerant to virus. |
| Angara ($F_1$) | Namdhari Seeds | 365.00 | Fruits long (14 x 1.4 cm), green, highly pungent, thin skinned, dark red at maturity, early fruiting. |
| NS 3074 ($F_1$) | Namdhari Seeds | 370.00 | Fruits long (13 x 1.2 cm), dark green, thin skin, highly pungent, wrinkled, mature in 70 DAP and deep red at maturity. |
| Goli ($F_1$) | Namdhari Seeds | 368.00 | Fruits long (11 x 1.2 cm), straight, dark green, glossy pungent, thin skinned, mature in 70 DAP, dark red at maturity. |
| Akash ($F_1$) | Namdhari Seeds | 395.00 | Fruits medium long (6 x 0.8 cm) light green thin skinned, very high pungent, solitary bearer, red at maturity and mature in 95 DAP. |
| Sudra ($F_1$) | Tropica Green Co. Ltd. | 100.00 | Strong growth habit. Many slim, large shape fruits. Fruit length 13 cm, width 0.7 cm and weight 5-6 g. Tolerant to Anthracnose, Potato Virus Y strain 1, 2 (PVY 1-2), |

Contd...

| | | | |
|---|---|---|---|
| | | | Phytophthora blight and Bacterial wilt, suitable for fresh and dry markets. |
| Salmon($F_1$) | Tropica green Co.Ltd. | 100.00 | It is vigorous and suitable for tropical cultivation. Bird type (*Capsicum annuum-Capsicum fruitescens*). Fruits, small, upright in clusters, very hot, 5-6 cm length, 0.8-1.0 cm in diameter. Pointed ends. Bright red colour and very hot flavour, plant erected and concentrated, suitable for dry and powder. Duration 120 days, good for autumn-winter cultivation. |
| Cayenne | Tropica Green Co.Ltd. | 95.00 | Plants vigorous, long slim wrinkled fruit, dark green to bright red colour, length 12-15 cm, dia 2 cm, wt. 15-16 g, medium thick, concentrated fruiting. Harvest 85-95 DAS. Resistant to Bacterial wilt and Anthracnose. |
| Red Hot ($F_1$) | Tropical Green Co.Ltd. | 100.00 | Plant very strong, vigorous, fruit length 12-13 cm, width 1.8 cm and wt. 17-19 g. Cultivation all the year round. Resistant to phytophthora and bacterial wilt. Harvest 90 DAS and during 3-4 months. |
| Fire Hot($F_1$) | Tropica Green Co.Ltd. | 98.00 | Late maturity, pendent fruits, 12-13cm long, 0.8 cm width, wt. 5-6 g. Flesh thin and hot. Suitable for fresh and dry. Resistant to phytopthora and bacterial wilt. |
| Grogol ($F_1$) | Tropica Green Co.Ltd. | 95.00 | Plant erect, upright, fruit 3-5 cm long, 0.8 cm diameter, wt. 3-4 g dark red colour. Very pungent. Harvest 50-60 DAP. |

Contd...

| | | | |
|---|---|---|---|
| | | | Suitable for fresh market and dry chilli. |
| Big Sun($F_1$) | Tropica Green Co. Ltd. | 98.00 | Plants semi-erect tall. Vigorous, prolific yielder, medium to late maturity. Fruits large (13 cm long), dia 1.8-2.0 cm, wt. 12 g, glossy, deep red, pungent, suitable for both artificial and sun drying. Tolerant to phytophthora and bacterial wilt. |
| Riki ($F_1$) | Tropica Green Co. Ltd. | 102.00 | Early maturity.Fruits long (14 cm), dia (2-2.2 cm), wt. 18-20g, pungent, excellent quality, glossy. Tolerant to phytophthora, Anthracnose and TMV. |
| Thailand($F_1$) | Tropica Green Co. Ltd. | 100.00 | Bird type chilli, plant erect, upright, moderately concentrated. Fruits long (3-5 cm), dia (0.8 cm), wt. 3-4 g, light green to bright red colour, very pungent. Harvest 50DAP. Well suited to fresh or dried market. |
| Super Hot($F_1$) | Tropica green Co. Ltd. | 98.00 | Plants strong, vigorous, fruits long (12-13 cm), width (1.8 cm), wt. 17-19 g and thick flesh. Resistant to Phytophthora. Harvest 35-40 DAP and during 3-4 months. |

## TOMATO

Tomato (*Lycopersicon lycopersicum* (L.) Karsten ex. Farw.; Syn. *Solanum lycopersion, Lycopersicon esculentum* mill.); 2n=24; syn. Tamatar, Vilayati baigan, Bilayati begun, Bapatala ghant, Ramamulaka, Seema Vankaya, Tambuta, Tabati, Welwangi, Takkali, Bilahi, Bhedra, Chhapperbhende, Tamattar; belongs to the family solanaceae, is the most important warm season fruit vegetable grown throughout the world. Among vegetables, tomato occupies 4$^{th}$ position in area and 2$^{nd}$ position in production in India. It is a herbaceous annual plant with bisexual flowers. The fruit is true berry.

Botanically tomato was called *Solanum lycopersicon* (lycos = wolf and persicon = peach). Later in 1978 cultivated tomato is renamed as *Lycopersicon esculentum* Mill. and wild form as *L. Pimpinellifolium.*

Genus *Lycopersicon* is divided into two sub-genera (a) *Eulycopersicon* (Red fruited and self compatible) (b) Eriopersicon (Green fruited and self incompatible).

*Eulycopersicon* includes two species *L. Esculentum* with large fruits and *L. Pimpinellifolium* with small fruits born in clusters. *Eriopersicon* mainly consists of wild sp. like *L. Hirsutum, L. Peruvianum, L. Pissisi, L. Glandulosum, L. Cheesmani.*

Rick (1976) divided genus *Lycopersicon* into following two groups based on their ability to cross with cultivated tomato :

Esculentum complex (crossable with cultivated tomato) → *L. esculentum, L. pimpinellifolium, L. cheesemani* and *L. hirsutum.*

Peruvianum complex (Not crossable with cultivated tomato) → *L.peruvianum, L. chinense*

Based on growth habit and fruit characters, five forms of *L. esculentum* have been identified :

| | | |
|---|---|---|
| Commune | - | Common tomato |
| Cerasiformae | - | Small fruited cherry tomato |
| Pyriformae | - | Pear shaped tomato |
| Grandifolium | - | Potato leaved tomato |
| Validum | - | Erect and upright tomato |

Tomato is a true diploid with 2n = 24 and self-pollinated crop due to hermaphrodite flowers, introvert stigma, internal and synchronized anther dehiscence, and stigma respectively. Self fertilization occurs when pollen grains are shedding during growth of style through anther corn. In warm regions of the country, some amount of crossing was observed when stigma protrudes out the level of anthers.

## Varieties and Hybrids

Quite a large number of varieties differing in their climatic requirements, growth habit, fruit quality, resistance to pest and diseases are developed for specific purposes like french market, processing, long distance transport etc.

Tomato varieties, broadly divided into two groups (1) determinate and (2) indeterminate. A brief differences is given below :

| S.No. | Determinate type | Indeterminate type |
|---|---|---|
| 1 | Main stem of tomato plant terminate in flower bud. | Main stems of tomato plant does not terminate in flower bud. |
| 2 | Main stem does not grow tall after flowering. | Main stems remains continue in growth throughout growing season. |
| 3 | Formation of more laterals. | Formation of less number of laterals. |
| 4 | There is self-topping of plant. | There is no self-topping of plant. |
| 5 | Flower cluster or truss forms in many cases at every internode. | Flower cluster or truss forms in many cases at every third inter node, separated by three leaves. |
| 6 | Comparatively having early maturity. | Comparatively requires long period for maturity. |
| 7 | Incidence of sun scalding is less because fruits remain hiding under leaves. | Incidence of sun scalding is more because fruits remain exposed to sun. |
| 8 | Plants do not require earthing. | Plants requires earthing and they are benefited with it. |
| 9 | Plants do not need staking. | Plants need staking. |
| 10 | Plants habit is bush type. | Plants growth straight i.e., 'non-bush' type. |
| 11 | Not much suitable for high planting density. | Suitable for high planting density. |

Some improved varieties are : Pusa Early dwarf, HS 101, Sweet-72, S-12, Punjab Chhuhara, Pusa Gaurav, Punjab Keshari, CO 3, La-Bonita, KS2, Sel-12, Roma, D.M.M., HS-102, Hissar Lalit, Hisar Arun, Co-1, Red Plum, Arka Alok, Hisar Lalima, Sonali, Pusa Ruby, Sel - 120, Pant $T_1$, Pant Bahar, Marglobe, Best of all, Punjab Tropic, LE-79, Sakthi, 'Pusa Uphar'. Pusa-120, Pant $T_3$, HS-110, Arka Saurabh, Arka Vikas, Sel-11, Arka Abha., Arka Ahuti, Arka Ashish, Arka Meghali, Pusa Sheetal, Kashi Vishesh, Kashi Amrit, Kashi Hemant, and Kashi Sharad, VC 48-1. Utkal Pallavi, Utkal Deepti, Utkal Kumari, Utkal Urbashi, Utkal Shrabani, Vishal, Vijaya, Maharaja. Some foreign varieties are : Avila, Ardon, Adit, Alican,Elgon, Jubulee, Tanza, Cora.

## Hybrids

*IARI New Delhi* : Pusa Hybrid-1, Pusa Hybrid- 2, Pusa hybrid-4, Pusa Divya, Pusa Sadabahar.

*IIHR Bangalore* : Arka Ahuti, Arka Ashish, Arka Vishal, Arka Vardan, Arka Shreshta, Arka Abhijit, Arka Meghali.

*Indo American Hybrid Seeds (IAHS)* : Mangala,Sonali,Naveen, Rohini, Baisakhi, IAHS-9, IAHS 88-1, IAHS 88-2, IAHS 88-3, IAHS-3121, IAHS-9501, IAHS-9502, IAHS-9503, IAHS-9504, Karnataka, Vaishali, Sheetal, Rashmi, Rupali, Rajani, Abhinas, Abinash-2, Nidhi, Ramiya, Meghana.

*Sungro Seeds* : Arjuna, Krishna, Karna, Nakul, SC-1, SC-3, Chiranjeev (621), Virat, Himlata, Himshiva, Sachin (OP), Saurav (OP).

*MAHYCO* : Sadabahar, Preeti, Gulmohar, S-28,MTH2, MTH 3, Priti, Sonali, MTH-6 (Gulmohar), S-26, S-68.Mandisen, Aswarya (S-82), Bhushan (S-83), Sadabahar, Cross-B, Gotya, S-78, S-72, MHTM-77, MHTM-84, Abhiruchi-202, MHTM-203.

*Century Seeds* : Century- 12, Rishi, Maitri, Manmohan, Supriya, Swarna, Himalaya-40, INDI-1,CTH-230,Yash, CTH-96, Neeladhari, Himroop, Lehar, Akash- 2, Maharishi, Yogi, Guru, CTH-315, Deva, Roopsi, INDI 11, Himalaya-40 and Best of all.

*Rallis India* : Madhuri, Mohini, Payal, Meenakshi

*Nath Sluis* : Nath series-63, 64, 65, 68, 76, 80, 84, 88, 89, 94, 95, 98, 250, Nath amrit-110, 239, 245, 255, 337, 601, 901.

*Nuzividu Seeds* : Priya, Vidiya.

*Sreemaa Seeds* : Sandhya, Abhishek, Upasana, Amarprem, Arti, Ashirbad, Viswajit, Amarjyoti, Arundhati, Cross-71, Jyoti, Anjali, Samaleswari, Saurav.

*Bejo Sheetal Seeds* : Meenakshi, Manisha (BSH-20), Megha, Madhuri, Tasty, BSS 36, Sultan, Charm, BSS 282, Sartaj, Tolstoi, BSS 320, Rakshak (BSS 368), BSS 369, Param, Rachit (BSS 419), BSS 350, Urvashi, Rishab (BSS 423), Nayak (BSS 422), BSS 211, Rahul, Ronco, Benito, Topspin, BSS 420,(BSS 98) Sheetal-4, Chetan.

*Ankur Seeds* : ARTH-3, ARTH-4, Ankur-308, Ankur-128, Ankur-501, Ankur-500, Ankur-210.

*Verma Seeds :* Shankar, Samrat.

***Suttons seeds :*** Sutton gram, Chhuhara, Suttongram Wonder, Sutton Sweet clusters, Lalmani, Maruraja, Sutton Gram Prolific.

***Nijjar seed :*** NH-12, NH- 24, NH-38.

***Sandoz India :*** Ratna, Tarica, Resika, Raki.

***Clause International :*** Top-21, ICO, Kristal, paltoon.

***Modi Agritech :*** Maharathi, Payal-1, Payal-2, Payal-4, Meth 40, Meth-315, Meth-412, Meth 711, Janki.

***Heritage Seeds :*** Paramamitra, Sarvodaya, Vidhata, Viswa Vidhata, Ganapati, Chaitali, C-17, Payal, S- 12, S-68, Sobhagya, Tyoti, Ajanta, Alora.

*NFCL :* Sundari, Lal, Imperial lady.

***AVRDC*** : Tiwan No 27, 35, 12, BARI- 2, BARI- 6.

***Foreign Materials*** : Tanza, Elgan, U.F., Avila, Alican, Cora,Ardon, Jubulee.

***Novartis Seeds*** : Rasika, Ratna, Rocky, TO 5975.

***Unicorn Seeds (Choice) :*** Kuber Geeta, Choice-108, Smitha, TMT-415, Sumerking, Snehalata, Pooja, Lekha, Protima, X-119, Jaiveer-1, Jaiveer-2, Nagmani, Nagmoti, PKM-1, Prestige, UTM-1370.

***Namdhari Seeds (Determinate Hybrids) :*** NS 815, NS 812, NS 816, Utsav, Amogh, NS 2535.

***Bacterial Wilt tolerant Hybrids :*** Swaraksha, Amar.

***Tomato Yellow Leaf Curl Virus tolerant hybrids :*** Ankush, Ananya, Uttam.

***Nematode tolerant hybrids*** : NS 709, NS 6138, NS 6141.

***Indeterminate hybrids :*** NS 5130, NS 443, NS 671, NS 679.

***Golden Seeds :*** Amar, GC-300, Subaru, Shivaji, Sumo, Ayusman, Praveen, Gokul, GC-1500, GC-100, Amulya, Lalmoti, Vijeta, GS-124, GS-14.

***Nunhems Seeds :*** S-230, S-176, Raja, S-7610, S-7712, Avtar (7711), NP-7730, NP-7725, NP-7704, 6109, Anant.

***Fortune Seeds :*** Kumari, Dipti, Vishal, Gayatri, Vijaya, Maharaja, Saravani, Preeti.

*Tokita Seeds :* Samurai.

*Tropica Seeds (Green CO.) :* T-01, T-02, T-03, T-05, T-09, T-12, T-26, T-27, T-29, T-35.

**Detail characteristics of Varieties and Hybrid are given below:**

*Pusa Divya :* The plants are indeterminate with profuse branching. Fruits are firm thick skinned, round to oval shaped, green with dark green stem end turn dark red on ripening. Due to thick skin, fruits have longer shelf-life with suitability for long distance transportation. Maturing period is 70-90 days from transplanting and 135-150 days from seed sowing in hills, however, this may be reduced in plains. On an average, 16 fruits weighs one kilogram with an average yield potential of 350-450 q/ha. Less severity to late blight and buck eye rot. Suitable for growing in Uttar Pradesh, Punjab, Bihar and Orissa.

*Pusa Sadabahar :* A new variety has been released from IARI, New Delhi. It is highly suitable for growing under wide range of (8-30$^0$ C) high temperature. Its plants are determinate dwarf. They can be accommodated more in number per unit area than other varieties. Very prolific-bearer with smooth oval to round attractive fruits with TSS 4.8%. This variety can be grown throughout the year in north Indian plains except rainy season (July to early Sept.). Pusa Sadabahar gives as mush as 40 tonnes /ha yield during main season. Ripening of fruits start 90 days after sowing.

*Pusa Red Plum :* It has been evolved from a cross of a cultivated tomato and the wild *Lycopersicon pimpinellifolium*. It is an early variety suitable for table use. It is suitable to grow in home gardens; habit is, vigorous and high yielding. The fruits are small, borne in cluster with red colour and high in vitamin C and sugar content.

*Pusa Rohini :* Plants are determinate, prolific-bearer, compact with green foliage and cut leaves. Fruits are red round, smooth,medium sized (70 g), thick skinned and uniform in ripening. It takes 105 days from transplanting to fruit ripening. It has a yield potential of 412 q/ ha, suitable for cultivationin spring-summer in north Indian plains. Its fruits are ideal for distant transportation. A new variety released by Delhi State Variety Release Committee.

*Pusa Early Dwarf :* A very popular variety from NBPGR, New Delhi grown for early crop in zones I, IV and VII. Plant typical dwarf, 50-55 cm in height with compact fruiting, fruits are slightly flattish in shape, medium large, uniform red, ribbed, obscure furrow, 5-6 locules, 30-40 fruits per plants. Maturity in 55-60 days after transplanting. Suitable for rainy season in plains and March sowing in hills.

***Pusa Ruby*** **:** A very popular variety from IARI New Delhi, derived from the cross, improved Meeruti x Sioux. Grown all over the country in plains. Plant height 80-85 cm, indeterminate spreading, less branched and hardy, fruits flattish round, small to medium, uniform red, slightly lobed (4-5 locules), slightly acidic, 25-30 fruits per plant. Early maturity (60-65 days). Suitable for autumn, winter and spring-summer seasons. Good for juice and ketchup.

***Pusa Uphar*** **:** Plants of Pusa Uphar are determinate prolific- bearer, upright, compact with dark green foliage. Fruits are borne in bunches of 4-5 each. They are attractive round, medium sized (60g), thick-skinned and uniform in ripening. It is slightly tolerant to fruit- borer under field conditions. It takes 100 days from transplanting to fruit ripening. The firm, mature and greenish-yellow tomatoes are ideal for distant marketing.

***Pusa Gaurav*** **:** It is developed from the exotic segregating generation of a cross between 'Glamour' and 'Watch'. The plants are dwarf, bushy with moderate foliage cover. The fruits are smooth, elliptical (egg-shaped), borne in clusters. The unripe fruits are 4 cm long and 3.5 cm in diameter. The ripe fruits are firm with thick flesh (0.6 cm) and to well filled locules that facilitate easy transportation to long distances. Pusa Gaurav is very suitable for processing, especially for whole fruit canning because it is without neck construction and has higher TSS (6 per cent) and better keeping quality at room temperature. Average fruit yield is 34.9 to 48.5 t/ha.

***Pusa Sheetal*** **:** It is evolved at IARI by combining the desirable attributes of the two cold set likes 'Balkan' and 'Jemnorrosuij'. The variety can set fruits even if the night temperature touches a low of 8°C. Plants are compact with moderate foliage cover and prolific bearing. Fruits are flattish round, smooth, attractive and develop uniformly red colour on maturity. This cultivar is very suitable for gaining as an additional early spring crop. It has distinct advantage for providing fruits during March in north India. On an average yield is 363.25 q/ha.

***Marglobe :*** Indeterminate, mid-season variety, plants large, vigorous, erect. Unripe fruits greenish with deep green shoulders, fruiting in bunches; mature fruit smooth, medium, scarlet red in colour, juicy, round, small, scar, thin skin, 3 locules, very heavy yielder.

***Roma :*** Plants semi-dwarf, good foliage, excellent bearing; unripe fruits whitish green with green stem end, borne in clusters of 4-5, mature fruits pear shaped, thick walled, two loculed with narrow and heavy yielder. Very good for ketchup and long distance transport, especially suited for hill.

***Sel 120 :*** A nematode resistant variety from IARI, New Delhi, identified in 1975 for zones I, IV, VI,VII and VIII. Semi-indeterminate, foliage dark green, good coverage, fruits attractive, round to flattish round, medium to large in size, smooth, uniform red in colour, less acidic and less seeded. Suitable for summer and winter.

***S-152 :*** The special feature of variety is that it can set fruits at high temperature ranging between 20.5 and 34.5°C, whereas commercial varieties failed to set fruits above 20°C.

***Sioux :*** Developed at IARI, Katrain, identified in 1977 for zones II, IV, VI and VII. Fruits round, smooth, medium to large attrative red on ripening, less seeded. Maturity 60-70 days after transplanting. Suitable for summer and spring crops.

***La-Bonita :*** A selection from NBPGR, New Delhi, identified in 1987 for zoens I and V. Plants are bushy. Fruits have yellow stem end, smooth, oblong shaped without neck, 2-3 locules, thick pericarp. Suitable for long distance transportation. Average yield 353 q/ha.

***Pusa Hybrid - 1 :*** Plants are determinate, compact with good foliage cover and prolific bearer; fruits are round, smooth and attractive; average fruit weight 63.0 g and fruit yield 32.0 t /ha.; can set fruits up to a temp of 28° C during night and has an added advantage of providing tomato from June to midJuly. Female Parent : Plants are determinate, compact with moderate foliage cover and prolific bearing; fruits are smooth, attractive and uniform red at maturity. Male Parent; Plants are determinate with very good foliage colour and prolific bearing ; fruits are yellow, round smooth attractive and uniform red at maturity.

***Pusa Hybrid 2 :*** The plants are compact , semi – determinate with good foliage cover and prolific bearing. Fruits are round to flattish - round, firm, smooth, attractive and develop a uniform red colour on maturity. It is highly tolerant to root-knot nematodes. Fruit size medium and average fruit yield varies from 44.94-79.67 tonnes/ ha. It is highly tolerant to nematode disease. Ripening is very uniform. Firm, mature and greenish-yellow tomatoes are useful for distant markets. Whereas fully ripe, red tomatoes are ideal for fresh market consumption.

***Pusa Hybrid - 4 :*** Plants are determinate, compact with dark green foliage, good foliage cover ; fruits are attractive, round, smooth, average weight of 70- 80 g ; three locules , thick pericarp and uniform ripening; field resistant to root knot nematode. Female Parent : Plants are semi-determinate, dark–green , medium foliage, cut leaves ; fruits are yellow, flattish round, smooth, average weight of 100g and uniform red at maturity. Male Parent Plants are determinate, dark green foliage, simple leaves; fruits are yellow, round to oblong, smooth, two locule, average weight of 70 g and thick pericarp.

***Pusa Hybrid 8 :*** Plants are determinate, compact with dark green foliage and prolific- bearing habit. Fruits are attractive, round, smooth, medium-sized (60.70 g) with thick pericarp and uniform in ripening. It takes 120 days from transplanting to fruit ripening in spring-summer crop and 105 days in autumn crop. The fruits are suitable for distance transportation. It has yield potential of 636.66 q/ha and suitable for cultivation in spring–summer in both Indian plains.

***HS-101 :*** A very promising variety bred at HAU, Hisar. Plant height 50cm, dwarf multi- branched sturdy and determinate in growth habit, fruits borne in cluster of 2-3, round, small to medium in size, 3-4 locules. Fruits ripen uniformly, 40-50 fruits per plant. Suitable for winter season for northern regions, identified in 1975.

***HS-102 :*** This is an early cultivar. Plants are determinate type. Fruits are with yellow stem-end, medium to small size, smooth, attractive and round in shape. Released by Haryana Agril. Univ. Hisar.

***HS-110 :*** It is a selection from exotic line. Determinte; late; potato leaf type; fruit large, smooth flashy, jointless; highly suitable for table purpose. Recommended for cultivation in western India.

***Sel. 32 :*** This variety was developed at HAU, Hisar and recommended for growing in II, VI and VII zone under AICVIP in 1996. Plants are determinate and fruits are flattish round, big sized and red coloured. The average yield is 30-35 tonnes/ha. This variety is susceptible to leaf curl virus diseases.

***Hisar Lalit :*** It has been developed from the cross of $R_2$ (Resistant Bangalore 7 x 'HS 101'). Back cross and pedigree method was adopted. It is highly resistant to the root-knot nematode, high yielding, produces fruits of a good size, appeal and flesh. Plants 48.7cm, semi-determinate early. Average fruit per plant 22.5, fruit weight 55.3 g, medium to large concentrated fruiting, higher yield, av. 229.3 q/ha, fruit colour deep red, shape round, days to fruit picking 65-70 days.

***Hisar Arun (Sel-7) :*** It has been developed by hybridization of Pusa Early Dwarf x 'K1' through pedigree selection in the $F_{10}$ generation.The tomato is extremely early bearing, in which fruits begin to ripen about 50 days after planting. Its plants are determinate, dwarf and bushy. Fruits round, medium to large, borne in cluster, deep red, firm with 4-6 locules and mostly meaty pulp. Average early yield is 153.9 q/ha and mean trial yield is 286.9 q/ha.

***Hisar Lalima (Sel 18) :*** It has been developed through hybridization of 'Pusa Early Dwarf' x 'HS 101'. It is an early maturing variety.The fruit is large, round, flat , attractive red, fleshy, with 4-6 locules, a thick pericarp, less prune to cracking and suitable for both (autumn and spring) seasons. Harvest index is very high because foliage is less and fruits more in number. It gives a total average yield of 28 tonnes/ha.

***Hisar Anmol (H 24) :*** Plant determinate, fruit medium size round red , fleshy, field resistant to tomato leaf curl virus. Recommended for cultivation in leaf curl disease infested season and area. It is a cross between Hisar Arun and *L. hirisutum f. glabratum*

*Sweet-72 :* A promising variety developed at Gwalior and identified in 1975 for zones I, IV and VII. Plants determinate type, fruit flattish round, green stem end, slightly furrowed. Uniform maturity and heavy yielder.

*Co.-1 :* It is a dwarf variety. Plants are semi-spreading type. Foliage is dark green in colour.The fruits are smooth and round in shape. They are borne in clusters of 6 to 8.The average fruit weight is about 125 g. It yields an average of 380 quintals fruits per hectare.

*Co-3 :* A mutant of Co-1 from Coimbatore. Tested at 23 coordinating centres and identified for cultivation in zones IV, V and VII during 1987. Average plant height (40 cm) erect, thick main stem with strong primary branches. Leaves normal and dark green. Determinate type of inflorescence. Fruit bright red, round, globular in shape and medium in size, born in clusters (4-5), 30-40 fruits/ plant. Crop duration 110-115 days. Yield potentiality 381 q/ha. Rich in Vit. C content (25 mg/100g) with acidity of 3.8%.

***Punjab Tropic :*** It is the outcome of the selection made from American introductions. It was released in 1973. Plants are determinate type. Fruits form in clusters of 4 to 6, deep red in colour. The average yield potential is 600 quintals per hectare.

***Keck-Ruth Ageti :*** It is selection from cross of EC 58055 and Punjab Tropic. It is determinate type. Fruits are pear shaped, medium sized, uniformly ripening having thick pericarp, 2-3 locules and a few seeds. It is tolerant to blight. TMV and wilt diseases. It yields an averge of 600 quintals fruits per hectare.

***S-12 :*** Developed at PAU, Ludhiana and identified for zone I,II, IV,VII and VIII. Plants are dwarf, bushy, vigorous, fruits are medium sized, compressed round. Uniform red coloured at maturity, Juicy, highly acidic. Average yield 175-280 q/ha. Suitable for fresh marketing for table purposes.

***Pub. NR-7 :*** Plants are dwarf. The fruits are flat round, medium-sized, uniform red, multilocular and juicy. This is tolerant to root knot nematodes. It gives about 400 g of seeds yield per quintal of fruit.

***Punjab Chhuhara :*** Suitable for long transportation. Developed at PAU, Ludhiana. Identified in 1981 for zones I, II, IV, Vi, VI and VIII. Plants dwarf. bushy growth habit with dense and luxuriant foliage. Fruits are medium sized, pear shaped, firm, fleshy, usually bilocular, less seedy, less sour, thick walled, uniformly red coloured at maturity, retain marketable quality for a week after harvesting under ordinary conditions during summer.

***Punjab Keshari :*** Developed through the hybridization of EC 55055 X Punjab Tropic followed by pedigreee selection at PAU, Ludhiana. Identified in 1987 for northern plains and hills. Average plant height 60 cm, bushy, no fruit cracking, early maturity, incidence of late blight and fruit borer moderate. Susceptible to root-knot nematodes. Average yield 286 q/ha.

***KS-2 :*** A kalianpur variety identified in 1985 for zones I, II, IV and VI. Foliage light green. Fruits flattish round, Slightly furrowed. Moderate, yield. Stem end is slightly green, number of locules 4-5.

***Kashi Vishesh* :** Plants are determinate with dark green foliage. The fruits are spherical, medium to large with an average weight of 80 g. The fruits are ready for harvesting 70-75 days after transplanting, the yield potential being 450-550 q/ha. Resistant to TLCV, can be cultivated in the TLCV prone areas. This is ideal for cultivation in Jammu and Kashmir, Himachal Pradesh, Uttaranchal, Punjab, Uttar Pradesh, Bihar, Jharkhand, Chhatisgarh, Orissa, Andhra Pradesh, Karnataka, Tamil Nadu and kerala.

***Kashi Amrit*:** Its plants are determinate, fruits large, flattish round (slightly indented at blossom end), attractive red with 5-6 locules. Its fruits are ready for plucking 75-80 days after transplanting with an average yield of 500-600 q/ha. It is suitable for Rajasthan, Gujarat, Haryana and Delhi.

***Kashi Hemant* :** Its plants are determinate, with attractive red colour and round shape fruit, concentrated fruiting. Fruit weight varies from 80-90 g, the average yield being 400-420 q/ha. This variety is suitable for cultivation in Chhatisgarh, Orissa, Andhra Pradesh, Madhya Pradesh and Maharashtra.

***Kashi Sharad*** : Its plants are indeterminate, with broad leaves (potato type). The fruits are attractive red, slightly oval and very firm, thick pericarp suitable for longer shelf life, number of fruits being 35-40, fruit weight 90-100 g. It gives an average yield of 400-500 q/ha. This variety has been released for cultivation in Jammu and Kashmir, Himachal Pradesh and Uttaranchal.

***Pant Bahar* :** Developed from line AC-238 at Pantnagar. Identified in 1985 for zones I, II, IV and V. plant height 93 cm, bushy and branched. Light green foliage, stem, relatively thin leaflets small in size. First ripening in 78 days. Fruit flattish round, medium in size, 5-6 locules, slightly ridged, red on maturity. Resistant to Verticillium and Fusarium wilt under field conditions. Good storage and processing qualities.

***Pant T-1* :** Identified in 1975 for zones I, II, IV, VII and VIII. Plant height 100-120 cm, tall erect in growth habit, foliage green in colour, fruit round. Slightly pointed at the stigmatic end, 4-5 locules, 4-5 fruits/truss, non- cracking, red on ripening. Susceptible to TMV.

***Pant T-3* :** Developed at Pantnagar by pure line selection. Identified for zones IV, V, VI, and VII on the basis of trails at 23 testing centres. Plant semi-determinate, thick, round hairy stem. Leaf medium sized with dark foliage. Most suitable for winter season. Fruits ripen in 60 days, weighing (65-78 gm/fruit), round, smooth surface, uniform red colour of skin. Suitable for processing. Average yield 192 q/ha.

***Arka Vikas :*** A selection from a variable population of American variety 'Tip Top, introduced at IIHR, Bangalore. Identified in 1987 for both the hills and plains. Fruits medium large (80-90 gm) with uniform deep red colour. Suitable for fresh market. Average yield 303 q/ha in 105-110 days after transplanting. Tolerant to moisture stress.

***Arka Saurabh :*** Selection made at IIHR, Bangalore from variable breeding line V-685 introduced from Canada. Plant semi determinate in growth. Fruits thick flesh, firm, round, medium large (70 gm), deep red and nipple tipped, suitable for both fresh market and processing. Resistant to fruit cracking. Average yield 308 q/ha in 105-110 days after transplanting. Good transport qualities and shelf life. Identified in 1987 for zones I, IV, V, VI and VII.

***Arka Meghali :*** Tomato 'Arka Meghali-cross of 'Arka Vikas' x IIHR 554 ' is a high yielding variety. Developed through a pedigree method of selection. It is a new variety suited to rainfed areas. It has semi determinate plant habit with oblate fruits. It is a superior to Pusa Ruby in a fruit size (63 g ), flesh thickness and fruit firmness, besides high-yield potential ( 0.75 kg/ plant ), total yield potential being 17.2- 18.2 tonnes / ha.

***Arka Shreshta :*** It has been developed at IIHR, Bangalore , plants are semi-determinate with light green foliage; fruits are round, firm, deep red, average weight of 75g, light green shoulder with jointed pedicel ; flesh is red ; shelf life 7 days under room temperature and excellent transport quality ; takes 55-60 days for first picking and continues up to 140days ; resistant to bacterial wilt; give an average yield of 70 t/ha. Female parent, plants are semi - determinate with green foliage , fruits are round, red with light green shoulders , average weight of 90g ; crop duration is 145 days with an average yield of 35 t / ha. Male parent ; plants are tall determinate with light green shoulder, average weight of 70 g ; crop duration is 135 days with an average yield of 40 t/ha.

***Arka Ahuti (Sel 11) :*** A pure line selection from Ottawa 60' (Canada). Selection for three generations followed by massing for 2 more generations helped to evolve a uniform high yielding variety with oblong, thick fleshed firm, bilocular and uniform ripening fruits.

It is semi- determinate and the yield potential is around 450 q/ha. TSS is 5.2% and suitable for processing.

*Arka Ashish (processing 674 SB, SB 1 HR lines)* : It is an improvement over 'UC82B' from California. It is determinate, uniform ripening, has very thick flesh, very firm bilocular oval fruits and concentrated fruit maturity. Ripened fruits store well on the vine. It is tolerant to powdery mildew, yields around 350-400 q /ha. Colour is excellent (lycopene 9mg). TSS is 4.8 %. Suitable for processing.

*Arka Vishal* : Suited for fresh market. Plants are indeterminate and requires staking and pruning. As fruit weight is around140g. Fruits are green shouldered, round, very firm , deep red on maturity, tolerant to fruit cracking and have good shelf life. Yields about 750 q / ha.

*Arka Vardan* : It is resistant to nematodes and bred for fresh market purposes. It is indeterminate in habit and requires staking and pruning. Average fruit weight is 140g. Fruits are flat round, firm thick fleshed , deep red, green shouldered with good shelf life. Yield 750 q / ha.

*Arka Abha* : Resistant to bacterial wilt, semi determinate, fruits are flat round, 70-90g, light green shouldered, yield 350 q / ha.

*Arka Alok* : Highly resistant to bacterial wilt. It is a determinate plant. Fruits are large square round, thick fleshed with light green shoulder. fruits yield about 350q/ha.

*Arka Abhijit* : It has been developed at IIHR, Bangalore. Plants are semi-determinate with dark foliage ; fruits are round, firm, deep red with green shoulder and jointed pedicel, average weight of 60g ; red flesh ; excellent transport quality with shelf life of 17 days at room temperature ; resistant to bacterial wilt and suitable for fresh market; takes 55 - 60 days to first picking and continues up to 140 days ; gives an average yield of 40 t/ha. Female Parent : Plants are semi determinate with green foliage ; fruits are oblate round , firm , red with light green shoulder, average weight 90g ; crop duration is 45 days with an average yield of 35 t / ha. Male Parent : Plants are determinate with dark green foliage ; fruits are round, firm, deep red with light green shoulder, average weight 80 g ; crop duration is 130 days with an average yield of 40 t / ha.

*Utkal Pallavi (BT-1)* : Determinate type, curly leaf. Fruit borne in cluster consisting of 6-12 in number, medium to small in size, pear shaped, nipple at blossom end, firm, fleshy. Matures in 85 to 90 days after sowing. Resistant to bacterial wilt and nematode wilt. It also perform well in saline soil. It is a national check variety for bacterial wilt resistant trial. Average yield is 375.75 q/ha.

***Utkal Deepti (BT-2)*** : Determinate type, slightly curling leaf. Fruit round shaped, small nipple present at the blossom end of the fruit, small and medium size, good keeping quality with 6-8 fruits per cluster. Matures in 80-85 days after sowing. Resistant to bacterial wilt and nematode wilt. It also performs well in saline soil. Average yield is 411.90 q/ha.

***Utkal Kumari (BT-10)*** : Indeterminate growth habit. Fruit borne in cluster consisting of 3-5 in no., medium large in size, round shaped, uniform red colour at maturity.Matueres in 90 to 95 days after sowing. Resistant to bacterial wilt disease. Average yield is 402.0 q/ha.

***Utkal Urbashi (BT-12)*** : Indeterminate growth habit, tall plant, good foliage cover. Fruit pear shaped, medium large size with slightly green shoulder, good keeping quality, 5-8 fruits per cluster. Matures in 100-105 days after sowing. It is resistant to bacterial wilt and tolerant to early blight. Average yield is 451.67 q/ha.

***Utkal Raja (BT-10-2-1)*** : Indeterminate growth habit, fruit round, medium large in size and deep red in colour at ripening. Fruits bear in clusters of 4-5 in numbers. Tolerant to bacterial wilt disease. Average yield is 449.0q/ha.

***Utkal Pragyan (BT 116-3-2)*** : Determinate in growth habit, fruit round, deep red in colour, medium in size with a blunt beak at distal end. Fruits bear in clusters of 5-6 in numbers, pulpy and tolerant to bacterial wilt disease. Matures in 90-95 days after sowing. Average yield is 412.22 q/ha.

***Utkal Shravani (BT-3)*** : Determinate in habit. Fruit flattish round, medium in size and juicy and matures in 85-90 days after sowing. Suitable for early planting. Tolerant to bacterial wilt.Average yield is 436.59 q/ha.

***Utkal Gayatri (BT-18)*** : Semi-determinate growth habit. Fruit oval in shape, medium small in size, 3-5 fruits per cluster. Matures in 95-100 days after sowing. Performs well in early planting. Resistant to bacterial wilt. Average yield is 412.62 q/ha.

***BT -17 :*** Determinate type. Fruit flattish round, medium in size and matures in 90-95 days after sowing. Resistant to bacterial wilt and nematode wilt.Average yield is 452.16 q/ha.

***BT-136 :*** Indeterminate type, fruit round to oval in shape, medium in size, 4-5 fruits per cluster, matures in 95-98 days. Tolerant to bacterial wilt. Average yield is 387.96 q/ha.

***Utkal Gautami (BTH-9) :*** Determinate type of hybrid. Fruit round to slightly oval in shape, medium large size, 3-5 fruits per cluster, very pulpy, matures in 96-98 days after sowing, low seed content. Resistant to bacterial wilt. Average yield is 661.69 q/ha.

***Solan Sagun :*** This is an early hybrid, the flowering starts 24 days after transplanting. It yields about 580 q/ha of fruits.This hybrid is quite suitable in low and mid hill of Himachal Pradesh and evolved from Horticultural Research Station, Kandaghat.

***Raja Shree :*** This hybrid is released in 1994 from Mahatma Phule Agricultural University, Rahuri. Fruits are medium (90-100 g), firm with pulp and ascorbic acid (42-50 mg/100 g) content, attractive orange coloured and oval round in shape. It is suitable for long transportation and comparatively less susceptible to the viral diseases.

***Sonali :*** It is a high-yielding, wilt resistant tomato, evolved from an exotic collection, sonali (sel-2), has short- statured compact plants bear oval fruits with firm skin. It matures in 120-140 days. The just-ripe fruits are ideal for picking. It gives as much as 50 tonnes of the fruit to a hectare. It is ideally suited to the bacterial wilt prone areas. The average, plant height 60 cm, number of cluster per plant 10, number of fruits per cluster 6, number of fruits per plant 65, weight/ fruit 70g, days to 50% flowering 40 days.

***Krishna Nagar (S-20) :*** This variety has been released by Department of Agriculture (WB). It is a medium to late variety. Fruits for first picking are available 70-80 days after transplanting. Fruits are medium-sized, round in shape and red in colour. It yields nearly 300-325 quintals fruits per hectare.

***Junagadh Ruby :*** Evolved through hybridization from a cross between inbred No.22 viz., Big Silani (Female) x Pusa Ruby (Male) during 1968. It bears medium – sized and ungrooved flat round fruits. The fruits are borne in clusters of 5 or 6.The fruit colour is deep orange red, fruit compact, pulpy and with well filled juice. It is tolerant to fruit borer. First picking starts 85-90 days after transplanting. It yields 240 quintals fruits per hectare.

*NDT 5 :* It is suitable for cultivation in Eastern Uttar Pradesh. Fruits oval, red, attractive medium to large in size and juicy. Average yield 250-300 q/ha. Mature in 80-90 days after planting.

*NDT 120 :* It is a suitable for cultivation in Eastern Uttar Pradesh. It is dwarf variety and mature in 75-80 days after transplanting. Fruits large, red fleshy and attractive. Average yield 275-300 q/ha.

*NS 815 (F1) :* Vigorous, determinate plant with strong disease resistance. NS 815 produces abundant crop of blocky square round fruits of 70-80g. Excellent appearance and firmness makes this hybrid most popular in Western India. Low pH with good colour and high soluble solids makes NS 815 very desireable for processing industry also. Wider adaptability makes it suitable for planting in Southern and Western plains during kharif and rabi seasons.

*NS 812 (F1) :* Jointless hybrid with good firmness and uniform fruits with smaller scar. NS 812 is vigorous, determinate plant with good foilage cover. Fruits are oval, of 80-85g, and are firm with good keeping quality and tolerant to cracking. This hybrid performs well in Southern, Central and Western India in kharif and rabi seasons.

*NS 816 (F1) :* Strong determinate plants with very good foliage cover. Medium to late maturity hybrid and yields large sized oval fruits of 100-120g. It is suitable for non staked culture. Excellent firmness, shipping quality, attractive colour make this hybrid suitable for both processing and fresh market. This hybrid has performed very well as rabi and kharif crop in Western and Northern India.

*Utsav (F1) :* This medium maturity, vigorous determinate hybrid is ideally suitable for Southern, Western and Central plains. The large sized oval furits (90g) are very firm having excellent crack tolerance. Wider adaptability with abundant yields makes this hybrid very popular in Southern and North Eastern India.

*Amogh (F1) :* Extra firm hybrid tomato has performed extremely well in Southern and Western India. Very high yields with excellent fruit qualities make merchants love this variety and fetches a premium to farmers. Early to medium sized fruits,(75-80g) and suitable for fresh market and processing industry.

***NS 2535 (F1)*** **:** Determinate early maturing hybrid with excellent vigour. Fruits are oval, weigh 80-90g with a very attractive red colur. Fruits are glossy and are attributed with superior firmness. This hybrid is a prolific bearer and well adapted for the kharif and rabi seasons through out India.

***Swaraskha (F1)*** **:** Hybrid with high tolerance to bacterial wilt. Bred specially for the bacterial wilt sick soils of Southern and Northern India. Vigorous determinate plants with excellent foliage cover and good yield potential. It produces attractive fruits, weighing around 90-100g of medium firmness. Recommended for early harvest at breaker stage. This hybrid has high acidity and very good flavour. Swaraksha has very wide adaptability including North Eastern India.

***Amar (F1)*** **:** High yielding vigorous tall determinate bacterial wilt tolerant hybrid. Fruits possess attractive square round shape and weigh 80-90g. Excellent fruit colour, superior firmness and good keeping quality are the special attributes of this hybrid, suitable for kharif and rabi seasons in bacterial wilt sick soils.

***Ankush (F1)*** **:** Vigorous determinate TYLCV tolerant hybrid with dark green foliage. Fruits are square round, weigh 90-100g with grossy red colour and very good firmness. This hybrid is very high yielding with prolonged period of harvest. Ideal hybrid for summer months in Central, Western and Southern India.

***Ananya (F1)*** **:** Very vigorous plants with good foliage cover. High yielding TYLCV tolerant hybrid with oval fruits weighing 85-90g. Fruits are glossy red and possess very good firmness. This hybrid is recommended for the summer months in Central, Western and Southern India. Good foliage and excellent yield are special features of this hybrid.

***Uttam (F1)*** **:** Vigorous determinate early TYLCV tolerant hybrid. Fruits are square round, weigh about 90g and possess very good firmness and are glossy red. High yield is the special attribute of this hybrid. The hybrid is suitable for summer in Central, Western and Southern India.

***NS 709 (F1)*** **:** Vigorous tall determinate hybrid with oval to square round fruits weighing 80g each. Fruits are endowed with excellent colour, firmness and very good flavour. Prolonged period of harvest and tolerance to nematodes are the notable features of this hybrid. It is a very heavy yielder, resistant to VFFN.

***NS 6138 (F1) :*** The hybrid has vigorous plants with tall determinate plant habit. Fruits are round to high round in shape, very firm, attractive and smooth. The average fruit weight is 140-150g, with deep red colour. Very high yielding hybrid with resistance to VFN.

***NS 6141 (F1) :*** Plants are vigorous with good foliage cover and determinate habit. Fruits are round to square round, firm, attractive and smooth. They weigh 130-140g each. It is a very high yielding hybrid with resistance to VFN.

***NS 5130 (F1) :*** Vigorous growth, continuous bearing habit and good foliage cover. The fruits are very firm with outstanding shelf life. Fruits have a mild green shoulder and are flat round, attractive and weigh 130-150g. Fruits are of excellent quality and can also be harvested at red ripe stage. Under good management the fruits can be continuously harvested for 5-6 months.

***NS 443 (F1) :*** A vigorous indeterminate hybrid with good foliage cover. The fruits are very attractive round to square round, each weighing 90-100 g in clusters. Early ness, high yield and excellent colur are striking features of this hybrid.

***NS 671 (F1) :*** An early maturing, indeterminate hybrid with good foliage cover. This hybrid bears oval to square round, firm fruits weighing around 90g each. Fruits are of very attractive red colour on maturity with good transportability, has superior yield potential.

***NS 679 (F1) :*** An indeterminate early maturing hybrid with good foliage. Fruits are flat round, weighing 120-130g with excellent colour and firmness. Suitable for long transportation. It has consisting high yields.

***Amar :*** Semi determinate, vigorous hybrid, producing large, deep red, round fruits with green shoulder weighing after 60–65 days of transplantation. The hybrid has very good tolerance to yellow leaf curl and cucumber mosaic virus diseases. The hybrid has a yield potential of 115–120 mt/ha.

***GC-300 :*** Strong determinate hybrid of medium early maturity. The first fruit harvest starts after 65-68 days of transplantation. Fruits are medium sized, 80-90 g, firm, square round and shining red and have very good keeping quality and suitable for long distance transportation. The hybrid has yield potential of 100-120 mt/ha and suitable for rainy and winter season cultivation.

***Subaru :*** Semi determinate hybrid of medium early maturity. First fruit harvest commences after 68 days of transplantation. Fruits are medium sized, 90-100 g, square round, deep red and possess excellent keeping quality and ideal for long distance transportation. Yields 110 -120mt/ha.

***Shivaji :*** A premium hybrid of golden seeds. Plants are of determinate habit and the first fruit picking starts 62-65 days after transplantation. Fruits weigh 85-90g, square round, shining red and have excellent keeping quality. The dark green foliage cover projects the fruits from sun scald. The hybrid has yield potential of 120-130mt/ha and higher yields can be obtained under intensive cultivation. Fruits are ideal for both fresh market and processing.

***Sumo :*** A semi determinate, medium early maturity hybrid, producing large round fruits weighing 120-130g, the fruits are round, deep red have very good keeping quality. Plants have good foliage cover to protect fruits from sun scald. Yields 100-150 mt/ha.

***Ayushman :*** A semi determinate hybrid, producing red round fruits with uniform shoulder. Fruits are firm, 80-100g and have good shelf life. The first fruits harvesting starts 55-60 days from planting. The hybrid has very good tolerance to bacterial wilt and recommended for cultivation in wilt sick prone areas. Yields 100-110 mt/ha.

***Praveen :*** Indeterminate hybrid of medium early maturity with attractive round, deep red glossy fruits and weigh70-80g. The fruits have thick skin, crack resistant and possess good shelf life. The first fruits harvesting starts after 60days and the good foliage cover protect fruits form sun scald. The hybrid has remarkable ability to resist higher temperatures. Yields about 150 mt/ha.

***Gokul :*** Determinate hybrid producing square round, deep red fruits with an average fruit weight of 60-70g. The fruits have good firmness and suitable for both market and processing. Yields 110-115mt/ha.

***GC-1500 :*** An outstanding early maturing hybrid, popular for its high heat tolerance. First fruit harvesting starts 50-55days after planting. The fruits are medium sized, firm, square round and weigh 70-80g and possess good keeping quality and suitable for long distance transportation. Yields100-110mt/ha.

*GC-100* : Determinate, vigorous hybrid of medium early maturity. Fruits square round, deep red, 80-90g and have good firmness. First fruit picking starts55-58 days after transplantation. The hybrid has excellent resistance to root knot nematode and ideal for cultivation in nematode sick soils. Yields 100-115 mt/ha.

*Amulya* : Multiple disease tolerant determinate hybrid of early maturity. First fruits harvest starts 55-58 days after planting. Fruits are square round, deep red and weigh 80-90g. The hybrid has very good multiple disease tolerance to yellow leaf curl and tobacco mosaic viruses, bacterial wilt, root knot nematode and fusarium wilt.

*Lalmoti* : A semi determinate, early maturing hybrid producing cherry type fruits. First fruit harvest starts 50-55 days after planting. Fruits are small, round, shining red and weigh 10-15g. very good flavour and taste makes it suitable for fresh market. About 30-40 fruits are found in each cluster. Good tolerance to bacterial wilt and Fusarium wilt (F-1,2).

*Vijeta* : Multiple disease tolerant semi-determinate hybrid of early maturity. First fruit harvest commences 58-60 days after planting. Fruits are round, deep red and weigh 90-100g. Strong tolerance to yellow leaf curl and tobacco mosaic viruses, bacterial wilt, root knot nematode and fusarium wilt (F-1).

*GS-124* : Determinate, vigorous hybrid of early maturity. Fruits are square round, deep red and have excellent firmness and weigh 70-80g. This hybrid has ability to flower and set fruits even under high temperature conditions. Suitable for long distance transportation. Yields120 mt/ha.

*GS -14* : This is a determinate hybrid. Plants are compact and sturdy. Fruits are square, pear shaped and uniformly deep red. Average fruit weight 70-80g. the hybrid has excellent resistance to *Verticillium, fusarium* and root knot nematodes. Fruits are very suitable for processing. Yield potential 95-100 mt/ha.

*Ankur-308* : Determinate growth habit.Fruit shape round with good taste and deep red colour.Average fruit weight ranges from 75-85 g. High yield potential.

*Ankur-128 (F-1)* : Indeterminate, medium spreading growth habit. Fruit is round in shape with good taste and medium red colour, Average fruit weight ranges from 45-55g. Good keeping quality. High yielding potential.

***Ankur-501 :*** High yielding hybrid. Determinate, medium compact growth habit with vigour. Attractive oval shaped fruit, deep red in colour. Average fruit weight ranges from 70to 80g.Good transport quality.

***Ankur-500 (F-1) :*** High yielding hybrid. Determinate, medium compact growth habit. Fruits are oval shaped, deep red in colour. Average fruit weight ranges from 70 to 80 g. Good transport quality.

***Ankur-210 (F-1) :*** Determinate, medium compact growth habit with good foliage. Fruits are round, deep red in colour. Average fruit weight ranges from 70-80g. Good keeping quality. High yield potential.

***Pallavi :*** Determnate, bushy early maturity and profused fruiting. Fruits are egg shaped, 70-80 gm. weight. Resistant to bacterial (*Pseudomonas solancearum*) and nematode wilt. Tolerant to fusarium and verticillium wilt. Suitable for long distance marketing.

***Deepti :*** Determinate, bushy early maturing, profuse fruiting, fruit are round shape, medium size, 60-70gm weight. Resistant to bacterial (*Pseudomonas solanacearum*) and nematode wilt. Tolerant to fusarium and verticillium wilt.

***Jyoti :*** Determinate, vigorous growth, medium maturity. Fruits are globe shaped 80-100 gm. weight, red colour, resistant to bacterial,(*Pseudomonas solanacearum*) nematode, and fusarium wilt.

***Anjali :*** Determinate, vigorous, early maturing, profuse fruiting Fruits are round 70-80 gm. weight, deep red colour, suitable to long distance transportation, resistant to bacterial (Pseudomonas solanacearum) and nematode wilt. Tolerant to fusarium and verticillium.

***Amarjyoti :*** Determinate, vigorous growth, medium mature fruits are globe shaped, 70-80 gms., red colour resistant to bacterial, nematode and fusarium wilt.

***Kumari :*** Vigorous, profuse branching, high yielding, medium maturity. Fruits are globe shaped, 80-100gm. weight, red colour, highly resistant to bacterial (*Pseudomonas solanacearum*), nematode and tolerant to fusarium and verticillium wilt.

***Samaleswari :*** Determinate, hight yielding suitable for cultivation in late rabi, fruits are egg shaped, 65-75 gm weight, red when matured. Resistant to bacterial, nematode and fusarium wilt.

***Saurav :*** Determinate, high yielding, suitable for cultivation in rainy days. Fruits are egg shaped, 80-90 gm. weight. Resistant to bacterial, fungal and nematode wilt.

***Tasty :*** Indeterminate, first harvest 70 days after transplanting, Fruit round, firm, skin, thick medium. Fruit weight-80gm, resistant to TMV, fusarium and verticillium, Tolerant to other virus complex. High yielding, good for fresh market, slightly sour in taste. Grown in both kharif, rabi season.

***BSS 366 :*** Plant indeterminate, 1st picking- 65 days after transplanting. Fruit round, firm-skin, fruit weight- 20gm. Tolerant to virus complex. Cherry tomato, fruits in clusters, very high yielding, good sweet taste. Suitable for both (kharif and rabi) season.

***Sultan :*** Plant determinate, first picking 75 days after transplanting. Fruit shape round, medium, fruit weight (180 gm). Resistant to fusarium and verticillium. Grown in Kharif and Rabi. It is having big fruit size and fairly tolerant to diseases.

***Charm :*** This is an indeterminate, oblong shaped hybrid. Fruit weight is approximately 80 gms. Having thick skin and good diseases tolerance. It is ready for first picking in 70 to 75 days from transplanting. It is good for processing.

***BSS 282 :*** Plant determinate, first picking, 65-100 days after transplanting. Thick skin, fruit weight, 90 to 100 gm. It is long industry type hybrid, having very high skin and suitable for long transportation. Grown in both (kharif and rabi) season.

***Sartaj :*** Plant indeterminate, first harvest 65 days after transplanting. Fruit round, medium, weight- 80-100 gm. Tolerant to virus complex. Suitable for kharif & rabi. Very early indeterminate hybrid, can be planted in rainy seasons, as it has good disease tolerance.

***Tolstoi :*** Plant indeterminate, first harvest 70-75 days after transplanting. Fruit square round, fruit weight 100 gm. Resistant to TMV, fusarium and verticillium. Tolerant to other virus complex. Round the year planting, except very hot summer. Excellent very high yielding hybrid, can bear upto 13 fruits per cluster. Plant is vigorous growing and tolerant to diseases.

***BSS 320 :*** Plant indeterminate, first harvest 75 days after transplanting. Fruit square round, weight 100-110 gm. Resistant to TMV, fusarium, verticillium. Tolerant to other virus complex. It has good shelf life (table keeping quality) of 4 weeks. It has very good yield potential. Suitable for both kharif and rabi season crop.

***Rakshak (BSS 368) :*** Plant indeterminate, first harvest 70-75 days after transplanting. Fruit square round. weight- 80-100 gm. Resistant to TMV, fusarium, verticillium, tolerant to other virus complex. This is tolerant to nematodes and good for fresh market. Grown in both kharif and rabi season.

***BSS 369 :*** Plant indeterminate, first harvest 75 days after transplanting. Fruit shape flattish round, weigh 100 gm. It is a good hybrid, having excellent keeping quality of 3 weeks. Resistant to fusarium, TMV and grown in kharif , rabi and also in mild summer.

***Param :*** Plant semi determinate, first harvest 60 days after transplanting. Fruit flattish round, weight 120gm. Resistance to fusarium and tolerant to virus complex. Round the year sowing excep very high temperature. This is a very vigorous growing, high yielding hybrid with sour taste and highly tolerant to disease complex.

***Rachit (BSS 419) :*** Plant determinate, first harvest at 65-70days after transplanting. Fruit square round. Very thick skin. Fruit weight 80-90gm. Resistant to fusarium, tolerant to virus complex. Fruit colou is dark red on ripening. Very high yielding hybrid with very stron skin & good for long transport. Suitable for kharif, rabi and mi summer season.

***BSS 350 :*** Plant determinate , first harvest 75days afte transplanting. Fruit square round, weight 120 gm. Tolerant to viru complex. Big fruit size, very high yielding , having good field resistanc to diseases. It is suitable for long transport and grown in kharif, rab and early summer.

***Urvashi :*** Plant determinate, first harvest 75 days afte transplanting. Fruit square round, weight- 120gm. Tolerant to viru complex. Big fruit size, very high yielding, having good field resistanc to diseases. It is suitable for long transport and grown in kharif, rab and early summer.

***Rishab (BSS 423) :*** Plant semi determinate, first harvest- 65 day after planting. Fruit square round, weight 100-120 gm. Tolerant t virus complex. Resistant to fusarium. Round the year sowing , except very hot summer. Good for summer sowing as well as winter and rainy season sowing. It is highly tolerant to diseases complex and so can be grown round the year.

***Nayak (BSS 422)*** : Plant semi-determinate, first harvest. 65 days after planting. Fruit shape square round, very thick skin, weight-100-120gm. Resistant to fusarium. Tolerant to virus complex. Round the year sowing, except very hot summer. Recommended for summer planting as well as rainy season planting. It is very tolerant to virus and other disease complex and so can be grown in rainy seasons and mild summer also.

***BSS 211*** : Plant determinate, first harvest-60 days after transplanting. Fruit square round very thick, weight-100 gm. Tolerant to virus complex. Grown in kharif, rabi and early summer. It has deep red coloured attractive fruits, with good skin quality and high yielding.

***Rahul*** : Plant determinate, first harvest 65 -70 days after transplanting. Fruit square round, weight 80-90 gm. Resistant to fusarium. Tolerant to virus complex and suitable for kharif, rabi and mild summer. Fruit colour is dark red on ripening. Very high yielding hybrid with very strong skin and good for long transport.

***Ronco*** : Plant determinate, first harvest 65 days after transplanting. Fruit shape square round, very thick, fruit weight 100-120 gm. Tolerant to virus complex. Resistant to fusarium. Grown in kharif, rabi and early summer. Good fruit shape, well covered with leaves, very high yielding hybrid. Suitable for long transportations.

***Benito*** : Plant determinate, first harvest 70 days after transplanting. Fruit square round thick, weight 100-120 gm. Tolerant to virus complex. Grown in kharif, rabi and early summer. It has excellent skin quality, plant is vigorous growing and fruits are well covered with leaves.

***Topspin*** : Plant determinate, first harvest – 60 days after planting. Fruit square round thick, weight-80 gms. Tolerant to virus complex. Resistant to fusarium. Grown in kharif, rabi and mild summer. It is very high yielding, early maturing hybrid which can be grown in all three seasons.

***BSS 420*** : Plant semi determinate, first harvest – 65 days after planting. Fruit round to square round, weight-100 gm. Resistant to fusarium, tolerant to nematode and to virus complex. This has very good vigour and heat tolerance. It can be grown well in rainy season as well as in mild summer. It is very productive.

*SUN 230 :* A full season hybrid which produces large vigorous determinate vines, combines with high yield of heart shaped bright red firm fruits, resistant to verticillium F1 and F2 and tolerance to early blight during wet season.

*SUN 145 :* An early maturing hybrid with medium sized determinate vines, produces round, firm and attractive red colour fruit with uniform ripening. It has verticillium. F1 and F2 resistance.

*SUN 176 :* A full season hybrid which produces medium large vigorous vines. Fruits are firm, heart shaped, smooth and bright and resistant to verticillium. F1 and F2. An ideal hybrid for growing during wet season if TLCV is not a problem.

*SUN 7610 :* Determinate bearing hybrid for a fresh market, that combines the unique qualities of early maturity, resistance to verticillium, fusarium 1 and 2 and high level of tolerance to bacterial wilt. Fruits are deep oblate in shape and deep red in colour at maturity. This is suitable for bacterial prone areas and post rainy season.

*7704 :* A full season fresh market hybrid with vigorous plant growth produces attractive oval shaped fruits (80-90 gms). Fruits are firm and suitable for long distance shipments, resistant to verticillium and fusarium race.

*7712 :* An early maturity hybrid (60-65 days) with determinate plant growth and prolific bearing. Fruits are oval to oblong in shape, 80-90gms in weight, firm, attractive red in colour at maturity.Suitable for long distance shipping.

*Avtar (7711 ) :* Indeterminate plant habit with dark green foliage. First picking starts 70-75 days after transplanting. Fruits are round with an average weight of 90-100 gms. Good tolerance to bacterial wilt. Good shelf life, attractive bright red colour and performance well in hilly areas.

*6109 :* This is mid-season hybrid and produces vigorous determinates vines combined with high yields of firm blocky fruits with good holding ability. this high viscosity hybird has very high solids.

***6200*** **:** Good for peeling and slicing. It has thick fruit walls and is extremely firm for high yields of peeled and diced product, it has high viscosity also, fruits are oval shaped weighing 80-90 gms, very firm and attractive red in colour.

***Kuber Geeta*** **:** A proven determinate hybrid adaptable to different growing conditions. Particularly dry and hot weather conditions. Produces blocky uniform deep red fruits weighing about 70 gms with excellent keeping quality. Resistant to VFFN.

***Hybrid 108*** **:** An excellent determinate hybrid with very good foliage cover and concentrated fruit setting. Fruits blocky, uniform, deep red, suitable for long distance transportation.

***Protima*** **:** A determinate vigorous early hybrid with good foliage cover produces square round blocky, very firm fruits of 70- 80 gms, turning bright on maturity, having good keeping / shipping quality.

***Summer King*** **:** An outstanding determinate hybrid with very good heat tolerance. Plants are vigorous with high foliage cover to protect fruit from sun scalding. Very firm fruits with bright red colour, weighing about 80 gms. Have very good keeping quality.

***Pooja*** **:** A determinate hybrid with wide adaptability. Strong plant with good foliage cover and profuse bearing produces square round, sweet and firm fruits weighing about 80 gms., with good keeping quality.

***Karuna*** **:** A very high yielding indeterminate hybrid. Strong plant with heavy foliage cover produces slightly oval square round fruits of about110 gms. with green shoulder attaining bright red colour on maturity. Prolonged flowering and fruiting gives an advantage of extending picking and higher yields. The very firm fruits are excellent choice for long distance shipping and keeping quality.

***Snehalatha*** **:** An indeterminate hybrid produces square round fruit of about 100 –200 gms, with excellent keeping quality. A large plant with good foliage cover prolific bearer with continuous flowering and fruiting. Fruits are green shoulders attaining bright red colour on maturity, can withstand long distance shipping.

***X 119*** **:** A second generation hybrid with determinate strong plant with good foliage. Fruits are square round, weighing about 80 gms. Heavy fruit setting ability with firm fruits makes it an excellent choice.

***Nagmani :*** A cherry type tomato hybrid, plant growth is indeterminate with sufficient fine foliage and spreading habit. Fruits are small oval, weighing 10 gms / fruit and each cluster has plenty of fruits. Very high yielding. Keeping quality of fruits is good. Highly tolerant to leaf curl virus.

***Nagmoti :*** A cherry Type tomato hybrid. Plant growth is indeterminate with sufficient fine foliage and spreading habit. Fruits are small round, weighing 15 gms / fruit and each cluster has plenty of keeping fruits. Very high yielding. Keeping quality of fruits is good. Highly tolerant to leaf curl virus.

***Pride :*** Determinate plant, wide adaptability, excellent foliage cover and profuse bearing. Fruits weighing about 80 gms. Firm fruits and good keeping quality.

***Prestige :*** A very strong determinate plant with good foliage cover produces square round fruits of 80 gms. Very firm and having attractive red colour on maturity.

***UTM -1370 :*** A determinate plant with excellent adaptability, attractive canopy and heavy fruit bearing are main features. Fruits are round, firm, weighing about 100 gms.

***Avinash-2 :*** Determinate hybrid with vigorous growth, dense green foliage. Profuse fruit setting. Most suitable hybrid for cultivation in TYLCV prone areas and season. Fruit is medium, thick, walled, round, glossy, dark red with average weight 70-90g, first picking starts in 50-55 days after transplanting. First TYLCV virus tolerant tomato hybrid in india. Yield 100-120mt / ha under optimal conditions.

***Tarica :*** Plant sturdy, determinate with dense green foliage providing fruit shelter. Fruit is oblang, frim, dark red , uniform with average weight 80-100 gm. First picking starts in 60-65 days after transplanting. Good shelf life. Ideal for long distance transportation. Tolerance to *fusarium and Verticillium wilt.* Yield 100mt/ ha under optimal conditions.

***Rasika :*** Plant early, determinate hybrid with vigorous growth. Good branching, dense green foliage providing fruit shelter. First picking starts 55-60 days after transplanting. Flowering / fruit setting is in clusters. Fruit is firm, round, deep red with average weight 80- 90 gm, uniform size till the last picking, fetch higher price in the market. Good shelf-life. Ideal for long distance transportation. Tolerant to *fusarium* and *virticillium* wilt. Yield 100 mt/ ha under optimal conditions.

***Ratna :*** Plant determinate of medium height. Dense green foliage providing fruit shelter. Flowering starts 35 days after transplanting. Fruit is round, firm, dark red with average weight 80-125 gm. First picking starts in 60-65 days after transplanting. Good shelf-life. Ideal for long distance transportation. Tolerant to *fusarium* and *verticillium* wilt. Yield 100mt / ha under optimal conditions.

***Rocky :*** Plant determinate of medium height. Strong, vigorous stem with green foliage. Flowering starts 35 days after transplanting, suitable for monsoon and winter cultivation. Fruit is round, firm, dark red with an average weight of 80-100g. Overlapping flushes. Ideal for long distance transportation. Yield 80-100mt/ha under optimum conditions.

***TO 5975 :*** Plant determinate of medium height. Strong, vigorous stem with green foliage. Flowering starts 35 days after transplanting, suitable for monsoon and winter cultivation. Fruit is oval- round, firm, dark red with an average weight of 80-100 g, overlapping flushes. Ideal for long distance transportation. Yield 80-100 mt / ha under ideal growing conditions.

***Lehar :*** The plants of this vigorous early dual purpose type hybrid are determinate having excellent foliage cover. The fruits are attractive, medium sized (av. 80gm) with excellent colour, flattish round in shape and concentrated maturity. A heavy producer.

***Neeladhari :*** An early, very high yielding determinate, self supporting bushy hybrid with good tolerance to bacterial wilt. Suitable for ground culture. Fruits attractive globe shaped with uniform shoulders, good colours and TSS 6%, small medium weighing 46gm on an average.Highly flavoursome.

***Maharishi :*** As the name suggests, Maharishi is an improvement over already popular Rishi. This new presentation is an early maturing type with better fruit weight of 80gm, round, firm and attractive medium sized fruits with uniform shoulders. Recommended for long distance shipping. An immense cropper.

***Roopsi :*** A dual purpose hybrid, determinate, vigorous with good cover. Fruits are uniformly globe in shape with uniform shoulder, most attractive deep scarlet-red in colour and exceptional beauty; firm with average weight of 70gm. Highly recommended for upmarket growers where grading of produce is of prime importance.

***Rishi*** **:** Rishi is vigorous and determinate hybrid suitable for fresh market and processing. An early uniformly maturity variety (115 days), produces most attractive medium sized (70g.) round and firm fruits, highly recommended for long distance transportation. Excellent material for higher temperatures.

***Maitri*** **:** Matri is a determinate and very high yielding hybrid suitable for fresh market and processing. This hybrid is well adapted to a wide range of environmental conditions thereby the availability of fruit is for a much longer period; off setting. Market fluctuations. An early-medium maturing (120-125days) having medium sized round and extremely firm fruits. recommended for long distance transportation.

***Century-12*** **:** An outstanding early, determinate hybrid. Fruits round, scarlet- red in colour weighing 62 g. on the average. Relatively more tolerant to extremes of temperature. Suits better to Indian taste. TSS 4.20, acidity-0.70, Lycopene content-5.90/100gm.

***Yogi*** **:** A recently developed determinate type with excellent fruiting potential. A vigorous and early maturing ; dual purpose hybrid. Fruits firm, squarish-deep globe with uniform shoulders, medium size (av. 70gm), 2-3 locules and good wall thickness. TSS 5.34%. Shelf life 1-2 weeks. Recommended for long distance transportation.

***Guru*** **:** Semi-indeterminate, early, vigororous and prolific, plants are tolerant to heat and moisture. Fruits medium sized, firm, oblongish-round with fainted green shoulders and average weight of 80 gm. Excellent uniform colour and attraractive, High yields with good shipping trait is make this hybrid an ideal choice for growers.

***INDI-1*** **:** A fresh market type, indeterminate plant with compact habit is highly recommended for poly-tunnels in area where the growing season is short. Highly responsive to high fertility conditions coupled with optimum moisture. Large firm fruits are deep oblate in shape with uniform shoulders and weigh 130gm. on an average. Excellent colour with good flavour.

***CTH-230*** **:** An excellent semi-determinate-early maturing hybrid suitable for distant markets. Very prolific yielder with firm medium size clear shoulder, round fruit with good colour and flavour. Average

80 g borne on clusters over a long period. Has marked tolerance to *Verticillium, Fusarium* and Bacterial wilt. A must for a progressive grower with wide adaptability.

***Himalaya 40 :*** Semi-indeterminate, early, vigours and prolific plants are tolerant to heat and moisture. Fruits medium sized, firm, oblongish-round with fainted green shoulders and average weight of 80 g, excellent uniform colour and attractive. High yield with good shipping traits is make this hybrid are ideal choice for growers.

***INDI 11 :*** A fresh market type. Indeterminate plant with compact habit is highly recommended for poly-tunnels in areas where the growing season is short. Highly response to high fertility conditions. coupled with optimum moisture.

***Gotya (S-41) :*** Plant type determinate, height at maturity 75-90 cm First marketable produce obtained 70-75 days after transplanting. Fruit oval, dark red with white shoulder and jointed pedicel. Average fruit weight 50-55g, TSS(brix); 4.0. Good for fresh market, good shelf life and transportability. Yield potential 80-90 t/ha. Recommended for Tamil Nadu, Maharashtra Karnataka, West Bengal, Andhra Pradesh and parts of Rajasthan.

***S-72 :*** Plant determinate, height at maturity 65-75cm. First marketable produce at 80-85 days after transplanting. Fruit blocky, dark red with green shoulder and jointed pedicel. Average fruit weight 80-85g. TSS(brix) 5.3. Yield potential 80-90 t/ha. Recommended for Karnataka, Tamil Nadu, Maharashtra and parts of Gujarat.

***S-75 :*** Plant determinate, height at maturity 55-60 cm. First marketable produce at 85-95 days after transplanting. Fruit dark red with white shoulder and jointed pedicel. Average, fruit weight 70-75g TSS (brix) 5.1. Good for fresh market, slow ripening character gives farmers an extended period for harvesting. Yield potential 80-90 t/ha, Recommended for Karnataka, Tamil Nadu and parts of Gujurat.

***Naveen :*** Plant indeterminate, firm oblong , average fruit weight 89 g; fruit firmness intermediate, average yield 57.5 tonnes/ha, TSS. 6.2 %, suitable for processing, tolerant to *verticillium* and *fusarium.*

***Karnataka :*** Plant indeterminate, fruit shape flat round (green shoulder), fruit weight 230g, fruit colour yellowish red fruit firmness intermediate, average yield 52.0 tonnes /ha. Tolerant to *verticillium, Fusarium,* root knot nematode, TSS 5.2 %.

***Baishakhi*** **:** Plants determinate with less foliage and very early hybrid. Fruits, medium size (80 g), red round, mature in 55 days, ideal for growing in summer season.The optimum temperature range is 25°-30°C, tolerant to *fusarium* and *verticillium* wilt.

***Meghana*** **:** Its plants are vigorous growing, fruits medium size (75 g), intense red, with round shape, fruits mature in 65 days. Suitable for both fresh market and processing. The temperature range is 25°-30°C and ideal for summer and rainy cultivation. Tolerant to late blight disease.

***Rohini*** **:** Vigorous growing hybrid, fruits medium size (90 g), red, slightly elongated, fruits mature in 65 days, ideally suited for long distance shipping, storage and processing. The optimum temperature range is 25°-30°C. Tolerant to fusarium, *verticillium* and nematode wilt.

***Ramya*** **:** Plant habit rigid, fruits medium size (80 g), red, round, suitable for long transportation; optimum temperature range is 25-30°C, ideal for rainy season growing. Fruits mature in 65 days. Tolerant to *fusarium, verticillium* and nematode wilt.

***IAHS 88-1*** **:** Plant indeterminate, fruits flat round (green shoulder), fruit weight 152 g, fruit colour orange red, fruit firm, average fruit yield 56.0 t /ha. Tolerant to *verticillium, fusarium, alternaria,* stem canker, *stemphyllium*. TSS 5.2%, very strong root system, good for the rainy season.

***IAHS 88-2*** **:** Plant determinate , fruit round, fruit weight 220 g, fruit colour deep red, firmness intermediate, average fruit yield 66.0 t/ ha. tolerant to *verticillium, fusarium,* root-knot nematode, *alternaria, stemphyllium,* TMV-1, TSS 5.4%, not suitable for hot season.

***Rajani*** **:** Plant determinate, fruit oblong (green shoulder), fruit weight 82.0 g, fruit colour orange red, fruit firm, average yield 72. 25t/ha, tolerant to *verticillium, fusarium, alternaria,* stem canker, stemphyllium. TSS 5.3 – 5.5%, suitable for processing.

***Rupali*** **:** Plant determinate, fruit oblong , fruit weight 87 g, fruit colour orange –red, firm, average fruit yield 79.5 t/ ha, tolerant to *verticillium, fusarium,* TSS. 5.6 –5.8 % and suitable for processing.

***Rashmi*** **:** Plant determinate, fruit oblong, fruit weight 65g, fruit colour deep red , firmness intermediate, average fruit yield 67.5 t/ha. Tolerant to *verticillium, fusarium, alternaria,* stem canker, TSS 5.5% suitable for processing.

***Vaishali :*** Plant determinate, fruit shape round, fruit weight 96 g, fruit colour deep red, firmness soft, average fruit yield 77.0 t/ha, tolerant to *verticillium, fusarium,* TSS 5.1%, good for rainy season and dry humid conditions.

***Sheetal :*** Plant determinate, fruit round, fruit weight 99g. colour orange red, firmness intermediate, average fruit yield 64.7t/ha, tolerant to *verticillium, fusarium,* TSS 5.4%, can set fruit at 15°C.

***Mangala :*** Plant determinate, fruit round, average weight 120g colour deep red, firmness intermediate, average fruit yield 57.8 t/ha, tolerant to *verticillium, fusarium,* root-knot nematode. TSS is 5.7%.

***IAHS 88-3 :*** Plant determinate, fruit shape pear shaped (nippled), weight 60-80g, colour orange-red, firmness intermediate , average fruit yields 82.5 t/ha. tolerant to *verticillium, fusarium,* alternaria, root-knot nematode, T.S.S 5.8 –6.0%, suitable for processing.

***Manisha (BSS-99) :*** It has developed by Bejo Sheetal Seeds Pvt. Ltd, Jalna. Plants are indeterminate ; fruits red with thick skin ; good shelf life and transport quality ; gives an average yield of 80-100 t/ ha. Female Parent (Rus-1) ; plants are indeterminate with dark green leaves ; fruits are round and appear in clusters. Male Parent (Rif-1) : Plants are semi -determinate with light green leaves and fruits are oval blocky.

***Megha (BSS – 98) :*** It has developed by Bejo Sheetal Seed Pvt.. Ltd, Jalna. Plants are determinate; fruits are red with thick skin ; sood shelf life and transport quality ; gives an average yield of 50-70 t /ha. Female Parents (Pbk 1) : Plants are determinate with green leaves and fruits are round-Male Parent (Rif-1) : Plants are semi determinate with light green leaves and fruits are oval blocky.

***Gulmohar (MTH-6) :*** It is developed by Maharashtra Hybrids Seeds Co. Ltd., Jalna. Plants are determinate ; fruits are red, square, average weight of 90g; TSS 4.17%; first picking starts from 60-65 days ; gives an average yield of 90- 100 t/ ha ; recommended for different regions.

***Meenakashi (BSS-20) :*** It is developed by Bejo Sheetal Seeds Pvt. Ltd.; Jalna. Plants are in determinate ; fruits are dark red , round global, average weight of 80 g ; resistant to *fusarium wilt*; gives an average yield of 60-100 t/ha. Female Parents; (Rus-1) : Plants are indeterminate with dark green leaves ; fruits are round and appears in clusters. Male Parents (HTR-1); Plants are determinate with green leaves and fruits are blocky.

***BSS-39 (Madhuri) :*** It is developed by Bejo Sheetal Seeds Pvt. Ltd., Jalna. Plants are determinate, early in maturity ; fruits are dark red , flat round with thick skin , average weight of 100 g ; crack resistant which can with stand long transportation ; gives an average yield of 50-70 t/ha. Female parent (Pot- 94) : plants are semi-determinate with pale leaves and blocky oblong fruits. Male parent (Olita-1) : plants are very early with light green leaves and flattish round fruits.

***ARTH-3 :*** It is developed by Ankur seeds Pvt.Ltd.,Nagpur. Plants are determinate with medium dark green leaves, fruits are equarish round, deep red ; flesh is red with 4.5 to 5.0%, TSS; moderately tolerant to drought ; harvesting starts after 80-85 days and continues upto 160-170days;gives an average yield of 88-95 tonnes/ha.

***Female parent :*** plants are determinate with medium to dark green leaves; fruits are dark red and flattish round. Male parent; plants are determinate with medium dark green leaves; fruits are dark red and flattish round.

***Male parent :*** plants are determinate with dark green leaves ; fruits are oval and dark red with thick wall.

***ARTH - 4 :*** Plants are semi spreading and indeterminate with dark green leaves ; fruits are deep red, flat round ; red flesh with 4-4.5 % TSS; moderately tolerant to draught ; harvesting starts after 80–85 days and continues up to 170-190 days ; gives an average yield of 100-110 t / ha ; Female Parents : plants are indeterminate with medium to dark green leaves; fruits are deep red, flat round and average weight of 40-45g. Male Parents : Plants are semi-spreading and indeterminate with medium green leaves ; fruits are deep red round and average weight of 50-60g.

***Sandhya :*** Medium stature, strong plant with medium foliage cover, fruits are round, medium large size (90–110 gm) with excellent deep red colour and profuse fruiting.

***Jajati :*** Vigorous compact plant with good foliage cover, medium maturity group, suitable for fresh market. Fruits are uniformly red coloured, round in shape and medium sized (75-95 gm per fruit). Tolerant to bacterial and root-knot nematode wilt.

***Abhishek :*** Strong, tall vigorous plant with good foliage cover and prolong fruiting period. Very high yielding. Fruits are medium sized (75-80 gm), oval shaped, firm and good shelf life. Resistant to bacterial wilt and tolerant to early blight diseases.

***Upasana :*** Vigorous compact plant with good foliage cover, fruits are medium large sized (90-120 gm), round to oval shaped and good keeping quality. Tolerant to bacterial, *fusarium* and *verticillium* wilt. Suitable for early planting.

***Amarprem :*** Indeterminate growth habit, strong plants, with prolong fruiting period. Fruits are medium sized (75-85 gm) good flavour, suitable for dual purpose i.e. both for fresh marketing and processing. Tolerant to early blight and resistant to fusarium wilt disease.

***Pratibha :*** Plant indeterminate, vigorous and compact, medium foliage cover, very good yielder and prolong fruiting period, fruits oblate, medium sized (75-80 gm), uniform deep red colour with very good transport quality. Tolerant to bacterial wilt and resistant to fusarium wilt.

***Arati :*** Strong plants with indeterminate growth habit prolong fruiting period. Developed for dual purpose both for fresh marketing and processing. Fruits are medium size (70–80 gm), round to oval in shape, deep red colour and good shelf life. Resistant to bacterial wilt and root knot nematode wilt.

***Ashirbad :*** A dual purpose hybrid for both fresh marketing and processing. Plant tall and profuse fruiting. Fruits are round, medium sized (70-80 gm) firm, uniform deep scarlet red colour with good shelf life. Tolerant to bacterial wilt and resistant to *fusarium* and *verticillium* wilt.

***Biswajit :*** Strong plant, vigorous, suitable for early planting and good foliage cover, fruits are round, medium (70-80 gm), uniform red colour with good keeping quality. Tolerant to bacterial wilt and root-knot nematode wilt.

***Amarjioti :*** A dual purpose hybrid both for fresh marketing and processing, plant tall and profuse fruiting. Fruits are round, medium sized (70-80 gm), firm uniform deep scarlet red colour with good shelf life. Tolerant to bacterial wilt and resistant to *fusarium* and *verticillium* wilt.

***Arundhati :*** Strong, tall, vigorous plants with good foliage cover and prolong fruiting period. Very high yielding. Fruits are medium sized (75-80 gm), oval shaped, firm and good shelf life. Resistant to bacterial wilt and tolerant to early blight diseases.

*Raja* : An early maturing hybrid which matures in about 60-65 days after transplanting, medium sized fruits are attractive bright red in colour and borne in cluster. Good for storage and long transportation.

*SUN 7610* : Determinate bearing hybrid for a fresh market, that combines the unique qualities of early maturity, resistance to *verticillium, fusarium* 1and 2 and high level of tolerance to bacterial wilt. Fruits are deep oblate in shape and deep red in colour at maturity. This is suitable for bacterial prone areas and post rainy season.

*7704* : A full season fresh market hybrid with vigorous plant growth produces attractive oval shaped fruits (80-90gms). Fruits are firm and suitable for long distance shipments, resistant to *verticillium* and *fusarium* race.

*7712* : An early maturity hybrid (60-65days ) with determinate plant growth and prolific bearing. Fruits are oval to oblong in shape, 80-90 gms in weight, firm, attractive red in colour at maturity. Suitable for long distance shipping.

*Avtar (7711 )* : Indeterminate plant habit with dark green foilage. First picking starts 70-75 days after transplanting. Fruits are round with an average weight of 90-100 gms. Good tolerance to bacterial wilt. Good shelf life, attractive bright red colour and performance well in hilly areas.

*6109* : This hybrid is mid-season and produces vigorous determinates vines combined with high yields of firm blocky fruits with good holding ability, this high viscosity hybrid has very high solids.

*6200* : Good for peeling and slicing. It has thick fruit walls and is extremely firm for high yields of peeled and diced product. it has high viscosity also. Fruits are oval shaped weighing 80-90 gms, very firm and attractive red in colour.

*Surabhi* : This early maturing diploid hybrid with a vigorous growth , produces more fruits and vigorous growth and protects fruits form sun burn. Globe shaped fruits are dark green in colour and good for long transportation and storage, flesh is firm, crispy sweet and juicy. Tolerant to high temperature and *fusarium* wilt race-1.

*Arjuna* : Plant determinate, first picking start 60-65 days after transplanting, fruit shape round, dark red, firmness good, average weight 100 g. The leaves provide good foliage cover to the fruit protecting them from sun scorching. Suitable both for processing and fresh market.

***Krishna*** **:** Plant determinate, first picking 60 days after transplanting, fruit square round, deep red, firmness very good, average weight 80g. ideally suited for rainy as well as Spring – Summer seasons. An early maturing hybrid suitable for processing.

***Karna*** **:** Plant semi-determinate, first picking 65 days after transplanting. Fruit round, deep red, firmness excellent, average weight 100 g, this hybrid has excellent shelf-life 20 days under room temperature. Highly suitable for long distance transportation.

***Nakul*** **:** Plant determinate, first picking 60 days after transplanting. Fruit square round, deep red, firmness very good, average weight 50g, highly suitable for processing.

***SC- 1*** **:** Plant determinate, first picking 65 days after transplanting, fruit round to slightly blocky, uniform bright red, firmness very good, average weight 100 g. tolerant to wilt, highly suitable for long distance transportation and processing, set fruit under high temperature.

***SC-3*** **:** Plant determinate, first picking 60 days after transplanting, fruit round, bright red, firmness very good, average weight 80 g, tolerant to wilt. Suitable for whole peeled fruit processing.

***Chiranjeev (621)*** **:** Plant determinate, first picking 55 days after transplanting, fruit square round, red, firmness good, average weight 100 g, suitable for early marketing grown during spring-summer season.

***Virat*** **:** Plant semi-determinate, first picking 60 days after transplanting, fruit round, deep red, firmness very good, average fruit weight 110 g, dark green foliage provides good protection to fruits from sun scorching.

***Himlata*** **:** Plant indeterminate, first picking 65 days after transplanting, fruit square round, uniform red, firmness good, average weight 100 g, the fruits are protected from sun scorching by vigorous foliage. Perform very well in Northern hills.

***HimShiva*** **:** Plant indeterminate, first picking 65 days after transplanting, fruit square round, red, firmness good, average fruit weight 90 g. very good for farmer in Northern hills.

***Sachin (OP) :*** Plant determinate, maturity 80 days after transplanting, fruit oval (slightly elongated), red, average weight 70 g, heavy yielder.

***Saurav (OP) :*** Plant determinate, first picking 75 days after transplanting, fruit oval, deep red, average weight 70 g, heavy yielder.

***BCTH 4 :*** Tomato BCTH 4 hybrid with semi-determinate growth, uniparous cyme with less serrated leaflet blade ; flattish-round, bright red, medium sized fruits, weighing 70-75 g with 4 locules and medium ribbing on fruit surface, BCTH 4 show ample scope of its commercial cultivation. Its fruits are medium-firm with appreciably thick pericarp (0.56 cm) which turn dark red on ripening. The ripe fruits contain appreciable TSS (4.0 ° Brix), lycopene (3.13 mg / 100 g fresh) contents and high vitamin C content (40.5 mg / 100 g fresh). On an average yield potential of 40-45 tonnes/ha. It is highly tolerant hybrid to tomato leaf curl virus and excellent for fresh consumption. It is suitable to grow in autumn-winter season with transplanting with first fortnight of October. Developed by BCKVV, Kalyani.

***BCTH 62 :*** It is a big-fruited processing quality hybrid with semi determinate growth habit, biparous cyme with oblong heart shaped greenish fruit shoulder which turn dark red on ripening. The fruits are big, weighing 110-125 g with 4 locules and light ribbing on fruit surface. Its fruits show somewhat slow ripening character and suitable for long distance transportation. Processing quality of the fruits is marked by its thick pericarp (0.62 cm), high TSS (4.4 ° Brix), lycopene (4.91 mg / 100 g fresh) and total carotenoid (7.26 mg / 100 g fresh) contents. On an average, 8-10 fruits weight one kilogram with an average yield potential of 45-50 tonnes / ha. Less severity to early blight and tomato leaf curl virus diseases and wide adaptability and can be grown in winter season successfully. It is developed by BCKVV, Kalyani.

□□□

# 6

# Legume Vegetables

## BROAD BEAN

Broad bean (*Vicia faba* L.); 2n = 12, 14 ; syn. Faba bean, Horse bean, Field bean, Tick bean and Windsor bean, belong to the family Leguminosae, sub-family Fabaceae ; commonly known as Makhan Sim, Falda papdi, Bakla. Chapparadavare, Avarakka, Simba, Avarai and Pedda Chikkudu ; is the fourth most important legume crop in the world after beans, peas and chick pea. It is the only bean and winter legume which can withstand sufficiently cold temperature.

Broad bean is one of the oldest cultivated plant and rich in carbohydrate, protein, minerals and vitamins and obility to fix atmospheric nitrogen. Pods are used as green, generally as cooked vegetable either singly or mixed with potatoes in curry form. It is also used in the preparation of mixed pickle with other vegetables. Boiled seeds of broad bean can also be consumed after applying salt and chilli or black pepper powder.

## Improved Varieties and Hybrids

*Germplasm resources* : Filippetti and Ricciardi (1993) reported that the International Board for Plant Genetic Resources (IBPGR) has developed a network of plant genetic resources centres, each with the responsibility of maintaining the world collection of a particular crop. The International Centre for Agricultural Research in the Dry Areas (ICARDA) has a world mandate for the improvement of faba bean. Since its inception, ICARDA has one of the largest faba bean germplasm collections for the continuous improvement of this crop. Faba bean germplasm is maintained at ICARDA in two collections :

1. International legume Faba Bean (ILB) collection (3305 accessions) containing the original germplasm accessions as received, usually heterogenous and heterozygous to a great extent.
2. Faba bean pure line (BPL) collection which has been derived from the ILB collection by a pre breeding process of cyclic single plant progeny rows.

The collection expeditions have also taken place in Cyprus, Egypt and Afghanistan. Collection of Vicia faba have also been made in the Mediterranean region by the Germplasm Institute, CNR, Bari, Italy and in the Ethiopia by Ethiopian Genetic Resources Centre.

The collection stored in Bari includes material from Algeria, Greece, Spain, Tunisia, Ethiopia, Italy, Egypt and Libya (Perrino, 1985). In Cyprus, the Agricultural Research Institute in collaboration with FAO and IBPGR collected 98 accessions of faba bean (Della, 1981). Afghanistan, Ethiopia, Germany, Turkey, Ciprus, Spain, Canada, UK, Morocco, Egypt, Iraq, Italy, Syria, and the former USSR are the main countries of origin of the major collections of broad bean germplasm.

There is no improved variety available in India, but several exotic varieties introduced from abroad are :

***Long pod type*** : Aquadule Claudin, Imperial White long pod, Masterpiece green long pod, Imperial green long pod, Red Epicure.

***Windsor type*** : Imperial White Windsor, Gaint Four seeded Green Windsor, Imperial Green Windsor.

***Local types*** : Bakla local- It is a selection from local population grown in North Uttar Pradesh.

***Synthetic varieties*** : A synthetic is a population of plants artificially produced by the breeder. A synthetic variety is an advanced generation of a seed mixture of selected genotypes (inbred lines, clones, hybrids etc.), maintained for a limited number of generations by hybridization in all combination among them.

***Throw's MS*** : It was the first synthetic faba bean to be grown commercially in the U.K.

***Maris Beagle*** : It is another English synthetic variety composed of five inbred lines and released in 1974.

***Soravi and Avrissot*** : These synthetic varieties were released in France; the first is an example of international cooperation between the Plant Breeding Institute of Cambridge (UK) and the National Institute for Agricultural Research (INRA-France). Now most of the North European faba beans are synthetic varieties (Filippetti and Ricciardi, 1993).

***Pusa Sumeet*** : Plants are 75 cm tall having on an average 10 branches/plant. A plant bears about 100 pods. The pod length and thickness are 6.0 cm and 1.3 cm respectively. It has attractive dark green pods and borne in cluster. The average yield potential is 180 q/ ha.

Besides above variety, 'Selection BR-1' and 'Selection BR-2' are black and yellow seeded varieties, respectively developed from Bihar state. 'Jawahar Vicia 73-81' is domant type, reported from Madhya Pradesh.

## COW PEA

Cow pea (*Vigna unguiculata* L.) syn. Southern pea, Black eyed pea, yard long bean, lobia, karamani, Baragudi, Avadai, Barbati, Chola, Mambayar, Chavli, is a diploid with 2n = 22, belong to the family Leguminoceae, sub-family Fabaceae and Genus Vigna. Three cultivated sub-species have been identified under the species *Vigna unguiculata*. These are :

1. *V. unguiculata* ssp. *unguiculata* = (dual purpose type).
2. *V. unguiculata* ssp. *cylindrica* (Syn. *V. unguiculata* ssp. Catjang, *V. sinensis* ssp. Catjang) = (grain type).
3. *V. unguiculata* var. sesquipedalis = (Vegetable type), known as yard long bean or asparagus bean is used as vegetable. It is

climbing or trailing. Pods are 30-90 cm long, pendulous, fleshy and tending to shrink when dry. Seeds are elongated and kidney shaped.

All the three cross each other. In a cross between *V. unguiculata* ssp. *unguiculata* (2n = 22) and *V. unguiculata* var. *sesquipedalis* (2n = 24), F1 had 2n = 23.

## Varieties and Hybrids

Varieties and cultivars respond differently to photoperiod. According to use, there are different varieties. However, the same variety may be raised for more than one purpose. Several cowpea varieties are grown in various parts of the country.

Cowpea varieties for cultivation in different states are :

| | |
|---|---|
| Jammu and Kashmir | C-152, F.S. 68. |
| Himachal Pradesh | C-152, Pusa-1, Sel 27. |
| Punjab | C-152, FS 68, Pusa Barsati. |
| Haryana | FS 68, L-1552. |
| Uttar Pradesh | Type 2, Pusa Barsati, Pusa-Do-fasli, Pusa Phalguni. |
| Bihar | C-152, FS 68. |
| Delhi | Pusa Rituraj, Pusa Komal. |
| West Bengal | 5269, Bidhan Bharati 1, Bidhan Bharati-2. |
| Rajasthan | JC-5, JC-10. |
| Madhya Pradesh | C-152, K-11, K-14, Pusa Barsati. |
| Orissa | C 152, FS 68, SEB-2, Utkal Manika, BCP-4. |
| Maharashtra | No.5, No.21, 19-14-1, C-152, IIHR 61-8. |
| Karnataka | C-152, Sel 61-B-1, S-288, IIHR 61-B, IIHR 16, Arka Garima, Arka Suman, Arka Samrudhi. |
| Tamil Nadu | Co-1, CO-2, Pusa-3, C-152, Pusa –Do-Fasli. |
| Kerala | Vyjayanthi, Lola, KMV1, Bhagyalakshmi, Kanakamani, Kairali Anaswara, Varun. |

Varieties and hybrids, developed by different institutions are mentioned below :

***IARI, New Delhi*** : Pusa Komal (Sel.1552)), Pusa Phalguni, Pusa Barsati, Pusa Dofasli, Pusa Rituraj.

***IIHR, Bangalore*** : IIHR 16, Arka Garima, Arka Suman, Arka Samrudhi.

***Kerala Agril. University*** : Vyjayanthi, Lola, KMV 1, Bhagyalakshmi, Kanakamani, Kairali, Anaswara, Varun.

***TNAU, Coimbatore*** : CO2, Vamban.

***NDAU & T, Faizabad, UP*** : Sel-2-1.

***PAU, Ludhiana*** : Sel-263.

***BCKVV, Kalyani*** : Bidhan Bharati-1, Bidhan Bharati-2.

***JNKVV, MP*** : Gwalior K 3B, Gwalior 11, Gwalior K-14.

***OUAT, Bhubaneswar*** : Utkal Manika, BCP-4.

***Pusa Komal (Sel.1552)*** : It is a product of P-82 and P-426 cowpea strains. It is compact, bushy, without vines or trailers. However, it put fourth short vines when light intensity and temperatures are relatively low. It has been bred at IARI, New Delhi. It has been recommended for release by the all India Coordinated Vegetable Improvement workshop, 1983. It is a dwarf and bushy cultivar and normal plant height is 30-45 cm. The peduncles are prominent. Each trifoliate base produces a peduncle, two to four pods of 20-22 cm long are produced on each peduncle simultaneously. The pods are bright green, non-fibrous, cylindrical soft and meaty, suitable for planting in spring summer and rainy seasons. Pods are light green, 25-30 cm long. It flowers in 45 days and produces green pods in 60-65 days. The bearing is almost synchronous and the pods can be harvested in three pickings. It is a photo insensitive variety and is resistant to bacterial blight.

***Pusa Phalguni*** : It is a selection from a promising introduction Dolique Du Tonkin (ex Canada). The plants are dwarf with bushy habit. It gives two flushes of dark green erect pods of 10-12 cm length with small cylindrical white seeds. Pods get ready for harvesting in about 60 days. Yield potential of green pods is 75 q/ha. Seeds small and white coloured. It is best suited for February-March sowing.

***Pusa Barsati*** : It is selection from exotic materials from Philippines. It is an early cultivar (45 days to first green pods picking), suitable for growing during the rainy season. It gives 2-3 flushes of light green pendent pods. The pods are 25-28 cm long and contain large green seeds. Seeds brown and white marked. Green pod yield potential is about 75 q/ha.

***Pusa Rituraj*** : This is a selection in germplasm by national Bureau of Plant and Genetic Resources, New Delhi. This is photo-insensitive, bushy variety and can be grown in summer and rainy seasons. First picking is possible in 40-50 days after sowing. Pods virtually cover the foliage. Pods are 20-25 cm long. It is a dual purpose variety as pods and seeds (brown) both can be consumed. Green pod yield potential is 80 q/ha.

***Pusa Do Fasli*** : It has been developed from cross of Pusa Phalguni and a long podded introduction from Philippines. It flowers in 30-40 days after sowing. Pods are light green, erect, about 18 cm long. Plants are bushy. It is photo-insensitive and is suitable for sowing in spring-summer and rainy seasons. Green pod yield are lower in summer planting than those in the rainy season. The green pod yield potential is about 80 q/ha.

***IIHR-16*** : IIHR-16 is an early maturing variety developed at IIHR, Bangalore, derived through pedigree selection from the cross Arka Garima x Pusa Komal. Plants are erect, bushy, 70-75 cm tall, photo-insensitive, pods are green, medium thick, medium long (15-18 cm), tender, fleshy without parchment, good cooking quality, gives an average yield of 19 t/ha in 70-75 days of crop duration. Recommended for cultivation and release in 1998 for the zones V1 (Rajasthan, Gujarat, Haryana and Delhi) and VII (M.P., excluding eastern areas and Maharashtra).

***Arka Garima (Sel-61-B)*** : This popular and pole type variety has been bred at IIHR, Bangalore, derived through backcross and pure line selection from the cross TUV-762 x *Vigna Unguiculata* sub. spp. *sesquipedalis*. Plants are tall, vigorous, bushy, spreading with small vines with light green leaves and purple flowers. Pods are light green, long thick, round, fleshy and stringless, tolerant to heat and drought; gives an average yield of 19 t/ ha in 90 days of crop duration. Recommended for release and cultivation in 1992 for the zones VII (MP excluding eastern areas and Maharashtra) and VIII (Karnataka, Tamil Nadu and Kerala).

***Arka Suman*** : This is an early maturing variety. It performs very well in Maharashtra, Karnataka and Kerala. The pod weight is higher (mean of 70 pods / kg). This variety is most suitable for rabi season. Bushy photo insensitive variety with medium long pods. Yield 15 t/ ha.

***Arka Samrudhi*** : Bushy photo insensitive variety with medium long pods. Yield 15 t/ha.

***Vyjayanthi*** : Trailing growth habit, long vine red coloured pods, brown seeds; average length of pod 50.62 cm, average fruit weight 16.17 g, no. of seeds/pod 20-22, productivity 12.6 t/ha.

***Lola*** : Trailing growth habit, smooth and extra long pods, glossy light green coloured long pods with purple tip, black seeds. Average length of pod 53.38 cm, average pod weight 22 g, number of pods/ plant 20-22, productivity 20.0 t/ha.

***KMV 1*** : Trailing growth habit, long light green pods with brown tip, immature seeds light green. Average length of pod 38 cm, number of seeds/pod 19, productivity 14.2 t/ha.

***Bhagyalakshmi*** : Early flowering, bushy growth habit, light green medium sized pods, mottled seeds, average length of pod 27.0 cm, aveage pod weight 7.13 g. Days to first harvest 48 days, productivity 6.48 t/ha.

***Kanakamani*** : Semi-trailing growth habit, pods-dark green, bold seeded and medium long, reddish brown seeds; average pod length 17.8 cm; average pod weight 2.0 g; seeds per pod 17; productivity 7.0 t/ha.

***Kairali*** : Semi-trailing growth habit, pink coloured medium long pods, reddish brown seeds; average length of pod 22.78 cm; average pod weight 7.08 g, days to first harvest 50 ; productivity 7.13 t/ha.

***Anaswara*** : Semi-trailing growth habit, pods-light green, bold seeded and medium long, purple flowers, av. pod length 28.12 cm, average pod weight 12.5 g, seeds/pod 19, productivity 12.5 t/ha.

***Varun*** : Tolerant to mosaic, semi-trailing growth habit, pink coloured medium long pods, reddish brown seeds. av. length of pod 27 cm, average pod weight 11 g, days to first harvest 50, productivity 8.4 t/ha.

***CO 2*** : A semi-spreading variety with 26 cm long pods having less fibre, yield 9.4 t/ha.

***Vamban*** : This variety has long green pods, yield 10.6 t/ha.

***Sel-2-1*** : This variety has been developed at NDUA & T, Faizabad. Plants are 70-75 cm of height. Pods are green, 25-30 cm length, black seeded; susceptible to pseudocercospora and viruses. Recommended for cultivation and release in 1993 for the zone IV (Punjab tara and Plains of U.P. and Bihar).

***Narendra Lobia 1*** : This variety was evolved at NDUAT, Narendranagar, Faizabad from a cross of L 1552 (now released as Pusa Komal) x Vranasi Local following pedigree method of breeding. It has been released by the U.P. state variety release committee meeting on 7.4.1995. It has determinate plant habit. Plant height is 40-45 cm. Foliage is green with large leaves. Green pods are 28-32 cm long with purple terminal end. Each pod contains 10-12 seeds. Seeds are bold (18 g/100 seeds) with black hilum. Edible pod maturity is 45-48 days and seed maturity 75-80 days. It is photo-insensitive and can be grown in summer and rainy both the seasons. Green pod yield potential is 90 q/ha.

***Sel.263*** : Sel-263 is an early maturing variety, developed through selection at PAU, Ludhiana. Plants are dwarf and can be grown in both spring and rainy seasons. Pods are green, thick fleshy, tender and medium (20 cm long), resistant to mosaic and golden mosaic viruses; gives an average pod yield of 22 t/ha.

***Cow pea-263*** : It is suitable for both spring and the rainy seasons. Its plant is dwarf, pods medium green, thick, meaty, tender and about 20 cm long. It is an early-maturing variety. It is free from the golden mosaic and comparatively resistant to other mosaic. Average green pod yield is 9.0 tonnes/ha and pod becomes ready 45-50 days after sowing.

***Bidhan Bharati 1*** : Its plants are compact and determinate with dark green foliage. It takes about 42-45 days from sowing to flowering and 55-60 days from sowing to first picking of pods. Colour of standard petal is white with light yellow at the base. Green, tender, thick, fleshy and appreciably long (25.2 cm) pods are borne on long and stout peduncle (18-19 cm). Seeds are bold, thick, slightly reniform, holstein coloured with buff and brown. In West Bengal condition, 13.4 tonnes of green pods can be harvested from a hectare during spring summer and autumn-winter seasons. Crude protein of green, tender pods is appreciably high (3.87 %). It shows resistance to cowpea mosaic and cowpea golden mosaic virus.

*Bidhan Bharati 2* : Its plants are semi-determinate and loose framed with yellowish green foliage. It takes 43-46 days to flower and first picking of pods can be picked 55-60 days after sowing. Colour of standard and wing petals are light purple violet with faint yellow at the base. Light green, tender, thick solid and fleshy pods of excellent size (25.8 cm) are borne on soft and short peduncle (14.8 cm). Seeds are flat, reniform and holstein coloured with buff and darktan. Average yield of tender pod is 15.9 tonnes/ha. Tender pods contain high (4 %) crude protein. It shows very low incidence of cowpea mosaic and golden mosaic virus diseases.

*Gwalior K 3B* : This variety can be grown in both the seasons and pods are ready to harvest 65-75 days after sowing. The seed colour is creamy with brown hilum. Average seed yield is about 12 q/ha.

*Gwalior K 11* : It can also be grown in both the seasons with the same duration as of above variety. The seeds are smaller than the above one.

*Gwalior K 14* : It can also be grown in both the seasons with the same duration as of above varieties. The seeds are medium in size, attractive, creamy white in colour with black hilum. The average seed yield is about 10 q/ha.

*Utkal Manika* : It is a dwarf, bushy, profused bearing all season variety. Pods are long (22 cm) and fleshy, green in colour, moderately tolerant to YMV. Matures in 50 to 55 days after sowing. Average yield is 69 q/ha.

*BCP 4* : It is a trailing type of cowpea. Pods are about 52 cm long, fleshy, light green in colour. matures in 60-65 days after sowing. It requires staking for better performance. Average yield is 108.0 q/ha.

*FS-68* : This variety is well suited for spring season. It is about a week late than Pusa Do-fasli. Its grains are attractive white. Pods are small and green with small white seeds. It is suitable for both grains and vegetable. It yields about 10-12 quintals of grain per hectare, suitable for cultivation in North India.

*Gomati* : Plants are semi viny, vigorous growing, first picking starts in 50-60 days, fruits two to three pods per inflorescence, pods smooth cylindrical, long, greenish white in colour.

***Kashi Sudha :*** It is a dwarf and bush type (50-60 cm height) early variety. It is sown in both Kharif and rabi seasons. It lakes 40-45 days from sowing to flowering. The pods become ready for harvesting in 50-55 days. On an average, it produces 20-25 pods / plant of 35-40 cm length. The pods are fleshy, green and free from parchment. It gives about 85-120 q/ha of green pods. It has been developed from Kashi Shyamal x NDCP - 13 at IIVR, Varanasi.

***Yard long Bean (V. Sesquipedalis) :*** It is commonly grown in home gardens, in Uttar Pradesh. It is not suitable for transportation as pods break easily. Pods are about 50 cm long. The plants are viny type and mature late (100 days). Green pod yield potential is 100 q/ha.

***Swarna (C-152) :*** It is medium duration variety suitable for rainy season plantings, for whole of the country, good for pure and mixed cropping. Though it is a grain variety, tender pods are used for vegetable purpose and yields 100-120 q/ha green pods.

***S-203 :*** A selection from a cross Sel-2 x Virginia, plant compact, short duration variety which can be harvested withn 75 to 80 days of sowing while the green pods can be harvested in 65 days. Because of its short duration, it is recommended for double cropping system in red soils of Karnataka. Green pod yield is 50-60 q/ha.

***S-488 :*** It is derived from Virginia x Iron Grey, yielding 30 % more than C 152. It matures about a fortnight earlier, shows good tolerance to *Xanthomonas vignicola* in the field and is of better cooking quality than C 152.

***SEB 2 :*** It is suitable for sowing both in autumn and spring (February-March) season. Crop duration is 90 days during both seasons. It is vegetable type variety, light irrigation is helpful in spring season for more yield.

***Philippines Early :*** Plant is trailing, early maturing variety, taking about 75 days. The first green pod picking is done after 55-60 days of sowing. The pods are long about 29 cm, creamy white, fibreless and bears about 17 seeds per pod. This is a vegetable variety popular in M.P. The average green pod yields 60-90 q/ha.

## CLUSTER BEAN

Cluster bean (*Cyamopsis tetragonoloba* L.) ; 2n = 14; syn. Guar, Thupi, Guara, Urahi, Koth Avarai, Gavar, Gor, Chikudu, Gorikaya,

Kothavara and Guvar, belongs to the family Leguminoceae; is a drought tolerant warm season crop grown for its tender pods for use as vegetable. African species, *cyamopsis senegalensis* is the ancestor of cluster bean. It is also used as a fodder purposes. It can be grown successfully in areas where average rainfall is 30-40 cm. It is cultivated mostly as rainfed crop in semiarid zones of northern India.

## Varieties and Hybrids

Varieties grown in India differ in height of plants. Cultivars grown in South India are vegetable types while those in North-West India are grown for seeds.

There are two well defined forms namely : Giant and Dwarf, Giant types are generally grown in Gujrat while dwarf are grown in Punjab and Uttar Pradesh. There are two dwarf forms, namely : smooth and hairy (fodder type). Vegetable types are mostly dwarf types with smooth appearance. Fodder types are mostly hairy.

Some improved cultivars are :

Pusa Mausami, Pusa Sadabahar, Pusa Navbahar, Sharad Bahar, Pardeshi, P-28-1-1, IC 11704, CP 78, 160-1, S299-7, Max No.1, 2, 3, 4 and 5; LPG 850.

**Detail characteristics of varieties and hybrids are given below:**

***Pusa Sadabahar*** : This variety is developed from the local cultivar 'Jaipur' of Rajasthan. It is a variety with single stem and high yield. It does well in the summer and in the rainy season. The tender and fibreless pods are green and more than 12 to 13 cm long. The first picking is done after 45 days of sowing during summer, whereas it takes about 55 days in the rainy season.

***Pusa Mausami*** : This variety is a selection from local cultivar of North India. It is another high yielding variety with smooth, bright green pods about 10 to 12 cm long. It is branched variety which is suitable for the rainy season. It is a late variety with first picking in about 65-80 days depending upon the weather and the region.

***Pusa Narbahar*** : It is a cross between 'Pusa Sadabahar' and 'Pusa Mausami. It is a new variety with the pod quality of Pusa Mausami nd the growing habit of Pusa Sadabahar and gives a good crop in both summer and rainy seasons. It is a unbranched, single stemmed variety which bears 10 to 12 cm long pods. The first picking of green pods starts 40 days after sowing.

***Sharad-Bahar*** : It is a single plant selection from I.C.-11704 the material having been obtained from Bombay (MS). A branching-type variety, that produces long succulent and attractive green pods. It can be grown in summer and the rainy season. It is a branched type producing 12-14 branches per plant. Its seed colour is light pinkish grey and it is flattish in shape; 100 seed weight 4.1 g.Average yield is 133 pods/plant.

***Pardeshi*** : A vegetable type variety grown in Gujarat, that produces long pubescent pods about 12-15 cm long.

***P.28-1-1*** : It is a single plant selection from the cross 'Pusa Navbahar and I.C.-11521. This variety is photo-insensitive and can be grown in both the season i.e., summer and rainy. It produce smooth, green and long pods. It grown with the break of rains.

***I.C. 11704*** : It is single plant selection from the material obtained from Namedo Umaji Co, Bombay. It is a branched type producing 12 to 14 branches per plant. The duration is four months. Green pod yield averages 190 to 200 q/ha in different environments.

***I.C. 11388*** : It is a single stem, selection from the material originally collected from Sukhpur area in Kutch.It is also a main season variety and can be grown in kharif season only. It is an early variety.

***PLG 850*** : This selection is originally from Sirohi in Rajasthan. It is also branched and smooth type producing 11 to 12 branches per plant. The yield ranged from 183 to 189 q per hectare.

***Brooks Guar*** : It is a high yielding, branching, glabrous variety resistant to major guar diseases. It is a potential dry land crop for sandy soil. The seed weight averaged 33 g per 100 pods.

## FRENCH BEAN

French Bean (*Phaseolus vulgaris* L.); 2n = 22; syn.Snap bean, String bean, Kidney bean, Haricot bean, Fresh bean, Pharash bean, Farasi Simba, Bangalore bean, Frans bean, Garden bean, Dry bean, Field bean, Navy bean, belong to the family Leguminoceae, is an excellent cool season legume vegetable grown for its tender pods, shelled green beans and dry beans (Rajmah). Genus phaseolus originated in New World.Following four species are under cultivation in New World :

1. *Phaseolus vulgaris* – Frenchbean, 2. *P. coccineus* –Runner or scarlet bean, 3. *P. Lunatus* – Lima bean, Butter bean or Madagascar bean, 4. *P. acutifolius* var. Latifolius – Tapery bean. Though it is a self pollinated crop, offers wide variability for plant growth (Bushy/climbing), colour of pod (green/waxy coloured), cross section of pod (flat/oval/round), pliability (stringed/stringless).

## Varieties and Hybrids

French bean varieties are of two types (i) dwarf or bush types and (ii) climbing or pole types. Each type bears green pods or wax pods, and varieties either string or stringless.

There are specific varieties for snap bean purpose, dry bean purpose and for processing. Processing varieties are very popular in the USA.

**Detail characteristics of varieties and hybrids are given below:**

## Dwarf or Bush Type Varieties

***Contender*** : This is an introduction from USA.This is an early variety, developed at Indian Agricultural Research Institute, New Delhi. Plants bushy with pink flower. The pods are stringless and dark green in colour. The seeds are oval in shape. This variety has been found to be resistant to mosaic and powdery mildew. Yield 20 t/ha.

***Giant Stringless*** : This variety was developed at IARI, New Delhi. Variety is early, pods are green in colour, medium size slightly curved and stringless. The seeds are glossy and yellowish brown. The yield potential is high.

***Jampa*** : This is a Mexican variety, suitable to grow in Maharashtra. This is very early variety, pods are available for first picking 50 days after sowing. The pods are flat, smooth, non-stringy and pale green in colour. The seeds are black, smooth, small in size and flat in shape. It is resistant to wilt and can grow successfully in warmer climate.

***Pusa Parvati*** : This is an outcome of mutation breeding (cv. Wax pod, and American Variety) done at IARI, New Delhi. The variety is early. Plants bushy with pink flowers. The pods are light green, attractive round. This is a high yielding, and most suitable for hills. The seeds are oval type. Resistant to mosaic and powdery mildew. Yield 22-25 t/ha.

***Bountiful*** : This variety is high yielding, suitable to grow from September to February in south India. On main stem, pods form in cluster. Pods are tender green, available for harvesting within a week of flowering. The pods remain for long period even when left unharvested The seeds are flat type.

***Premier*** : This variety is suitable for the plains of Tamil Nadu. This is less susceptible to wilt and mosaic. The seeds are black in colour and oval in shape.

***Pant Anupama*** : This variety was developed by GBPUA & T, Pantnagar. This is an early maturing variety, suitable for both hills as well as plains of Uttar Pradesh and elsewhere too. Plants bushy and dwarf with concentrated fruiting. The pods of this variety, are round, smooth tender, nonstring, translucent and attractive in appearance. This is resistant to angular leaf spot, tolerant to blight, common mosaic and sclerotinia mould. Yield 9 t/ha.

***Arka Komal (Selection-9)*** : This variety was developed at Indian Horticultural Research Institute, Bangalore Introduced bushy variety from Afghanistan. The plants are erect and bushy. The pods are straight and flat. The pods can tolerate long distance transportation. It has better yield performance than Pusa Parvati.

***Arka Subidha (IIHR 909) :*** Plants bushy and photosensitive. Pods straight and oval, light green, stringless and fleshy. Yeilds 9 t/ha in 70 days. Developed at IIHR, Bangalore.

***Arka Anoop :*** Its plants are bushy and photo-insensitive. It can be grown during all the three seasons. Its pods are ready for picking 45 days after sowing. Pods are flat, smooth, long (17-18 cm) and crisp with less parchment, the yield potential being 20.0 tonnes/ha in 70 days. It is resistant to both rust and bacterial blight diseases.

***TKD1*** : It is a hybrid derivative of a cross between two pole types, viz., Sel.1 and PV 118. A pole type suitable for growing in hills.The green tender pod yield ranged from 50-60 q/ha in 90-100 days. It is developed by TNAU, Tamil Nadu.

***Selection-5 :*** This is also developed at IIHR, Bangalore. The pods are round stringless. It is suitable to grow in kharif season.

***SVM-1*** : Pods green round stringless, 13-14 cm long. Ready for picking in 65-75 days. Average pod yield is 105-125 q/ha. Resistant to angular leaf spot both in leaves and pods. A dual purpose variety and thus can be used as vegetable and seeds as pulse. Recommended for mid hill areas.

***Laxmi*** : Pods are formed in cluster of three, 13-14 cm long, stringless, green round and attractive. Pods ready for picking in 65-70 days. Seeds are white in colour with light yellow scar. Average green pod yield is 120-140 q/ha. Tolerant to angular leaf spot disease.

***VL-Boni-1*** : This variety has been developed at VPKAS, Almora, plants are dwarf (30-40 cm) having white flower with purple tinged. Pods are round, light green, attractive, stringless and fleshy; first picking starts 45-60 days; moderately susceptible to foliar diseases; gives 100-120 q/ha pod yield. Suitable for multiple cropping and can be planted in the hills from March to July.

***YED 1*** : Bushy, dual purpose variety suitable for kharif season. Pods slightly flat, 15 cm long. It can give a green pod yield of 97.5 q/ha. Seeds are bold, attractive and dark-purple in colour. Tolerant to root rot, rust, yellow mosaic and anthracnose. Yield 9.75 t of pods or 6.3 t of grains/ha in 105 days. This is developed at TNAU, Tamil Nadu.

***Ooty 1*** : It has yield potential of 340 q/ha of green pods in a duration of 80 days. The variety is moderately resistant to leaf spot, anthracnose and pod borer. It is released by TNAU, Tamil Nadu.

***Phule Surekha*** : Phule surekha have been developed from Jampa improved variety by employing pure line selection. It is a semi-determinate variety. Its pods are straight flat with average length of 9.25 cm and breadth 1.07 cm. It is suitable for kharif, rabi and summer seasons (Photo-thermo-insensitive). It is specifically bread for small straight the edible pods and resistant to anthracnose, leaf crinkle, bean yellow-mosaic virus and wilt (caused by stem fly attack). It gives highest yield of (153-39 q/ha) which was 53.68 % higher than contender. It is released by MPKVV, Rahuri.

***Kashi Param*** : Its plant determinate in height, grow 70 cm high and are bush type, with dark green leaves. Its pods are attractive, fleshy, average length 14.7 cm, round and dark green. Its green pods yield is 120-140 q/ha, suitable for cultivation in Jammu and kashmir, Himachal Pradesh, Uttaranchal, Madhya Pradesh and Maharashtra. This variety is developed by IIVR, Varanasi.

***NDVP 8*** : This variety was developed at NDUA & T, Faizabad, UP. A mid season variety with pod yield of 105 q/ha.

***NDVP 10*** : This variety was also developed at NDAU & T, Faizabad, UP. A mid season variety with pod yield of 100 q/ha.

### Pole Type or Twining Type Varieties

***Kentuky Wonder*** : This is a late maturing variety, developed at IARI, New Delhi. The pods are green in colour, large curved, round and stringed at later stage. The seeds are light brown in colour. This variety has high yield potential. Average pod yield is 100-125 q/ha.

***KKL 1*** : It is a pole type best suited for elevation of 800-2,400 m. Pods long (28 cm) with low fibre. Seeds white and flat. It has a potential yield of 70 t/ha of pods or 30 q of grains per hectare. It is released by TNAU, Tamil Nadu.

***Pusa Himlata*** : It is a climbing or pole-type, french bean. Vine is medium in height (2.5 m long). The internode distance is small, each nodes bears 3-4 flowers. Pods are medium-sized (14 cm), round, meaty, stringless and light green the early maturing, ready for picking in 60 days. Seeds are white and suitable for processing as green-shelled bean. Mature seeds also have good cooking quality and are suitable for use as dry bean (Rajmash.). It is developed by IARI, Regional Station, Katrain.

***Other varieties and genotypes*** : Swarn Priya, Black Prince, VL Bean 2, VL Bean 3, Aparna, IVFB-1, IVFB-2, IVRFB-1, VL 5, Sel 2 and Sel-4.

## HYACINTH BEAN

Hyacinth bean (syn. Indian bean) *Lablab purpureus* L. (syn. *Dolichos lablab, D. purpureus*); 2n = 22, 24, also known as field bean or dolichos bean, is grown throughout tropical regions of Asia, Africa and America. In India it is grown as a field crop in Tamil Nadu, Andhra Pradesh, Karnataka, Madhya Pradesh and Maharashtra. In Kerala, the photo sensitive pole type are grown in homesteads by trailing to bower for its tender fruits which are used as cooked vegetable. Dry beans are also used in various vegetable preparations.

Indian bean belongs to the family Leguminosae; commonly known as Sim, Valpapdi, Sem, Avare, Moang, Avara, Baragudi,. Mochai and Chikkudu ; is supposed to havre originated in India as wild forms. From India, it might have been introduced in China, Western Asia and Egypt. The original ancestral species of the crop is considered as *Dolichos purpureus*.

Indian bean is good source of protein (3.8 g), minerals and vitamins. Amino acid composition of Indian bean is comparable to other legume proteins. Methionine is the most limiting amino-acid in Indian bean. Seeds of Indian bean also contain trypsin inhibitor, lectin, phytic acid and polyphenol. The toxicity of lectin can be eliminited by autoclaving (heating) the meal. It is primarily grown for green pods, which are cooked as vegetable like other beans. The dry seeds are also used for various vegetable preparations. The foliage of the crop provides hay, silage and green manure.

Plants are robust twiner, lasting more than one year and 15.6 m long. Leaves are pinnately trifoliate, leaflets ovate or ovate-rhomboid; flowers are shortly pedicelled, violet or white, in long stalked. Pods are subsessile, tuberculate on sutures, 3-6 seeded, glabrous when ripe, 5-20 cm long and 1.2-2.2 cm broad. Seeds variable in colour as white, ochreous with black dots or black with white dots, uniformly brown or uniformily \black.

There are two cultivated types viz., (1) *Lablab purpureus* var. typicus is vegetable type and grown for its soft and edible pods. (ii) *Lablab purpureus* var. lignosus; is the field bean cultivated for dry seeds as pulse. Both varieties are cross compatible.

## Varieties and Hybrids

Hyacinth bean has great variation in terms of plant habit, pod characters, grown all over the country. It is very popular in north eastern India where large endemic forms and promising varieties have been identified. Similarly in south and north India, certain varieties are grown. A brief description of the improved varieties is given below:

| Developing institution | Variety | Special features |
|---|---|---|
| IIHR, Bangalore | Arka Jay | Photo-insensitive bush variety developed by selection at F7 from Hebbal Avare 3 (Bush) x IIHR 99 (Pole –recurrent parent). Pods long, light green slightly curved. Yield 12 t/ha in 75-80 days. |
| | Arka Vijay | Photo-insensitive bush variety developed by selection at F7 from Hebbal Avare 3 (Bush) x Pusa Early Prolific (Pole). Pods |

Contd...

| | | |
|---|---|---|
| | | short, dark green with characteristic aroma. Yield 12 t/ha in 75-80 days. |
| UAS, Bangalore | Hebbal Avare-1 | Bush and photo insensitive variety. Pods small and soft. Yield 0.8 t/ha in 90-100 days. |
| | Hebbal Avare-3 | Bush and photo insensitive variety. Flowers white. Pods green, 2-3 seeded. Seeds brown, round and small. Yield 8-10 t/ha in 90-100 days. |
| | Hebbal Avare-4 | This variety is a determinate in habit and is very early. Bush and photo insensitive variety. Pods soft and harvested in 5 pickings, yield 6 t/ha. |
| IARI, New Delhi | Pusa Early Prolific | It is an improved vine variety. It bears early. The pods are thin and form in bunches. It is suitable for sowing in early spring and early autumn in the northern Indian plains. Pole type. Pods flat, green, narrow, cycle shaped.Pod length 9.3 cm, width 1.5 cm, weight 3.5 g. Yield 14 t/ha in 200-215 days. |
| | Pusa sem 2 | Pole type. Pods semi-flat, dark green, fleshy and stringless. Pod length 15-17 cm. Yield 13-22 t/ha in 200-215 days. Tolerant to anthracnose, yellow bean mosaic virus, aphids, pod borers and frost. |
| | Pusa Sem 3 | Pole type. Pods flat, green, fleshy and stringless, pod length 15 cm. Yield 17-27 t/ha in 200-215 days.Tolerant to anthracnose, yellow bean mosaic virus, aphids, pod borers and frost. |
| KKVP, Dapoli | Wal Konkan 1 | This was developed from the cross between Wal 2-K2 x |

Contd...

| | | |
|---|---|---|
| | | Wal 125-36. Bushy, photo-insensitive, resistant to yellow mosaic virus. Yield 9-10 t/ha in 110-115 days. |
| | Konkan Bhushan (DPLD1) | Bushy, photo-insensitive, pods green, tubular, dual purpose, yield 8 t/ha in 100-110 days. |
| Gujarat | Val-125-136 | This variety is selection from local germplasm. Plant height 60-75 cm, foliage green, pods are borne in cluster. Seeds are milky white. The crop duration is 110-120 days. |
| Tamil Nadu University | CO 1 | Pole type, pods green, fleshy with slow fibre development. Pod weight 9.7 g. Yield 18-20 t/ha in 160-180 days. |
| | Co 2 | Pole type. Pods flat green with purple margin. Pod length 9.3 cm, width 2.1 cm, weight 6.17 g. Yield 11.8 t/ha in 210-220 days. |
| | Co 3 | Pole type. Pods fleshy, green with purple tinge.Pod length 10.6 cm, width 4.8 cm. weight 11.77 g.Seeds black. Yield 10.0 t/ha in 230 days. |
| | Co 4 | Pole type. Pods deep purple throughout and fleshy. Pod length 10.2 cm.Width 3.3 cm, weight 7.43 g. Seeds black. Yield 13.5 t/ha in 215-220 days. |
| | Co 5 | Pole type. Pods long, narrow, light green to white in colour, tubular, curved with serrated margin. Pod length 13.4 cm, breadth 1.5 cm, weight 5.26 g. Seeds chocolate brown. Yield 6-7 t/ha in 235 days. |
| | Co 6 | Bush variety.Selected from cross of DL 3169 x Co 5. Pods long, |

Contd...

| | | |
|---|---|---|
| | | succulent, flat, greenish white, broad, slightly curved and bloated. Yield 12 t/ha in 240 days. |
| | Co 7 | Bush variety. Selected from cross of DL 3169 x Co 5. Pods long, succulent, flat, greenish white, broad, slightly curved and bloated. Yield 12 t/ha in 240 days. |
| | Co 8 | Bush variety. Pods green tubular and fleshy. Yield 6-8t/ha in 120 days. |
| | Co 9 | Bush variety. Pods and grains are used. Yield 7-8 t/ha in 120 days. |
| | Co 10 | Bush variety. Induced mutant from Co 6 by gamma ray (24 krad). Pods greenish white, tubular and curved.Yield 5-6 t/ha in 120 days. |
| | Co 11 | Bush variety. Hybrid derivative of Co 9 x a pandal type. Compact plant type. Pods flat and light green with purple margin. Yield 9-10 t/ha. |
| | Co 12 | Bush variety. Hybrid derivative of Co 9 x Co 4. Pods deep purple.Yield 10-12 t/ha in 110 days. |
| | Co 13 | Bush variety. Hybrid derivative of Co 9 x a trailing type.Pods long green. Yield 10 t/ha in 110-120 days. |
| CSAUA & T, Kanpur | Rajani | Pole type.Pods narrow oval in cross section, shining green. Pod length 10.4 cm, width 1.2, cm, weight 1.78 g. Yield 7.8 t/ha in 200-210 days. |

Contd...

| | | |
|---|---|---|
| | KDB 403 | Pole type. Pods long, narrow, shining green.Pod length 12.9 cm, width 1.2 cm., weight 2.0 g. Yield 5-6 t/ha in 180-210 days. |
| | KDB 405 | Pole type. Pods medium long, narrow, dark green band in the middle and borders light green.Pod length 9.6 cm,width 1.3 cm, weight 1.1 g. Yield 3-4 t/ha in 180-200 days. |
| Kalyanpur-T-1 | CSUAT | This variety is pole type and good yielder. |
| MPKV, Akola | Dasarawal | This variety is pole type and good yielder. Pods dirty green with purple tinge at both borders. Pod length 7.8 cm, width 2.0 cm, weight 3.2 g. Yield 7.8 t/ha. |
| | Deepaliwal | Pole type. Pods extra long (18.4 cm), white, not smooth due to bulging at each seed. Pod width 2.7 cm, weight 1.5 g. Yield 6.8 t/ha in 200-210 days. |
| JNKVV, Jabalpur | JDL 37 | It is a high yielder. |
| | JDL 79 | Pole type. Pods flat, broad, whitish green with parrot green border along the line of seed attachment. Pod length 11.8 cm, width 3.6cm, weight 1.5 g. Yield 5.6 t/ha in 200 days. |
| | JDL 53 | Pole type. Pods flat, small narrow, dull whitish green with purple tinge along the boarder. Pod length 7.2 cm, width 1.8 cm,weight 3.75 g. Yield 10-12 t/ha in 200-220 days. |

***AHDB3 :*** It is a selection from landraces collected from tribal area of northern Gujarat. Developed through single plant selection for highest quality and early yield under arid agroclimate conditions of Rajasthan, its tender pods are 13.85 cm long and 1.05 cm wide.

Average weight of a tender pod is 8.52 g. At marketable stages, weight of pod pulp and seed kernel are 7.03 and 1.51 g, respectively. Pods are soft and shining green in colour. Average pod yield is 1.768 kg per plant. Plants are medium in growth habit and about 3.0 m tall. Flowers are purple in colour. It is developed by Central Institute for Arid Horticulture, Bikaner, Rajasthan.

***AHDB16 :*** It is a selection from a population collected from tribal area of South Rajasthan. It is developed through single plant selection for earliest harvest as early as in mid october and higher early yield under arid conditions. Tender pods are 10.23 cm long and 1.58 cm wide. Average pod weight at tender stage is 5.63 g. The weight of pod pulp and seed kernel is 3.76 and 1.85 g, respectively. Medium-sized pods are whitish green and sickle-shaped. Pods are smooth, soft and shining. Average pod yield is 1.760 kg / plant. Plants are medium in growth habit and 2.75-3.25 m in height. Flowers are white in colour.

Both (AHDB 16 and AHDB3) varieties have viniy growth habit. They require support or trellis system for commercial cultivation.

## JACK BEAN

Jack bean (*Canavalia ensiformis* L.) 2n = 22; belons to the family leguminosae; is hardy drought resistant tropical legume vegetable grown for their young pods and immature seeds which are used as a vegetable. It is also known as Horse bean.

It is bushy erect annual herb, 1-2 m high, tips of branches tend to twine; particularly in shade. Lvs trifoliate, shortly hairy; petiole usually longer than leaflets, grooved above, stout, with large pulvinus at base and at base of each leaflet; stipels present, leaflets elliptic to ovate, rounded at base and apex, 6-15 x 5-10 cm, terminal leaflet long-stalked, side leaflets short-stalked with unequal base, entire. Inflorescence is a curved raceme with 10-15 flowers in groups of 3-5. Pods stout, straw-coloured when ripe, pendent, 20-30 x 2.0-2.5 cm, ribbed near upper suture; 8-20 seeded. Seeds white, about 2 x 1.3 cm, somewhat flattened; hilum less than half length of seed, 8 mm long, pale brown with orange margin. Bagged flowers set pods and seeds. Hand crossing is difficult as the emasculated flowers tend to drop, but it can be done and should be confined to the lowest 2-3 flowers of an inflorescence. Under natural condition the flowers are visited by bees and 20 per cent or more cross-pollination occurs.

The immature fruits and beans contain about; water 75.2 per cent; protein 6.9 per cent; fat 0.5 per cent; carbohydrate 13.3 per cent; fibre 3.3 per cent; ash 0.8 per cent. The ripe dried beans contain,; water 11.0 per cent; protein 23.4 per cent; fat 1.2 per cent, carbohydrate 55.3 per cent; fibre 4.9 per cent; ash 4.2 per cent.

The crop is grown through seeds.Seed germination is epigeal. Seeds are sown with the onset of south west monsoon during May-June in pits or mounds at spacing of 60 x 60 cm. Some times seeds are broadcasted for manure crop. The seed rate is 50-70 kg/ha. On an average 2.0-2.5 kg pods are harvested from a plant. The yield of green matter is 40-50 tonnes/ha. and the dried bean yield varies between 12-15 tonnes/ha.

### Varieties and Hybrids

There is no improved varieties in this crop. The commercial crop is still grown from locally available materials (genotypes available).

## PEA

Pea (*Pisum sativum* L.) is a diploid with 2n = 14; syn. Mattar, Bataneelu, Mattri, Motor, Mah, Bara, Pathani, Watana, Muttar, Batani is a popular vegetable and pulse crop of India. Field pea with colour flowers belong to *Pisum arvense* and the garden pea with white flowers belong to *Pisum sativum*. Pea is a herbaceous annual plant with tap root system.

There are two types of peas, generally cultivated all over the world. One is the garden pea and the other is field pea.

I. ***Garden pea (Pisum sativum var. hortense)*** : It is also called table pea. In this type, young, green seeds are used mostly in vegetables and also for canning purposes. Seeds are bold and wrinkled. The plants are generally white flowered. Leaf axils are generally green. Seeds are yellowish, whitish or bluish green.

II. ***Field pea (Pisum sativum var. arvense)*** : In this type, the ripe, mature seeds are used as pulse (dal). They are also grown as a forage or green manure crop. Field peas are hardy plants and grown on a large scale without irrigation. These plants are also able to withstand frost. Seeds are round and little angular. Generally they have coloured flowers. Leaf axils are often pigmented. Seeds are grayish green, grayish brown or grayish yellow.

## Varieties and Hybrids

Pea cultivars grown in different parts of the world exhibit wide variation in height of stem, branching, pod size, seeds per pod, shelling percentage, smoothness of seeds (smooth/wrinkled) etc.

The cultivars/varieties are grouped on the basis of various characters as given below :

### Based on maturity period

a) ***Early types*** : Green pods will be ready for harvest by 65 days after sowing.

b) ***Mid season types*** : Pods will be ready for harvest by 85-90 days after sowing.

c) ***Late main season types*** : Pods will be ready for harvest by 110 days after sowing.

### Based on height of plant

a) ***Bush or dwarf types*** : Usually this types are early.

b) ***Medium tall*** : In this, mid season varieties are generally fall.

c) ***Tall*** : In this group, late types/ varieties are included and need support.

Some important varieties cultivated in different placed of India are given below :

***Madhya Pradesh*** : Kalangini, Khapan, Kheda

***Uttar Pradesh*** : Kap, Bonia, Local yellow Batri, Lucknow Bonia, Hara Bonia.

***Amritsar*** : Asauji, Hoshiarpuri, Kanawari.

The improve varieties are given below :

***Early season varieties*** : Arkel, meteor, Asauji, Early Badger, Jawahar Mater-4, Hara Bona, Jawahar Matar-3, Early December, Pant Mater 2, Pant Sabji Matar 3, Hisar Harit (PH 1). Alaska, Early Superb, Little Marvel, Matar Ageta 6, VL Mater 7, Harbhajan (Ec 33866).

***Main season and late varieties*** : Bonneville, Lincoln, New Line Perfection, Jawahar Mater 1 (JM 1 or GC141), Jawahar Mater 2 (JM 2 or GC 477), Pant Uphar (IP3), VL 3, P 88, Azad Pea 1 (KS 123), Azad

Pea 4 (KS 225), Narendra Vegetable Pea 1 (NDVP-4), Pant Matar 2 (PM 2).

***Early smooth seeded*** : Asaugi, Lucknow Boniya, Alaska, Early Superb, Meteor.

***Early wrinkled seeded*** : Arkel, Early Badger, Little Marvel, Kelvedon Wonder, Early December.

***Wrinkled-seed main season and late season types*** : Bonneville, T-19, Lincoln, Delwiche commands, Khapar Kheda, NP 29, Perfection New, Line Thomas Laxton, Alderman, GC-14, GC 195, Early Giant.

***Smooth Seeded main season type*** : Kanwari, Rachna.

***Edible podded cultivar*** : Sylvia, Aparna, Matar Ageti-6, Mithi Phali.

**Varieties developed by different institutions/organizations:**

***IARI, New Delhi*** : Arkel, Bonneville, Sylvia, NP 29, Asauji, Meteor

***IARI, Regional Station, Katrain*** : Lineoln.

***IIHR, Bangalore*** : Arka Ajit, UN 53-6.

***IIVR, Varanasi*** : VRP 2, VRP 3, Kashi Nandini (VRP 3), Kashi Shakti (VRP 7), Kashi Udai, Kashi Mukti.

***TNAU, Coimbatore*** : Ooty 1.

***NDAU & T, UP*** : NDVP 8, NDVP 10.

***PAU, Ludhiana*** : Punjab 88, Matar Ageta 6.

***CSAUA & T, Kanpur*** : Azad P-2 (PRS4), Azad P-3 (PARS 3)

***JNKVV, Jabalpur*** : Jawahar Matar 1 (JM 1, GL 141), Jawahar Matar2, Jawahar Matar-3 (Early December), Jawahar Matar 4 (JM 4, GL 195), Jawahar Matar 15, Jawahar Matar 54, Jawahar Pea 71, Jawahar Pea 83, Harbhajan.

***GBPUA & T, Pantnagar*** : PM 2, Pant Uphar (IP 3), Pant Sabji Matar3.

***HAU, Hisar*** : Hisar Harit (PH1).

***VPKAS, Almora*** : VL Matar 3, VL Agethi Matar 7 (VL 7), VL 8, Vivek (VL Matar 6).

***HPKV, Palampur*** : Palam Priya.

## Hybrids

*Pro Agro* : PGS Hybrid Green Pearl

**Detail characteristics of varieties and hybrid are given below:**

***Arkel*** : Plants are dwarf, first flower appears in about 30 days after sowing. It takes 60-65 days to first picking. Pods are well filled and attractive. Three harvesting is done. The average green pod yield is about 50-60 q/ha.

***Meteor*** : Plants are 35-40 cm tall, dark green, flowers borne generally singly, pods dark green, 8 cm long, well filled with 7 seeds, having shelling percentage 45. It is a round-seeded early variety introduced from England. It is also sweet and gives the first harvest in about 70 days in Delhi.

***Asauji*** : It was selected at the IARI for sowing the early crop around the middle of September. It is a dwarf plant growing to a height of about 37.5 cm. The pods are dark green, curved, about 7.5 cm long and well filled with 6 to 7 smooth seeds per pod. The pods are ready for harvesting in about 60 days.

***Early Badger*** : It is an American variety suitable for sowing in early October. It is a dwarf wrinkled seeded variety which gets ready for the first picking in 60 to 65 days. The pods are well-filled and the seeds are bold and sweet.

***Bonneville*** : This mid season variety is an introduction from America; introduced by IARI, New Delhi. Plants are medium-tall, flowers in 55-60 days after sowing with two pods per peduncle. Pods are light green, straight, 9 cm long, having 6-7 ovules per pod; gives an average pod yield of 85 q/ha with 45 % of shelling. Seeds are green and wrinkled.

***Sylvia*** : Introduced edible podded variety suitable for kitchen garden. Pods curved, yellowish green without parchment.

***N.P. 29*** : It is developed from IARI, New Delhi. It is wrinkle-seeded variety, very tasty and pods are ready to harvest after 100 to 110 days of sowing. It is suitable for dehydration.

***Lincoln*** : An old exotic introduction recommended by IARI, katrain. It was identified in 1987 for zones i.e., I and IV where it gave 50 % increased yield over control. Plants are medium tall double podded, pods are large, dark green, grains are sweet in taste, 8-9 cm long. It takes 85-90 days to first picking. Shelling percentage is 46. Good for canning and transport, having long shelf life.

*Perfection New Line* : It is a heavy yielding mid-season variety introduced from America. The plant is medium tall and the pod is about 8 cm long. Dark green, sweet and well filled with wrinkled seeds. It is ready for first picking in about 80 to 85 days.

*VRP 2* : Plants 50 cm tall. Pods straight and medium sized. First harvest 55-58 DAS, yield 10 t/ha.

*VRP 3* : Mid season variety which flower between early and mid season i.e., 45-50 DAS. Yield 8 t/ha.

*Kashi Nandini* : This is earliest variety and can be sown from 25 October to 25 November in northern plains. Plants are erect, 60 cm high, dark green with 7-8 pods/plant. Pods are 8-9 cm long, attractive with well filled 8-9 ovules, shelling quality being 47-48 %. The pods are ready for picking 55 days after sowing. Total green pod yield being 110-120 q/ha. It is tolerant to leaf miner and pod-borer.

*Kashi Udai* : An early-maturing variety, its plants grow 60 cm high. It is sown from October end to 25 November in plains. Plants with dark green foliage have short internodes and 8-10 pods/plant. Pods are 9-10 cm long, attractive with 8-9 bold ovules/pod with shelling quality of 48 %. The pods are picked up 60-65 days after sowing, total green pod yield being 100-110 q/ha.

*Kashi Mukti* : An early maturing garden pea, it is resistant to powdery mildew. Last week of October to November is ideal sowing in plains. Plant height is 50-53 cm. The plants are slender with shorter leaves, pods are 8.5-9.0 cm long, attractive filled with 8-9 bold and soft-textured seeds, with 48-49 % shelling and yield of 110-120 q/ha.

*Kashi Shakti* : This is a high yielding, medium-maturing variety. It can be sown in November in plains. Plant height is 90-98 cm. Plants have dark green foliage with 11-12 pods/plant. Pods are 10-10.5 cm long, well-filled with 7.5-8.5 bold seeds, shelling quality of 48-49 % and yield of 140-160 q/ha.

*Ooty 1* : A dwarf variety with a yield of 11.9 t/ha in 90 days, resistant to white fly.

*NDVP 8* : Mid season variety with a yield of 10 t/ha.

*NDVP 10* : A mid season variety with a yield of 10 t/ha.

*P-88* : A derivative of cross Puss-2 x Morrasis-55, bred at PAU Ludhiana and identified in 1985 for zones I, IV, V, VI. Vine length 75.5 cm, 1 or 2 flowers per axil. Days to flowering 72, leaves dark green, average number of seeds per pod 7.5, less sweet Shelling percentage 47.3. Mature seeds green wrinkled. Susceptible to powdery mildew.

*Matar Ageta 6* : It is an early and high yielding garden pea; the plants are dwarf, quick-growing, erect with green foliage and ready for first picking within 7 weeks after sowing. Pods are long, 12-15 in number, borne singly and in pairs, well filled, containing six grains on an average, the shelling percentage being 44. The grains are sweet and taste is better, the seed is smooth, dented and light green with high dry matter, chlorophyll and crude protein contents and a low value of peroxidase activity (OD/min/mg/protein) compared with "Arkel" and 'Harabona", lending to processing. More than 50 per cent of the total green pods can be harvested in the first picking in 47 days in this variety. On an average yield is 4.8 tonnes/ha.

*Azad P-1* : plants are tall. First flower appears in about 45 days after sowing. Pods dark green in colour and borne in clusters. First harvesting can be done 75 days after seed sowing. 3-4 picking are done. Average green pod yield is about 85-90 q/ha.

*Azad P-2(PRS 4)* : Resistant to powdery mildew. Plants tall (130-150). Straight and smooth pods, yield 12t/ha in 90-95 days. It is a medium duration variety. Pods are of medium size with 6 to 7 ovules/ pod.

*Azad P-3* : Plants are early, erect medium tall, dark green in colour. First flower appears in about 40 days after seed sowing. Pods are well filled, bold, green and attractive. First picking can be obtained in 70 days after sowing. 3-4 picking are done. Average green pod yield is about 65-70 q/ha.

*Jawahar Matar-1* : A derivative of the cross T-19 x Greater Progress, developed at Gwalior. It was identified in 1975 for zones I, IV, VII. Plant height 82 cm, leaves normal shaped, green in colour. Flowers of white colour, two flowers per axil. Pods straight with beak like outgrowth at the lower end. Average size of pod is 8.8 x 1.4 cm. 8-9 seeds per pods, shelling 52 %, mature seeds green in colour, wrinkled.

*Jawahar Matar-2* : Pods dark green, big, curved with 8-10 sweet ovules, wrinkle seeded, susceptible to powdery mildew.

*Jawahar Matar-3* : Early season variety developed through selection from cross between T 19 x Early Badger. First picking in 50 DAS, pods 7 cm long, light green and round oval ovules.

*Jawahar Matar-4* : An early variety developed from cross of T 19 x Little Marvel, by JNKVV, Jabalpur and recommended by the 3rd workshop on vegetable improvement for zone number IV and VII. Its average yield in the coordinated trials was 123 q/ha which was 38 per cent higher than control. Plant height 70 cm, green pod, stem and foliage. Pods are well filled. Mature seeds wrinkled green.

*Jawahar Matar 15* : Resistant to powdery mildew and fusarium wilt. Plants dwarf, yield 13 t/ha.

*Jawahar Matar 54* : Powdery mildew resistant variety with big incurved pods enclosing 8-9 big wrinkled seeds. Yield 7 t/ha.

*Early December* : Also developed by JNKVV and recommended for zones IV, V, VII on the basis of its superior performance at 13 centres under all India Coordinated Vegetables Improvement Project in 1977. Plants are bushy 70 cm tall with white flowers. Pods are round to oval shaped 5 cm long and 1 cm broad. Numbers of seeds/pods is five and shelling percentage is 45.

*Jawahar Peas 71 (JP 71)* : A progeny of double cross (Arkel x JP 829) x (JP 501 x JM 1) having powdery mildew resistance belongs to mid-season group. The plants are 50 cm in height and the pods are medium sized (7 cm) with 6-7 ovules. The yield of green pods varies around 120 q/ha. The seeds are green, wrinkled and bigger (100 seed weight 17 g).

*Jawahar Peas 83* : Mid season powdery mildew resistant variety developed through double cross (Arkel x JP 829) x (46C x JP 501). Plants dwarf. Pods big and curved with 8 green and sweet ovules, yield 12-13 t/ha.

*Harbhajan* : Early variety resembling to field pea. Susceptible to powdery mildew. Average yield 3 t/ha.

*Ageta* : Plants are small, erect and green in colour. It is suitable for early sowing. First flower appears in about 25 days after seed sowing and it takes about 6 weeks for first picking. Two picking are done. Average green pod yield is 50-55 q/ha.

*Pant Matar-2 (PM-2)* : PM-2 is an early maturing variety, developed at GBPUA & T, Pantanagar; derived through pedigree selection from the cross Early Badger x IP-3 (Pant Uphar). Pods are relatively smaller than Arkel; gives an average green pod yield of 100 q/ha.

***Pant Uphar*** : Developed at Pantanagar from line IP-3 and identified in 1985 for Sutlej and Gangetic plains. Plant height 75-80 cm, relatively thin small leaflets, foliage light green. Flowers white, two buds per axil, length of pod 7-8 cm. Shelling 52 %. Susceptible to powdery mildew. Maturity 75 to 80 days.

***Pant Sabji Matar 3*** : Early season variety with long curved pods with 8-9 ovules. Picking starts 60-75 DAS. Yield 9 t/ha.

***Hisar Harit*** : It is a mid-early and high yielding. Bulk pedigree method of selection was followed in Bonneville x 023 for developing this variety. This variety is semi dwarf, early and takes 60 days for first picking; foliage green, single to double-podded; pod large green, well-filled; seed green dimpled after drying. The plants have small internodes, the number of pods per plant is more than the other varieties and harvest index high.

***VL-3*** : Developed from a cross of old sugar x early wrinkled dwarf 2-2-a, followed by a pedigree selection at VPKAS, Almora. Identified in 1987 for zones I, IV, VII. Determinate plant types, average height 67 cm, light green, white flowers, pod light green. Average pod length 6.8 cm. Number of seeds per pod 5, shelling percentage 46. Maturity 145 days, green pod yield 100 q/ha. Resistant to powdery mildew and wilt.

***VL Ageti Matar-7*** : It is a cross between Pant Uphar (IP-3) and Arkel. The plants are dwarf (65-70 cm) with green foliage, white flowers and double pods/ peduncle. It can be easily distinguished from the 'Arkel' as it grows at extremely faster growth rate and light green colour and early vegetative phase. The pods are light green, smooth, attractive, slightly incurved towards suture at the distal end. The pods are medium-sized (8 cm) with 6-7 grains/pod and filled completely. Green grains are dimpled, bold, very sweet with high T.S.S. (16.8 %) and dry seeds are wrinkled and light green colour. The average green pod yield is 10-12.5 tonnes/ha.

***VL8*** : Mid season variety with a yield of 10 t/ha.

***Vivek 6*** : Its plants are dwarf and vigorous with dark green foliage. It is across of Uphar and VL Matar 3, a medium sized variety. It matures in 135 to 140 days in mid hills, whereas it takes 80-85 days in plains. The pods are smooth straight, 6-7 cm long, light green and completely filled. The pods contain 6 grains in each having tolerance to cold and moisture stress conditions. Average pod yield is 79-103 q/ha.

***Palam Priya*** : It is a medium maturity (110-days), medium tall (60-67 cm), powdery mildew resistant garden pea. Its plants are vigorous with dark green foliage and branching habit. The pods are borne in double. These are smooth, straight, light green, well filled with average length of 8-9 cm. The average seeds/pod are 7-8 with about 60 % shelling. Its green peas contain 15-16 % TSS. The seeds become wrinkled upon maturity with light bluish green colour. On an average it gives 4-5 green pod pickings in hills with an average pod yield of 125-150 q ha.

***Arka Ajit*** : Resistant to powdery mildew and rust. Yield 10 t/ha in 90 days.

***Arka Karthik*** : It is a mid-season variety growing to a height of 65 cm. It does not require any staking. It has combined resistance to powdery mildew and rust. Its pods are ready for picking 60 days after sowing. They are 10.5-11.0 cm long, seeds are bold, green and sweet with 58% shelling. It has a yield potential of 11 tonnes / ha in 90 days.

***Vivek Matar 10*** : Plants are quite distinct in both growth and yield attributes. The average green pod yield is 90-110 q/ha which mature in 120-130 days. Its plant height 50-60 cm, branches 3-4, first flower appear at 7-8 node, green foliage, leaf normal, two white flower/ peduncle, dark green slightly curved pod, mostly double pods / peduncle, 13-15 pods / plant, 143 pods / kg, 8-9 cm pod length, 45% shelling and greenish white wrinkled seeds. It shows 50% flowering in 80-85 days in hills and 60-65 days in plains. It is recommended for sowing from first week of November to last week of November in mid hills as well as in plains. It is resistant to white rot, wilt, leaf blight and moderately resistant to powdery mildew under field conditions and tolerant to pod-borer.

***UN 53-6*** : A snap pea where the whole pod is edible. Yield 8-9 t/ ha in 90 days.

***Type 19*** : It is a selection from a sample of Varanasi district of UP. It matures in 120 days.It has dark green foliage and white flowers. Seeds are wrinkled and greenish white. Pods are ready for picking in about 75 days. It yields about 70-100 q/ha of green pods.

***Swarna Rekha*** : It is an early maturing selection from local material of Bihar. It matures in about 120 days. Plants are semi-spreading type with white flowers. Seeds are round, smooth and creamy in colour with black hilum. It is suitable for green pods and dry seeds as well. Its average yield is 80-100 q/ha of green pods and 15-20 q/ha of dry grains per hectare.

***Khapenkheda*** : It is a local selection from Maharashtra state. Pods are small and seeds are wrinkled. Average yield of green pods is about 50-60 quintals and grain yield about 10-12 q/ha.

## RUNNER BEAN

Runner bean (*Phaseolus coccineus* L.) 2n = 22; also called multiflora bean, belongs to the family Leguminoceae (Fabaceae), is a climbing perennial, but is grown as an annual vegetable for the immature green pods. Plants are long day twiny vines, suited for the summer season. Plants are generally vigorous, indeterminate to determinate depending upon the variety. Plants are frosted at low temperature but underground tap roots remain viable and generate with the rise in temperatures. Immature pods, green seeds mature dry seeds are used as food.

*Phaseolus coccineus* subsp. *coccineus* and *P. coccineus* subsp. *darwinianus* are grown in most part of Central and South America. In Europe,Africa and the USA, *P. coccineus* subsp, *coccineus* is grown for its green pods. *Phaseolus coccineus* subspecies *coccineus* formosus glabellus, *griseus* and *darwinianus* produce fertile hybrids when hybridized with *P. coccineus* varieties. It was absorved that *P. coccineus* x *P. acutifolius* and *P. coccineus* x *P. vulgaris* hybrids were sufficiently fertile to produce advance generation. Resistance to bean yellow mosaic virus was transferred from *P. coccineus* to *P. vulgaris*.

Flowers are grouped in biflorous fascicles on elongated pseudoracemes. The flowers of subsp. *coccineus* are red or white and those of subsp. *darwinianus* are white or lilac-purple. The degree of natural cross-pollination usually varied from 20 % to 50 % for cultivar in Britain, and for a wild population from 45 % to 90 %. Natural cross-pollination is performed by bumble-bees (*Bombus* spp.) and honey bees (*Apis mellifera*). *Phaseolus coccineus* was the only self incompatible species out of 41 investigated taxa. The gametophytic system of self-incompatibility is involved in *P. coccineus*.

## Varieties and Hybrids

There are some varieties have been developed in runner bean. These are; Tender Pod, Kelvedon Marvel, Bianco de Spagna, Montecarlo, Monterosa. Some lines have identified suitable for cultivation and have good resistance to different diseases.These lines are : PI 28764536 and PI 417603.

# SWORD BEAN

Sword bean (*Canavalia gladiata* (Jacq.) Dc.; 2n = 22, 44 belongs to the family Lequminosae; is hardy drought, resistant tropical legume vegetable grown for their young pods for use as cooked vegetable. It is also grown as green manure crops, forage crops and also cover crop. The pods contain high amount of protein, minerals and fibre. It also contains globulins like canavalin and a thermo labile toxic principle causing hemorrhage of the mucus living of the stomach and intestine. Usually the water left after boiling are removed to avoid a slight astringent taste while preparing dishes. The ripe seeds should be eaten with caution as they may be somewhat poisonous; they should be boiled in salt and the water changed.

Sword bean (*C. gladiata*) is of old World origin and is probably derived from *C. virosa* wight and Arn, which grows wild in tropical Asia and Africa. It is widely cultivated in the East, particularly in India and has now spread throughtout the tropics. It has become naturalized in some areas.

Sword bean has climbining vine growth, broader and thicker pods and scarlet colorued seeds compared to jack bean; the vegetative part and flowers are similar to *C. ensiformis*, but the petioles are shorter than the leaflets; the leaf lets are larger, 10-18 x 6-14 cm, glabrescent, and are acuminate with a short point at apex; fls are larger, 3.5 cm long, and white or pinkish; the pods are more curved and with more strongly developed ridges, broader than that of jackbean and is 15-25 cm long, 3.5-5.0 cm broad and 8-16 seeded. The seeds are larger, typically scarlet coloured, 2.5-3.5 cm long, with a dark brown hilum, 2-2.5 cm long, extending almost the entire length of the seed.

The unripe fresh beans contain about : water 88.6 per cent, protein 2.7 per cent; fat 0.2 per cent; carbohydrate 6.4 per cent; fibre 1.5 per cent; ash 0.6 per cent.

Sword bean bagged flowers set pods and seeds. Hand crossing is difficult as the emasculated flowers tend to drop, but it can be done and should be confined to the lowest 2-3 flowers of an inflorescence. Under natural conditions the flowers are visited by bees and 20 per cent or more cross-pollination occurs.

The crop has a twining habit and is usually grown in the homesteads by trailing over small tree or walls.

### Varieties and Hybrids

There is no named variety in this crop. Some local strains/varieties are popular with the growers. They are used for cultivation. Very recently one variety has been developed by Central Institute for Arid Horticulture, Rajasthan ; which detail is given below :

***AHSB 1 :*** It has been developed by selection. It is very early for harvesting under extremes of arid conditions. Plants are medium in growth with dark green leaves and white flowers. Flowering starts in 68-72 days, where first picking is from 90-95 days of sowing as a rainy-winter season crop (July-March). It produces uniform pods which are used for vegetable at tender stages.

Light shining green, tender pods of 20-28 cm length, 2.8-3.5 cm width and 50-60 g weight are better in quality and ready for picking 12-18 days after setting. The number of marketable tender pods ranges from 22.4 to 30.5 with a yield potential of about 1.732 kg / plant. For seed production, about 20 mature pods gives 400 g seeds / plant. The creamy white seeds are large and kidney-shaped. The seeds are about 2.52 cm in length and 1.61 cm in width. The weight of 1,000 seeds is 2.152 kg.

## VEGETABLE SOYBEAN

Vegetable soybean (*Glycine max* (L.) Merrill); 2n = 40; belongs to the family Fabaceae (syn. Leguminosae), is an amazingly versatile crop, popular in countries like China, Japan, Korea, Taiwan, Philippines and USA. It provide 35 % of the world's vegetable oil and is a major (11 %) protein source for over one billion Asians. This is also cultivated for green shelled bean just like peas and for soybean milk. Delayed harvest results in hardiness and change of colour. Taste is related to sucrose content and glutamic acid in seeds.

Soybean is an annual plant with bushy growth habit, self-pollinated diploid legume vegetable. An erect and productive crop i is believed to have been domesticated for food from its viny wild relative, *Glycine soja* Sieb and Zucc in eastern China more than three thousand years ago. Spread first throughout eastern China and Korea in prehistoric times, was exported by sea to Japan around the sixth century AD and at the same time westward down the silk road to Nepal and northern India.

The genus *Glycine* Wild, is divided into the subgenera *Soja* (Moench) F.J. Herm and *Glycine*. Cultigen *G. max* alongwith its ancestral species *Soja* belongs to subgenus *Soja*. Sub genus *Glycine* at present is composed of 15 species, most of which are diploid (2n = 40) and are indigenous to Australia.Wild species belonging to the sub genus *Glycine* possess many valuable traits such as tolerance to salt, heat, cold, drought or other adverse growing conditions, and resistance to some viral and fungal diseases.

## Varieties and Hybrids

Quite a large number of varieties differing in their climatic requirements, growth habit, pod quality, seeds types and for specific purposes have been developed.At least 15 countries maintain permanent collections of soybean germplasm. Among them, China, Japan, USA, Korea are important. The green cotyledon traits in soybean is usually an indication that the variety is used for edamame (vegetable type). Varieties with small, flat, black seeds are used extensively to make a spicy fermented garnish for vegetable dishes in southern China.In Japan, yellow, round, small seeds are used primarily for natto preparation and larger seeds for tofu. Varieties with large black seeds are used primarily to produce holding treats in Japan.

Some important commercial varieties released by leading soybean producing countries are given below :

### 1. Soybean varieties released by USA

a) ***Large seeded varieties*** : Beeson, Disoy, Corsoy 79, Emerald, Grande, HP 201, HP 202, HP 203, HP 204, IA 1002, IA 1003, IA 2009, IA 2010,Kanrich, Kim, Magna, Marion,Merrimax, Prize, Proto,Vinton 81, Verde, Zane.

b) ***Small-seeded varieties*** : Camp, Chico, dare, Essex, IL 1, IL 2, Minnatto, SS 201, SS 202, Vance.

2. Soybean varieties released byJapan

Toyosuzu, Kitamusume, Okushirome, Yamashiratama, Shirosennari, Enrei, Tamahikari, Tamahomare, Suzunari, Nakasennari, Gindaizu, Shirotae,(All are large seeded), Akasaya, Tambagura, Hourei, Ootsuru, Shin 4, Tousan 149, Tousan 155, Tachinagaha.

## WINGED BEAN

Winged bean (*Psophocarpus tetragonolobus* L.); 2 n = 18, 22; family: leguminosae; syn. Goa bean, four angled bean, Manila bean, Princess pea, asparagus bean, Chathur lobia (in Hindi), is wonder crop where in all parts are edible and highly nutritious. Synonyms 'Soyas rival', 'God sent vegetable', and vegetable of twentieth century are attributed due to this. This is also known as a super market on a stalk', because the plant combines the desirable characteristics of common bean, pea, spinach, mushroom and plants.

It is tropical herbaceous perennial grown as an annual with winy and twining stem. Leaves are trifoliate. Inflorescence is an auxiliary raceme and flower colour ranges from white to deep reddish purple. Pods are four angled and vary in length, shape and colour of seeds Storage roots are small (50 g), spindle shaped or irregular.

A somatic chromosome number of 2n = 26 has been reported. Chromosomes, polymorphism is found to exist in winged bean indicating a possible polyploidy nature of the crop. The related species of winged bean are *Psophocarpus pallustris* and *P. longepedunculatus*. These species have a common progenitor. The other species are *P. scandens*, *P. lancifolius*, *P. grandiflorus* and *P. lecomtai*.

### Varieties and Hybrids

The winged bean is a genetically diversified species that exhibits a wide variety of easily classified or continuous characters. Among the first 26 accessories introduced at the Mayaguez Institute of Tropical Agriculture (MITA), 64 types were found, however, many of these differed only in pimple characteristics, such as, flower or seed colour, that might be found in any segregating population.

The varieties from Papua new Guinea, and to a lesser extent from Indonesia are the most diverse and the least improved. Several classes of varieties can be distinguished as given below :

(i) Wild varieties, with small fibrous pods.
(ii) Improved varieties with large, course pods.
(iii) Dual-purpose varieties or varieties weed chiefly as a source of tubers.
(iv) Highly improved varieties with thin pods and narrow wings.
(v) Varieties that produce chiefly foliage and therefore are useful as forage.

Of-course, there is a continuity among varieties and there is no strict division into classes. The varieties seen in Java and Thailand are principally of the highly improved types and are excellent vegetables.

**Varieties from Malaysia institute of Tropical Agriculture**

***Bogor*** : It is a standard type seen in the markets of South East Asia, produces winged bean over a long season, provided proper care is taken during spring and summer. It is perennial in habit and shows continued production after two years.

***Ribbon*** : This variety produces well but chiefly during winter.

***Butterfly and Mariposa*** : These varieties are typical of the heavy bearing, square-podded winter types.

***Lunita*** : It is a similar to butterfly and mariposa but with a flattened pod.

***Always and Siempre*** : These two varieties are year-round varieties in Puerto Rico.

***Alipasto*** : It produces sufficient foliage and also produce enough seeds for maintenance.

***Chimbux*** : It is a good varieties for seed production, although its purple pod is less than acceptable as a green bean.

***Tinge*** : It is the best variety for use as a dried seed. When seeds are soaked in a solution of sodium bicarbonate, more than 95 per cent softened on cooking and are very palatable.

## Collection of Asian Countries

The collection include more than 100 in Indonesia. Some accessions i.e., from PT 1 to PT 16 were tested under Coimbatore conditions.

***PT 2 and PT 16 :*** These two varieties performed well under coimbatore conditions, giving more than 1.0 kg vegetable pods per plant.

***IIHR, Bangalore*** : IIHR, Bangalore has promising selections.These are Sel.21, Sel. 60 and Sel. 71 give very high yield and quality pods.

***WBC 2 :*** It is a promising selection from Meghalaya and is found very suitable for north eastern region.

□□□

# 7

# Cucurbitaceous Vegetables

## ASH GOURD

Ash gourd *(Benincasa hispida* Cong.), 2n = 24; syn. Hairy melon, winter melon, ash pumpkin, white pumpkin, chinese preserving melon, wax gourd, white gourd, cultivated for its immature as well as mature fruits. Fruits are used as a cooked vegetable and also used in confectionary and ayurvedic medicinal preparation, variability in ash gourd is limited except for size and shape of fruits.

### Varieties and Hybrids

In order to classify interspecific variation in *Benincasa hispida* four major categories are proposed as horticultural or cultivar groups.

1) ***Ridged winter melon group*** : Very similar to the unridged winter melon group except that seeds have ridged margins and leaves have shallow lobes.

2) ***Unridged winter melon group*** : Seeds unridged, fruits large (0.5-1.0 m long), cylindrical, rind dark green with little or no waxy bloom, mature leaves with shallow to moderate lobes.

3) ***Fuzzy gourd group*** **:** Seeds ridged, fruits small (20-25cm in length), narrowly cylindrical, rind green, covered with white, soft hairs without bloom, mature leaves shallowly lobed.

4) ***Wax gourd group*** **:** Seed ridged,fruits small to large (10-60cm in diameter), round and flattened at both ends to oblong, rind light green, covered with hairs or hairless with a white, waxy bloom; mature leaves deeply lobed. In this group, three cultivars types are described based as fruit characters : -

(a) Fruits nearly round and essentially hairless. (b) Fruits nearly round and hairy and (c) Fruits oblong and hairy.

As per colour ash gourd is two types viz.

(a) purple green coloured (b) green coloured.

Importlant improved varieties and hybrids are : KAU local, Indu, Co1, Co2, Shakthi, Karikumbala, Boodikumbala, IVAG-502, Sel-1, Thudilyalur, Okinawa No.1, Kurokawa Early.

## Hybrids

***Mahyco*** : MAH 1

***Tropica Green Co*** : Fat Boy

***Others*** : BH1 and BH 2 (Big type)

*Ashgourd* is graft compatible with a muskmelon scion. Such a scion could be rescue the muskmelon crop in areas prone to wilt caused by *fusarium oxysporum* sp. *melonis*. The ash gourd lines are resistant to the wilt.

**The detail characteristics of varieties and hybrids are given below:**

***Co-1*** : It is a medium early variety. Fruit is attractive, globular, bigger in size (5 to 6 kg) with less seeds and more flesh. Yield 204 q/ha. Duration is 140 days.

***CO-2*** : Early maturing (120) days, fruits small weighing 3.0 kg, long spherical shape, less seeds (around 200-300 seeds), light green coloured flesh, higher yielder than CO1.

***KAU Local*** : Medium sized oval to oblong fruits with high flesh thickness, fruit length 45-55 cm, L : B ratio 2.05, fruit weight 6.1-8.1 kg, productivity 28.2 t/ha.

***Indu*** : Medium sized round fruits, average fruit length 24.3 cm, average fruit breadth 23.78 cm, L : B ratio 1.02, weight 4.82 kg, productivity 24.5 t/ha. Tolerant to mosaic disease, Released by KAU, Thrissur.

***Shakthi*** : Fruits long and cylindrical, yield 30-35 t/ha in 140-155 days. Developed by APAU, Hyderabad.

***Karikumbala*** : Local cultivar having dark green fruits without ashy coat. Released by UAS, Bangalore.

***Boodikumbala*** : Local cultivar where the fruits are covered with ashy coat. Developed by UAS Bangalore.

***IVAG-502*** : Fruits oblong with average weight of 12-13 kg. Yield 30-35 t/ha. Developed by IIVR Varanasi.

***Okinawa No.1*** : It is a high yielding and popular variety of Japan.

***Kurokawa Early*** : It is a high yielding and early variety. It is a popular variety of Japan.

***S1*** : It is a selection from local collection and has been recommended for Punjab.

***Thudiyalur*** : It is a local cultivar and popular in Maharashtra.

***MAH-1*** : The fruit is whitish wax like layer found on the surface after maturity of fruits. Fruits little longer. Average fruit weight 4-5 kg, fruit matures after 110-130 days.

***Fat Boy*** : It is suitable for all tropical growing condition. Elongated light green fruits, mature in sixty days, size 30 cm width, 17 cm long, weight 4 kg and tolerant to sun burn.

## BITTER GOURD

Bitterground (*Momordica charantia* L.); 2n = 22; syn. Bitter cucumber, Balsam pear, Karela, Karavella, Cherbal, samapfel, Kerula, Uchche, Quisaulbarri, Kakara, Pavakka, Karathi, Momordique Charandia, Kareli, Tita, Kakarakaya, Pakarkai, Balsamper, belongs to the family cucurbitaceae, is an important cucurbit fruits vegetable grown mainly in tropics. The alkaloid momordicasoides imparts the bitter taste to the fruit. The crop is grown for its immature tuberculate fruits which have a unique bitter taste.

Genus *Momordica*, to which bittergourd belongs, has the following four species in India.

*Momordica Charantia* (Cultivated bittergourd).

*Momordica Charantia* var. *muricata* (Highly bitter and small fruited wild bittergourd).

*Momordica dioica* (Kakrol, spine gourd).

*Momordica cochinchinensis* (Sweet gourd of Assam).

*Momordica balsamina* : Immature fruits are used as vegetable or pickled.

All species have 2n = 2x = 28.

Wild species are *M. subangulata* Bl., *M. denudata* and *M. macrophylla*. The crop is highly cross-pollinated due to monoecy.

Bittergourd is a monoecious annual climber with duration of 100-120 days, propagated through seeds. Fruit is pendulous, fusiform, ribbed with numerous tubercles. Consumers preferences in bittergourd vary from region to region depending on size, colour, presence or absence of tubercles/ridges and bitterness (momordicin) of fruits.

## Varieties and Hybrids

The varieties can be broadly classified into small fruit type (7-10 cm long) and large fruit type (15-18 cm long). Small type generally grown in summer and large type in rainy season. According to consumer choice, bitter gourd varies from region to region depending on size, colour, presence or absence of tubercles/ridges and bitterness of fruits. Some improved varieties are given below:

Pusa Do Mousmi, Kalyanpur Barahmasi, Pusa Vishesh, Kalyanpur Sona, Co-1, Coimbatore long, M.D.U.-1, M.D-4, Priya, Arka Harit, UK-1, Punjab 14, MC 23, Kashi Harit and Kashi Urvashi.

## Hybrids

***T.A.R.I., New Delhi :*** Pusa Hybrid-2.

***Indo American hybrid Seeds :*** Hybrid-49.

***Mahyco :*** White Long, Green Long, MBTH-101, MBTH-102, MBTH-104, MHBI-7, MHBI-10, MHBI-13, MHBI-15.

***Tokita Seeds :*** Midori long.

***Modi Agritech :*** MBGH-51.

*Sungro Seeds :* Vivek.

*Century Seeds :* Tijarti.

*Bejo Sheetal Seeds :* Prachi, Priya, Sonam, Poonam.

*Unicorn Seeds (Choice) :* Bhagya, Heera, BIT 124, BIT 094.

*Open Pollinated Selection :* Green Long, White Long, Pusa Do Mausami, Hirkani.

*Tropica Seeds (Green CO.) :* Comet, Mari, Solok

*Sreemaa Seeds :* Nakhara (Improved), Green Long.

*Nunhems Seeds :* Summer Green, Chaman, Kiran.

*Ankur Seeds :* Karishma, Hema, Reshma, Shreya, Seema.

*Golden Seeds :* Vishesh, Vinay, Shubra.

*Namdhari Seeds :* NS 431, NS 433, Shashank.

*Annapurna Beej :* Green Long, Annapurna Express, Nakhara Improve, Patnai, Gangajali, Uchha Bolder, Jhumka.

**Detail Characteristics varieties and Hybrid are given below :**

*Pusa Do Mausami :* This variety has been obtained through selection from local germplasm at IARI and released by the same organisation. It is suitable for spring- summer and rainy both the seasons. The fruits attain edible stage in about 55 days from sowing. The fruits are dark-green, long (18 cm), medium thick, somewhat club shaped having 7-8 continuous ridges. At edible stage, 8-10 fruits weigh one kg. Average fruit yield 178.08 q /ha.

*Pusa Vishesh :* It is dwarf-vine variety suitable for planting at higher plant density than that of long-vined other varieties. The fruits are glossy-green, medium long, thick and suitable as vegetable type, for stuffing and dehydration. It has been released by IARI release committee, New Delhi. It is particularly suitable for Spring-Summer season in Northern plains. It takes about 55-60 days to first harvesting. The average fruit weight is 115 g. Fruit yield varies from 112.66-195.00 qt/ha.

*Kalyanpur Barahmasi* : This has been developed at CSAUAT-Vegetable Research Station, Kalyanpur, Kanpur. It is a vigorous creeper. Fruits are long, thin and tapering. Yield potential is 100-125 q/ha in about 120 days. It is reported to be tolerant to fruitfly and mosaic. It is recommended for Northern plains.

***Kalyanpur Sona :*** This variety has been developed at CSAUAT-Vegetable Research Station Kalyanpur. The vine is creeper with green foliage. Fruits are medium long, thick and suitable for stuffing. The yield potential is 110-125 q/ha in 120 days crop duration. It is also reported to be tolerant to mosaic and fruitfly. It is suitable for UP. This has been duly notified by the Central seed committee in 1996 after its release by the UP state variety Release Committee.

***Coimbatore long :*** It is developed from Agriculture Research Institute Coimbatore. Fruits long thin, colour white, surface ridging broken, fruit weight 70g. It takes 58-62 days for first picking. Vines prolific, heavy yielder. Average yield is 110q/ha. Suitable for rainy season crop.

***Preethi (MC-84) :*** It is developed at KAU, Vellanikara, vine medium, fruit long very thin green colour, broken surface ridging. Average fruit weight 60 g, takes 60 days for first picking, average fruit yield 60 q/ha.

***VK-1 (Priya) :*** This variety came as a result of selection at Kerala Agricultural University, Vellanikkara from the local germplasm. It is characterized by extra-long fruits (40 cm) and heavy fruiting (about 50 fruits/plant). The plants vigorously, trained on a pandel, the main branches upto 6 m long. Putting out more than 36 primary branches. Female flower appears 40 days after sowing on the 24 th node, about 68 female flowers /plant are produced, 56 mature into fruits, each weighing about 240 g. the fruit is 40 cm long and 17 cm in girth with a flesh thickness of 5.9 mm and TSS 3.12 percent. Each fruit contains 24 seeds weighing 62g (100 seed weigh 24 g). plants start bearing 40-45 days after sowing. Fruits are ready for harvest 2 weeks after fruit set, producing upto 30 tonnes of fruits per hectare. Priya may be cultivated both in kharif and rabi season.

***Long white :*** Fruit long, tender white.

***Pride of Gujarat :*** Fruit small, roundish, whitish green with a few white dots on thin skin.

***Punjab-14 :*** Fruits light green, smaller vines, suitable for spring summer and rainy season. Released by PAU, Ludhiana.

***Arka Harit*** **:** This was developed through selection from local materials of Rajasthan at IIHR, Bangalore and also released by the same institute as back as in 1973. The fruits are spindle shaped, attractive, glossy green with smooth regular ribs and thick flesh. Fruits are ready for harvest in 55-60 days. Average fruit weight 90 g. The yield potential is about 120 q/ha in 120 days. It grows well both in summer and rainy season but maximum yield is obtained in rainy season.

***Pusa Hybrid -1*** **:** Its vines are medium in growth, the leaves are broadly ovate, lobed, surface soft, medium-sized and dark in colour. The fruits are attractively green, medium-long and medium-thick (Fruit length, 13.5 cm and breadth, 5.0 cm) with irregular smooth tubercles on the surface. On an average a fruit weigh 100g. Fruit flesh is 7.3 mm thick. The fruits are suitable for making vegetable, pickles and also for dehydration purpose. It is early maturing (50-55 days) and superior in fruit quality to Pusa Do Mausami and Pusa Vishesh. Its average yield is 218 q/ha. It is ideally cultivated in spring-summer season.

***Pusa Hybrid-2*** **:** The fruits are dark green, medium-long and medium thick (fruit length 12.5 cm and breadth 4.5 cm) with irregular smooth tubercles on surface. On an average a fruit weight is 90 g. The fruits are suitable for making vegetable, pickles and also for dehydration purpose. It is early maturing (50-55 days) and superior in fruit yield to pusa do mausami and MC 84. It also gave higher yield than hybrid, PBIGH 1. The average yield is 215 q/ha. It has been recommended for cultivation in spring-summer in northern India, Orissa and Andhra Pradesh.

***Co-1*** **:** It is a selection from T.N.A.U., Coimbatore. It is of medium duration, with high yielding potential. The fruits are dark green, spiny, 20-25 cm long, fusiform and single fruit weight is 100-120 g. It is highly suitable for early harvest. The average fruit yield is 144.0 q /ha.

***MDU 1*** **:** This variety evolved by irradiatin of MC 103 by γ-rays (50 kr) and was released by T.N.A.U., Coimbatore. The fruits are 40 cm long and creamy white. Individual fruit weigh 400 g and the yield per hectare is 320 quintals.

***MBTH -101*** **:** Plant flowers upto 50% after 40-45 days, early hybrid. Average fruit weight 65-70 g; fruits are spindle shape, 15-20 cm long; colour deep green; big ridges; high yielder and has high yielding potentiality.

***MBTH-102*** : It flowers after 40-45 days upto 50%, medium duration hybrid; average fruit weight 100-200 g ; fruit colour white; 35-45 cm long; thick skin with less number of ridges. Less seeds and high yielder.

***MBTH-104*** : It flowrs upto 50% after 40-55 days; medium duration variety; Average fruits weight 100-200g fruit 20-30 cm long, fruit colour snow white thick skin; more ridges, less seeds and high yielder.

***MBTH-107*** : It is little earlier hybrid than others medium, colour green, 25-30 cm long, regular ridges. Fruits uniformly slender. Average fruit weight 60-65 g. high yielder.

***Nakhara Improved*** : Early variety with heavy yielde, fruits are 12-13 cm long, medium size fruit with dark green colour, surface mining and more fruiting. Average fruit weight 60-70 g. Can be cultivated throughout the year. It has consumer pereference.

***Green small*** : It has 15 cm long fruits, average fruit weight 100-120 g with dark green colour, fruits bears at each node. Heavy yielding variety.

***Priyanka*** : Large white spindle shaped fruits with smooth spine, thick, flesh and less seeds.Average fruit length 25 cm, girth 20 cm, weight 300 g. Productivity 28.0 t/ha. Released by KAU, Thrissur.

***Coimbatore long white*** : Extra long fruits (60-65 cm) with white colour, yield 15 t/ha. Released by TNAU, Coimbatore.

***CoBgH1*** : $F_1$ hybrid developed by crossing MC 84 x MDU 1. Fruits are light green in colour, plumpy with more warts, each weighs 200 g – 300 g. Yields 44.40 t/ha in 115-120 days. It is rich in momordicin (2.99 mg per 100 g). Released by TNAU, Coimbatore.

***Konkan Tara*** : Released by Konkan Krishi Vidya Peeth, Dapoli. Fruits green, prickly, medium long (15-16 cm) and spindle shaped. Yield 24 t/ha.

***Hirkani*** : Fruits dark green, 15-20 cm long, spindle shaped with warts and prickles, yield 14 t/ha in 160 days. Developed by MPKV, Rahuri.

***Phule Green*** : Fruits dark green, 25-30 cm long, prickled, tolerant to downy mildew, yield 23 t/ha in 160-180 days. Developed by MPKV, rahuri.

***BBG 4 :*** It is a good creeper. Fruit medium in size (length 12.8 cm, girth 10.4 cm), green in colour, solitary bearing habit, sometimes two fruits per cluster. Performs well throughout the year. Average yield is 83.66 q/ha. Developed by Orissa University of Agriculture and Technology, Bhubaneswar.

***Bhagya :*** A proven hybrid. Medium long green fruits are 12 –20 cm and weighs 50 –100 gm. First picking starts in about 63 days from sowing. Harvesting continuous for long period.

***Heera :*** An excellent small fruited hybrid variety with vigorous vine and good foliage. High yielding fruits are spindle shaped with medium sharp teeth of length 8 –12 cm. Fruits weighing 40 –60 cm can be harvested in about 65 days. Good for continious picking.

***Ruchi :*** An outstanding hybrid with strong vine and good foliage. Fruits are long about 20 –30 cm with pure white colour and medium sharp teeth, weighing around 80 –150 gm. first picking begins in 60 – 65 days.

***BIT-094 :*** It is a hybrids of vigorous creeper, medium long fruits with prominent teeth. Dark green in colour with white patches at blossom end.

***BIT 124 :*** An early maturity hybrid with strong vines and profuse bearing. Long dark green fruits of length 25–30cm, tapering with prominent teeth. Ideally suited for continuous and prolonged harvest.

***Vivek :*** First fruit picking start in 63 days after sowing, fruit shape straight, dark green colour and 22cm long, average weight 80 g, very strong vine, gives profuse early bearing and give more yield and better quality fruits when vine is train on trellis.

***MHBI-10 :*** Fruits snow white, slender with good teething, length 20-25 cm, weight 120-150 g. Fruit harvest at 55-60 days after sowing; yield potential 60-65 t/ha.

***MHBI-13 :*** Fruits green with medium teething, length 30-35 cm, average fruit weight 125-150 g. First harvesting starts 55-60 days after sowing. Yield potential is 60-65 t/ha.

***BHBI-15 :*** Fruit short, green with dense teething, length 16-18 cm, average fruit weight 70 g. First harvesting starts after 55-58 days of sowing. Yield potential 60-65 t/ha.

*Prachi :* Fruit dark green, medium long with very good teething, length 20-25 cm. First harvest after 70 days of sowing. It is very good hybrid with excellent dark green teething.

*Priya :* It is green colour hybrid, which gives very good yield and can be grown in summer, rainy and winter season. Fruit is thick and medium long with good teething, length 25 cm. First harvest 70-75 days after sowing.

*Sonam :* It is small, very high yielding hybrid, which can be grown in summer, rainy season and mild winter. Fruit short with good teething, length 8-10 cm. First harvest starts in 75 days from sowing.

*Poonam :* It is very productive, medium early and fruits are attractive white, fruits are medium long with good teething, length 25 cm, white colour fruits. First harvest starts at 75 days after sowing.

*NS 431 (F1) :* An early,medium vigorous prolific and continuous bearing hybrid. Fruits are attractive dark green, thick, short with spines. Each fruits weighs 100-150 g. Flesh is white, tender with excellent keeping and cooking quality. Slow seed development and less number of seeds per fruit.

*NS 433 (F1) :* An early vigorous, prolific and continuous bearing hybrid. Fruits are medium long straight (20-30 cm) attractive white with spines, each weighing 150-180 g. Flesh is white and tender with slow seed development and good keeping quality.

*Shashank (F1) :* This is a very high yielding hybrid, Asiatic type vigorous, with large vine; good tolerance to diseases. Fruits are glossy, attractive pale green in colour and of good texture. Fruits are cylindrical in shape, with straight shoulder and rounded at blossom end, with each fruit weighing 300-400 g. It is an early hybrid coming to first harvest in 40-45 days after sowing.

*Summer Green :* This hybrid is early prolific bearer with partial resistance to diseases. Fruits are shining dark green in colour and spindle in shape. Excellent crispy and tender fruits makes it ideal for both fresh vegetable and frying.

*Chaman :* This early maturing (60 days) hybrid has vigorous plant growth with profuse fruit bearing. Long (15-16 cm) fruits are attractive green in colour with spines weighing about 75-80 g. Good for long distance transportation.

***Kiran :*** Fruits of this high yielding hybrid are dark green in colour, straight long with spines weighing about 70-80 g. Crispyness and tender taste makes this hybrid an ideal choice for fresh vegetable and frying.

***Vishesh :*** Early maturing vigorous hybrid with excellent fruit setting habit. Fruits are straight, 20-25 cm long, uniformly dark green and shining and have prominent and non continuous tubercles (teeth).Average fruit weight 100-125 g. Inside flesh white and seeds mature slowly. Continuous fruiting allows prolonged harvesting period and fruits are excellent for export market.

***Vinay :*** Early maturing vigorous hybrid and a prolific yielder. Fruits are uniformly shining dark green, attractive, 30-35 cm long and have prominent and non-continuous tubercles. Average fruit weight 200-250 g. Inside flesh pure white and bitter. The hybrid has good yield potential and performs extremely well under bower system of cultivation.

***Shubra :*** Early maturing vigorous hybrid with excellent yield potential. First harvesting commences 50-55 days after planting. Fruits are attractive, uniform, long, pure white, 25-30 cm in length with prominent non-continuous tubercles and weigh 200-250 g. Inside flesh pure white and less bitter.

***Seema (F1) :*** Strong plant vigour produce excellent quality fruits for longer period. Compact, fleshy, spindle shaped and attractive green coloured fruits. Flowering starts in 50-56 days and fruits are ready for first picking after a week. High yield potential. Good keeping quality and transport.

***Shreya [White (F1)] :*** Vigorous growth habit. Fruits are medium long, white coloured with greenish tinge. Fruiting starts in 50-52 days after sowing. High yield potential.

***Reshma (F1) :*** Strong plant vigour. Fruits are medium long, green coloured weighing about 120-130 g. Fruiting starts in 55-60 days. Fleshy fruits with high yielding potential.

***Karishma (F1) :*** Very strong plant vigour with continuity. Elongated small green coloured fruits. Picking starts in 52-57 days. Compact fleshy fruits.

***Hema (F1) :*** Strong plant vigour. Fruits are cylindrical, medium long, green coloured with continuous smooth ridges. Picking starts in 52-67 days. High yielding potential and good keeping quality.

***Midori Long (F1) :*** Vigorous growing, prolific, dark green vine, continuous bearing hybrid. Fruits are shiny dark green with dark green spines. Average fruit length is 18 cm and weight 80 g. First picking after 60-65 days of sowing. Tolerant to downy and powdery mildews.

***Comet (F1) :*** It is a local Vietnamese type, mature in around 50 days (early type), fruit size 5.5 cm width, 22 cm long, weigh 180 g. Medium green colour. High yielding early hybrid.

***Mari (F1) :*** It is Indonesian type hybrid, fruits 30 cm long, 5 cm width, around 180 g fruit weight, mature in 50 days. Pale green skin and high yielding early hybrid.

***Solok (F1) :*** It is Indonesian type, fruit 25 cm long, weigh 250-300 g each, mature in 55-60 days. Medium green skin with large shoulders and long shelf life.

## BOTTLE GOURD

Bottlegourd (*Lagenaria siceraria* (mol.) Standle. (syn. *L. Ciceraria*, *L. leucantha* Rusby, *L. vulgaris* Ser) 2n = 22 ; is also known as white flowered gourd, Lauki, Chia, Doodhi, Jatilao, jatilas, Tikatalana, Lau, Anapa, Sorakaya, Surakai, Dudhya, Bhopla, Sorekayi, Sorakaya, Surakai, Alabu, Dhudhya, Halagumbla, Ghaikadu, Kashisose, Gurde, Flaschenkurbis, Bitter bottle gourd, white pumpkin, Lauka, Daani, Aal; belongs to the family cucurbitaceae. The genus *Lagenaria* to which bottle gourd belongs is characterized by key characters-fruits fleshy, vary in shape and size and many seeded pepo, flowers solitary and chalky white opening at night.

Bottlegourd is an important warm - season fruit vegetable. It is grown throughout India and its fruits are available in the market throughout the year. The name bottlegourd is due to bottle like shape of fruit and its use as container in the past. Fruits at tender stage are used as a cooked vegetable and for preparation of sweets and pickles. Hard shells of mature fruits are used as water jugs, domestic utensils, floats for fishing nets, etc. Bottlegourd is a typical warm season vegetable. Though crop tolerates cool climate better than muskmelon and watermelon, it can not tolerate frost. Crop is grown during summer and rainy season. In places where water is not scare, it is grown throughout the year. Being a monoecious crop, bottle gourd is strictly cross-pollinated. Bees are the pollinators.

## Varieties and Hybrids

Thousands of years of cultivation of this cross-pollinated crop have resulted in a large variation in the shape and size of the fruits. These vary from very long and slender to thick and round. There are different varieties especially suited for summer and the rainy seasons. Some improved varieties are given below:

Pusa Summer Prolific Long, Pusa Summer Prolific Round, Pusa Naveen, Pusa Sandesh. Arka Bahar, White Surat, Hot Season Long White, Sarmiwali Safaid Lambee, Doodhi Long White and All Season Long, Hot Long Green and Hot Round Summer Season. Mansoon Queen (long), Calcutta Charmer (round), Pocha's Long White (summer), Sutton's Selected (Club shaped; Rainy), Suttons Long White Lauki (winter), Sutton's Round Lauki (spring), V. Sadashiv's Extra Long, Punjab Long, Punjab Round, Punjab Komal. Jagtial Long, Co1, Kasi Ganga, Kasi Komal.

## Hybrids

***IARI*** **:** Pusa Meghdoot Pusa Manjari, Pusa hybrid 3.

***Mahyco*** **:** MGH- 1 (Surekha), MGH-4 (Warad), MGH-5, MGH-7, MHBG-8, MHBG-10.

***Century Seeds*** **:** Harit, C.B.H.-4 (Gutka Type).

***Nath Seeds*** **:** N.B.G,H-22

***Nuzividu Seeds*** **:** N.B.G.H.-405

***Sandoz Seeds*** **:** Mulayam.

***Haritage Seeds*** **:** Heritage long, Ganesh Gupta.

***Sungro Seeds*** **:** Swati.

***Bejo Sheetal Seeds*** **:** Shramik, BSS 333, Akash.

***Unicorn Seeds (Choice)*** **:** Anand, Angad, Suraiyya, All Rounder.

***Nunhems Seeds*** **:** Gauri, Nargis.

***Ankur Seeds*** **:** Vivek, Pratik, Vikrant.

***Golden Seeds*** **:** GSH-1, Gola.

***Namdhari Seeds*** **:** Kaveri, NS 421, Shantala, Mridula, Manmath, NS 6255.

***Tokita Seeds*** **:** Naveen.

***Malav Seeds*** **:** Suman.

***Annapurna Beej*** **:** P.S.P.L., Round Star.

**Detail of characteristics varieties and hybrids are given below :**

***Pusa Summer Prolific Long :*** It was developed at IARI through selection from local germplasm, vine growth large, prolific, fruiting, fruits green, days to first harvest is 70 days. It is particularly suitable for growing as summer crop, although it can be grown in riny season also, fruits are 40-50 cm in length and 20-25 cm in girth. Fruits shapes long, club-shaped, curved, thin neck, pointed blossom end with small scar medium thick, light green. Total fruits yield varies from 15-18 q/ha, spring summer to rainy.

***Pusa Summer Prolific Round :*** This was developed through selection from local germplasm at IARI and has been released by the same institution. It has vigorous growth, days to first harvest from 70-72 days. Number of fruits/plants 5-7, weight 735-740g. fruits shape round, flat, medium and green, round fruits of 15-18 cm girth. It is prolific bearer and heavy yielder. Fruits yield 131-172 q/ha. Suitable for both summer and kharif season.

***Pusa Naveen :*** This is the latest variety developed and released by IARI, New Delhi. More than 75 collections were purified and evaluated under the renewed programme which was taken-up in 1984-85. Sel-48 was released as Pusa Naveen by IARI, New Delhi, variety release committee. It is high yielding (300q/ha) and takes about 60 days for first harvesting. Fruits are perfectly cylindrical and straight without any crook neck or curve. Average fruit weight is 850 g.

***Pusa Sandesh :*** Vines are small to medium- long, with short internodes, leaves medium-sized, green and softly pubescent. Fruit is attractive green, round, deep oblate, medium-sized, weighing on an average 600g each. It is early in maturity. The fruits are ready for picking 60-65 days after sowing in summer and 55-60 days after sowing in kharif season. The average yield is 29.8 and 32.8 tonnes/ha in summer and kharif seasons.

***Pusa Samridhi :*** It has medium vegetative growth (av. vine length 3.5 m). Leaves are medium-sized broadly reniform with shallow undulating margin and softly pubscent. The fruit is attractive green, smooth with normal pubescence, straight, slightly clubbed and without neck, 25-30 cm long, medium thick (6.7-7.5 cm diameter). The average fruit weight is 1.0 kg. It matures 50-55 days for first harvesting. It gives an average fruit yield of 272 q/ha in summer and 307 q/ha in Kharif, 18 and 13% more than Pusa Naveen during summer and Kharif season, respectively. It is superior in Vitamin C (8.07), Phosphorus

(20.39), Calcium (19.35) and iron (1.35 mg / 100 g edible portion) than other varieties.

***Thar Samridhi* :** It has been derived from segregating F2 population of a cross combination, Banswara Local-1 x Gujarat Local-1. It is uniform high yielding and early. Its vines are medium in growth habit, attaining an average length of 2.42-3.65 m and 5.5-6.2 branches. The leaves are medium-sized (15-18 cm in length and 19-23 cm in width). The first male flower anthesis is after 40-43 days on 10-13 nodal position and first female flower appears on 12-14 nodal position 46-49 days after sowing. The fruit picking starts 50-55 days after sowing.

The fruits are attractive, oblate-round, weighing 450-750 g each. Tender fruits of 'A' grade in 9-11 days after fruit setting (weighing 400-500 g, 12-14 cm in length and 28-30 cm in girth) can be harvested on alternate day. The tender fruit shining light green in colour have sparse pubescence. The internal fruit flesh is white and solid. On an average, its yield is 4.83-5.78 kg / plant, and 240-300 q/ha.

***Punjab Komal* :** It is the advance generation of a cross between LC 11 (round) and LC 5 (oblong). It is early maturing, medium sized, oblong fruited variety developed and released by PAU, Ludhiana. Marketable fruits are available in about 70 days after sowing. The fruits are light green with pubescence. There are 10-12 fruits per vine. The fruits are tender and borne on medium long, thin pedicels from the fourth or fifth node onwards. It is tolerant to cucumber mosaic virus (CMV). On an average each fruits weighing 623.21g and yield 50 tonnes/ha of tender fruits. The yield potential is 400 q/ha and it has nerrow sex ratio 1:14

***Punjab Round* :** It is a sister line of 'Punjab Komal'. It is late by 10 days from 'Punjab Komal'. The average fruit weight is 525.82 g. The number of fruits per vine is 10.42 and total fruit yield is 42.67 tonnes/ ha and sex ratio 1 : 23.

***Punjab Long* :** Fruit small, light green, first picking starts 70-80 days after sowing.Yield 40 t/ha. Tolerant to CMV.

***Narendra Rashmi* :** Fruits bottle shaped having shallow neck, average weight 1.0 kg. Yield 30 t/ha.

***BBOG 3-2* :** It is a trailing type. Fruit round in shape medium to small in size and light green in colour. Average fruit weight is 1.0 to 1.5 kg. Matures in 60-65 days after sowing. Average yield is 524 q/ha.

Developed by Orissa University of Agriculture and Technology, Bhubaneswar.

***Co.1*** **:** It is developed from different collection of varieties, in breeding and individual plant selection at Coimbatore. Plant growth is moderately vigorous, light green, hairy stem and round, slightly deep, lobed, green leaves (length 25-30 cm and breadth 15-20 cm). The fruits are medium sized (length 45-50 cm and breadth 20-30 cm), bottle shaped with prominent centrally depressed neck and light green in colour Fruit weight 2.0-2.5 kg on an average and is medium seeded (400-500 seeds, weighting 40-50 g). It is an early maturing variety and takes 50-55 days for first flowering, 85-95 days for first harvesting. Crop duration is 135 days. Yield potential is 36.6 tonnes/ha.

***Samrat*** **:** Fruits cylindrical without crook neck, greenish white, 30-40 cm long, weighing 700-800 g, yield 43 t/ha. Released by MPAU, Rahuri.

***Pant Sankar Lauki 1*** **:** This $F_1$ hybrid has medium long fruits released by Central Variety Released Committee in 1999, developed by GBPUAT, Pantnagar.

***Azad Nutan*** **:** The plants are early in flowering by virtue of appearance of female flower at lower nodes and at short internodes. The stature of plant is tall. The fruits are medium long, shining, light green, smooth, soft but firm and attractive, weighing on an average 1.0-1.5 kg each. The bearing is heavy. The glut in fruiting comes around November with every node bearing a fruit. The fruits are without significant neck which enables easy packaging for distant marketing. Azad Nutan can be successfully grown during both spring- summer and rainy seasons. Yet it does relatively better during rainy season. It is early in maturity and its fruits are ready for picking 55-60 days after sowing.

***Arka Bahar*** **:** This was developed and released by IIHR, Bangalore. This is a selection in local cultivars of Karnataka. Fruits are straight, devoid of crook neck, medium in size, each weighing about 1 kg. The skin is light green with white tender flesh. The flesh is shining, tender with pleasant aroma. The yield potential is about 400 q/ha in 75 -120 days in both season.

***Kalyanpur long Green*** **:** This variety was developed at vegetable research station, Kalyanpur of CS Azad University of Agriculture and

Technology, Kanpur. The vines are vigorous and long. Fruits are long with tapering and some what pointed blossom end. The yield potential is 300 q/ha in 120 days.

***Pusa Meghdoot*** : This is an F1 hybrid cultivar between Pusa Summer Prolific Long and Sel-2, developed and released by IARI, New Delhi in 1971. Early and high yielder, fruits are long, light green and attractive. It is relatively early and suitable for cultivation in spring-summer season. It has shown considerable yield heterosis over Pusa Summer Prolific Long. Vine large, fruits mature in 70-72 days fruit weight 1100-1250 g, number of fruits per plant 4.5-7.2, fruit long, club-shaped, curved, thin neck, pointed blossom end with small scar, medium thick green. Suitable for both season. Average total yield 145.0-250.0 q/ha.

***Pusa Manjari*** : This is a round fruited F1 hybrid cultivar developed and released at IARI, New Delhi in 1971 from a cross of Pusa Summer Prolific Round and Sel-11. It has given 48% higher early yield and 106% total yield over Pusa Summer Prolific Round. The vine growth medium to large. Fruit maturity 65-68 days, fruits weigh 750-760g., number of fruits per plant 4.8 to 6.0. Fruits shape round flat, medium blossom end scar, medium size and green. It is early and give 48 per cent higher yield over PSPR. Average Fruits yield. 165-179 q/ha. Suitable for summer and rainy season.

***Pusa hybrid 3*** : Its plants are medium-long with internodes medium, leaves medium sized, green and dotly pubiscent. Fruits are attractively green, long and straight without any neck, weighing, an average one kg each. The Fruits are highly suitable for easy packing in card-board boxes for distance marketing. It can successfully be grown both in spring-summer (Feb-June) and kharif (Jul-Nov) seasons. The fruits become ready for picking in 60 days in spring-summer season and in 50-55 days in khsrif season.

***Azad hybrids*** : Its plants are medium- long with medium internodes, leaves medium-sized, green and softly pubescent, fruits are attractively green, long and straight without any neck, weighing an average 1 kg each. The fruits are highly suitable for easy packaging in card board boxes for distant marketing. It can successfully be grown both in spring- summer (Feb-June) and Kharif (July-Nov) seasons. The fruits become ready for picking in 60 days in spring summer season and in 55-58 days in kharif season.

***MGH-1 :*** Flowers upto 50% after 40 days. It is early hybrid variety, high yielder; fruits flattish round, colour light green, can be grown both in kharif and Rabi.

***Warad (MGH-4) :*** It gives 50% flowering after 40-50 days of germinations, medium duration. Grow luxuriantly, high yielder. Fruits long and round, 40-45 cm long. Easy to send distant market, colour light green; can be grown both in kharif and rabi.

***MGH-7 :*** Early hybrid variety, high yielder. Fruit colour light green, just like bottle, 40-45 cm long, easy to transport distant places, can be grown in both season i.e. kharif and Rabi.

***MHBG-8 :*** Fruits cylindrical, shiny green, tender with attractive lining and soft flesh. Average fruit weight 600-750 g, length 40-45 cm. First harvest 50-55 days after sowing. Ideal for long distance transportation. Yield potential 65-70 t/ha.

***Akash :*** Fruits flattish round, first harvest start in 60-65 days from sowing and can be harvest in small size as well as big size. A very high yielding hybrid can be grown in summer, rainy season and mild winter.

***Shramik :*** It is very vigorous growing and high yielding hybrid, having light green colour. Suitable for all three seasons. Fruit cylindrical, 1.5 ft, long and mature in 65 days after sowing.

***BSS 333 :*** It is very vigorous growing and high yielding hybrid, having light green colour. Fruits cylindrical to bottle shaped, weighing 3.0 kg each, length 3-4 ft, mature in 75 days after sowing. Grown in all three seasons.

***Anand :*** A strong vined are highly prolific bearer hybrid. Fruits are cylindrical in shape. Apple green coloured fruits are 40 –45 cm long with medium girth and are very tender.

***Angad :*** An excellent high yielding gutka type hybrid. Cylindrical shape fruits are light green with length of 30-45cm. Crop comes to 50% flower after 45-50 days and harvesting begins 65-70 days after sowing. Good for long distance transportation.

***Suraiyya :*** An outstanding performer. PSPL type hybrid produces long slender, light green fruits of length 35–50 cm, attains 50 % flowering in 45–50 days, harvesting begins in 65–70 days after sowing. Shipping quality is good.

***All Rounder :*** A high yielding hybrid, fruits are oblate- round shaped, light green colour. Fruits are ready for harvesting in 60 -70 days from the time of sowing.

***Swati :*** First picking start in 50 days after sowing, fruit shape cylindrical, light green in colour, 45– 60 cm length, average fruit weight 2.0 kg, suitable for planting throughout the year except in severe winter season.

***Vivek (F1) :*** Flat round, compact, light green fruits. Shining skin and small hairs makes the fruit attractive. High yield, longer duration and earliness gives higher returns. Picking starts in 60 to 65 days.

***Pratik (F1) :*** Vigorous uniform growth, fruits are cylindrical in shape, light green in colour, weighing 950-1200 g per fruit. Picking starts in 68-72 days. High yield potential.

***Vikrant (F1) :*** Vines are vigorous in growth habit. Fruits are bottle shaped, light green coloured weighing 1000-1250 g per fruit. Picking starts in 65-70 days. High yielder.

***GSH-1 :*** Early vigorous and prolific growing hybrid, bearing uniformly cylindrical, medium long (40-50 cm), light or pale green fruits with fine pubescence. Fruit flesh pure white and tender with soft seeds. Average fruit weight 500-800 g. Fruits have excellent keeping quality and suitable for long distance transportation. Yields 12-13 mt/ha.

***Gola :*** Very early maturing hybrid. Prolific fruit bearing habit. Fruit round, green shining and tender. Average fruit weight 0.5 to 0.75 kg. First harvest start 40-45 days after sowing. Prolific fruit setting habit. Very delicious.Slow seed development. Yield 60-70 mt/ha under optimal conditions.

***Gauri :*** This high yielding and early maturing hybrid bears long cylindrical fruits of light green colour. The first harvest is possible after 42-45 days of sowing with an average fruit length of 30-35 cm and average weight of 0.8 to 1.0 kg. The straight fruits are good for long distance transportation.

***Nargis :*** This predominantly female flowering hybrid has vigorous plant growth and high yielder. Light green fruits are long, cylindrical, straight weighing approximately 1.0-1.25 kg per fruit. Good for long transportation.

***Kaveri (F1) :*** An early and vigorous hybrid with continuous and prolific bearing habit. Medium to long fruits (25-30 cm) are uniform green and perfect cylindrical shape without neck. Each fruit weigh 300-400 g and is of excellent cooking and keeping qualities.White and tender flesh is embedded with few slow maturing seeds.

***NS 421 (F1) :*** An early, vigorous and regular bearing hybrid. Straight, uniform green, cylindrical shaped fruits weigh 350-400 g and measure 30-35 cm.Flesh is white and tender with good cooking quality. Slow seed maturity and less number of seeds per fruit makes this superior.

***Shantala (F1) :*** This extra early hyubrid is a high yielder with uniform green, cylindrical, extra long (75-80 cm) fruits. Vigorous plants bear fruits continuously with tender seeds of slow maturity.

***Mridula (F1) :*** It is a vigorous, prolific and continuous bearing hybrid. Fruits are shiny green, round in shape with 300-350 g in weight. This hybrid is early and superior in yield.

***Manmath (F1) :*** Plants are very vigorous and continuous bearer. Fruit are flat round, green in colour with300-350 g in weight. Fruits heavy very good uniform shape. It is an exceptional yielder.

***NS 6255 (F1) :*** Vigorous plants, prolific and continuous bearer. Fruits are round with pronounced neck, uniform green in colour and weigh 300-350 g. It is heavy yielder.

***Naveen (F1) :*** It is a early vigorous hybrid with continuous bearing, fruits are perfect cylindrical without neck, length 30-40 cm. Average fruit weight 500 g, white and tender flesh with few immature seeds.

## CANTALOUPE

Cantaloupe or Cantaloupe melon (*Cucumis melo* var. *cantaloupensis*); is a tetraploid (2n=48), cantaloupe (Europe) has thick, scaly, rough rind, often deeply grooved. It is used as negatable in immature stage and as dessert fruits after ripening.

### Varieties and Hybrids

**The detail characteristics of varieties and hybrids and given below:**

***MHC-5 :*** It is developed by mahyco. Fruits round, yellow skin with good netting, flesh yellowish, orange with very good texture, good flavour and aroma. Average fruit weight 0.8-1.0 kg; TSS (brix) 14. Fruit mature in 75-80 days. It has yield potential of 25-30 t/ha.

*MHC-9 :* It has also been developed by Mahyco. Fruit oval and has yellow skin with good netting, flesh orange with very good texture, good flavour and aroma. Average fruit weight 1.0-1.5 kg; TSS (brix) 10-11. Fruit mature in 80-85 days.

***Early Dew :*** It is an early Honey Dew type cantaloupe. The rind is creamy green at maturity with lime green colour. This hybrid has round (14 x 14 cm) fruit with 1.1-1.4 kg weight, without netting, creamy green rind at maturity, flesh lime green. Tolerant to fusarium wilt (race) and fruit slips when matures.

***Gallicum :*** A very sweet hybrid, produces round, slightly netted fruits with luscious green flesh. Its fruits have an intense sweet aroma. Average fruit weight is about 1.0 kg.Fruits round (12 x 12 cm), weight 1.0-1.1 kg with fine net very light, rind colour greenish yellow at maturity, flesh green with very sweet aroma. Tolerant to powdery mildew (race 1).

***Magnum-45 :*** A vigorous, very early hybrid produces medium netted fruits with higher sugar content. The fruits are good for long distance shipping. The flesh is deep orange and the seed cavity small and dry. It has a concentrated set for early harvest. Fruits oval, weigh 1.1-1.4 kg each, medium net light sutures, yellowish rind, deep orange flesh. Good for shipping and early hybrid, tolerant to powdery mildew (race 1).

## CHOW-CHOW

Chow-Chow (*Sechium edule* Swartz); 2n = 24, is also known as Choyote, Christophine, Askas; belongs to the family cucurbitaceae. It is indigenous to southern Mexico and Central America and was common vegetable among the Aztecs prior to the Conquest. It has now spead throughout the tropics. A monotypic genus, monoecious and consequently cross-pollinated in nature. It is a single seeded fruit vegetable, perennial rooted vine crop and viviparous in nature. It produces large numbers of edible fruit, tuberous roots and tender leaves and tendrils. It is an important part of dietary system of tribal of the region. The fruits as well as tuberous roots are used as vegetable. The young leaves and tender shoots are some times eaten as spinach.

Chow-chow is a monoecious, robust, climbing or sprawling, sparsely hairy, herbaceous, perennial with annual vine growth. Length upto 12m, with tuberous root. Stems longitudinally furrowed, tendrils large, branched. Leaves broadly ovate, angled; petiole 3-15 cm long. Flowers small, axillary; male flowers in small peduncled clusters. Female flowers solitary usually in same axil as male. The colour of both male and female flowers is light yellow and cream colour. The fruit is fleshy, pyriform, usually longitudinal furrows. The colour of the fruit is dark green to whitish green. The surface of fruit is prickly wrinkled with 5 to 7 ridges, 10-12 cm long flesh whitish, single seeded, weighing 100-350 g. Each fruit has deep fissure at the blossom-end, which is the main and only site for the seed. The seed is large, flat, white, 3-5 cm long present at fissure junction. When fruit reaches to maturity, in most of the cases seed emerges out and start germinating due to very short dormancy period, if fruits are not harvested. The sprouted fruits are immediately plucked and planted.

The edible portion, which constitutes about 80 per cent of the fruit contains : water 89.8 per cent; protein 0.9 per cent; fat 0.2 per cent; carbohydrate 7.7 per cent; fibre 0.4 per cent, ash 1.0 per cent; vitamin A 50 (IU); vitamin $B_1$ 0.3 mg; Vitamin $B_2$ 0.04 mg; niacin 0.5 mg; vitamin C 11.0 mg; calories 26 (per 100 g of fresh edible portion). The roots contain 79.0 per cent water and 17.8 per cent carbohydrate.

The chow-chow grows best at a altitudes above 1,000 ft in the tropis in areas of moderate rainfall. The whole mature fruit is planted horizontally and thinly covered with soil. They are usually placed 4-6 ft apart along a fense or trellis, but if allowed to sprawl will need much more space. Harvesting of the fruits begins 3-5 months after planting and continues for many months. They are picked when ½ to 1 lb in weight and can be stored for several weeks. Fruits can easily be shipped and stored for 2 to 4 months at moderate temperature.

### Varieties and Hybrids

Of course, named varieties of chow-chow are not available, but on the basis of variation in fruit characters, the following distinguished types have been reported. They are : Round white, long white, pointed green, broad green and oval green.

## CUCUMBER

Cucumber (*Cucumis sativus* L.); 2n=14; syn., Gherkins, Khira, Dosakaya, Sukasa, Kakri, Vellarikkais Kakrikai; is a warm season fruit vegetable used as salad and for pickling purpose. It belongs to the family cucurbitaceae, genus. *Cucumis, C. sativus* L. The key characters of the genus are fruits fleshy, many seeded pepo, monoecious in nature. Flowers are solitary and yellow in colour. All other sex forms - andromonoecious, gyno-monoecious, hermaphrodite, androecious, gynoecious are seen in cucumber. Being monoecious, it is highly cross-pollinated. The genus cucumis has over 40 species of which 8 are Indian. These species are either in wild or in cultivated forms. The important among are *C. callosus, C. hardwicki, C. melo, C. prophaterum* and *C. dispaceus*.

Boss (1959) classified *C. sativus* into sub-species based on origin and fruit shape.

*C. sativus* spp. *rigidus* : The english long fruited cucumber of East Asiatic type.

*C. sativus* spp. *agrestis* : The Soviet-far-East type.

*C. sativus* spp. *gracillis.*

*C. sativus* spp. *sphaerocarpus* (syn. *C. sphaerocarpus*).

Whitaker and Davis (1962) classified cucumber on the basis of edible quality into : slicing types i.e. used for salad purpose; and pickling types which is used after cooking.

Purseglove (1974) classified cucumber on the basis of season of cultivation into : hot weather cultivars; and rainy-season cultivars.

## Varieties and Hybrids

There are two form of cucumbers. Hot weather form known as Gherkins, comprise creeping plant. The rainy seasons have much larger fruits commonly grown in throughout the country. Cultivars differ in colour of spines, shape, size, colour and shape of fruits at tender stage and mature stage. Cultivars usually classified on the basis of their use, fresh market (slicing) and pickling. In general, fruits of slicing cucumber are larger than pickling and they develop a darker and heavier skin with uniformly cylindrical shape.

Present day pickling cultivars tend to develop shorter vines than fresh market ones and they are more prolific. Balam khira of Saharanpur region is akin to pickling type of Wests. The fruits of pickling cultivars are smaller and the seed cavity develops more slowly, an important characteristic affecting suitability for storage in brine. In pickling cultivars, a thin skin with a lighter colour is preferred.

Cucumber cultivars have also been classified according to the spine colour. Generally, slicing cucumbr have white spine colour and pickling cucumber have black spine colour.

Varieties of Cucumber have been also classified into four groups on continental basis :

(1) *European-American* : Some of the English cultivated varieties belong to this group.

(2) *West Asiatic* : A very Xerophytic species.

(3) *Chinese* : Very large, long fruits with semi glossy rind.

(4) *Indo-Japan* : Himalayan and hermaphrodite type.

***Some improved varieties :*** Japanese long green, Straight Eight, China, Poinsette, Balam Khira, Khira poona, Mundosa, Long spine green type, Spineless long green, Sikkim cucumber, Sheetal.

***Varieties from USA :*** Marketer, Straight eight, Colorado, Palmetto, Niagara (These are slicing varieties).

***Gherkins :*** It is known as pickling cucumber.Varieties are Ajax and calypsopus.

Besides improved cultivars, hybrids have been developed by different Institution and seed companies.

***I.A.R.I :*** Pusa Sanyog, DCH-1

***Indo American Hybrid Seeds :*** Priya, Hybrid No 1, 9531

***Mahyco :*** M.C.U.H 1/2/3/101/401, MCTH-101.

***Century Seeds :*** Hybrid No.1, Laila (Longmelon).

***Nath Sluis :*** Caulipse - (F1)

***Clause international :*** Rigal, Royal

***Indo-American Export :*** Click, Lucky, Strike, P.S - 64487.

***Ankur Seeds :*** Mohini, ARLMH-1 (Longmelon).

***Unicorn Seeds (Choice) :*** Greenstar, Sneha, Shruti, Champion, Lenyadri, Zeenat, Calypso.

***Open pollinated selection :*** Himangi, Poinsett, Mysore Green Long, Barsati.

***Norvatis Seeds :*** Octopus (Pickling Type).

***Sreemaa Seeds :*** Supriya, Rainy Special, Nirmalya. (all are O.P. varieties)

***Namdhari Seeds :*** NS 601, Himangi (OP).

***Nunhems Seeds :*** Ajax, Stimora, Tripti (19), Rani (50).

***Tropica Seeds (Green CO.) :*** L-04, L-05, L-06, L-07, Mekong (L-08), L-09, Panak, Storm, Eclipse, L-999.

***Annapurna Beej :*** Barapata, Rajmata, Chaitali, Seven Star (imp), Baishakhi (imp), Barsa Mangal, Rainy Season Frame, Green Sona.

***Golden seeds :*** Rajdhani, Uphar.

**Detail characteristics of varieties and Hybrids are given below:**

***Japanese Long Green :*** This is a temperate variety specifically suited to hills and lower hills. It is extra- early with 45 days maturity. Fruits are 30-40 cm long, with green skin, white spines and light green crisp flesh. It is an introduction, released by IARI, Regional Station, Katrain.

***Straight Eight :*** This is a variety of cucumber having cylindrical very symmetrical fruit introduced by Ferry-Morse Seed Co., Detroit, Michigan, U.S.A, in 1935. It was introduced in India and released by IARI, Regional Station, Katrain, Kullu Valley. It is early, suited for hills and dark green leaves, 20-25 cm long thick, straight fruits with round ends. The fruits are light green and heavy yielder.

***China :*** Medium late, straight variety, very hardy and prolific. Fruit very long (50cm), slender, with deep green skin, white spined, flesh is white, firm and crisp.

***Poinsett :*** This is an American varierty initially introduced and multiplied by National Seeds Corporation of India. Fruits are 20-30 cm long, dull dark green in colour. It is resistant to downy mildew, powdery mildew, anthracnose and angular leaf spot. It has been originally bred at Charleston, South Carolina, U.S.A.

***Sheetal :*** It is developed from local collection of germplasm at KKV (Konkan Krishi Vidya Peeth) Dapoli, M.S. Sheetal is an early-maturing and high-yielding cultivar. Its fruits are green of medium length, each fruit on an average weigh 300-350 g. The vine measure up to 2.7 m with as many as 10 branches to a vine. It is suitable for cultivation on the hill slopes receiving high rain fall during Kharif. The fruits are ready for harvest 50-55 days after sowing, which may continue up to 70-85 days after sowing, regularly every 3-4 days. A well managed crop yield, 30-40 tonnes/ha in Kharif and 20-23 tonnes/ha in summer.

***Himangi :*** It is developed by selection from the cross Poinsett x Kalyanpur Ageti. Fruits are white in colour and resistant to bronzing. Duration 100-110 days. Average yield 19 tonnes/ha. Developed byMPKV, Rahuri (Maharashtra).

***Phule Shubhangi :*** Developed by selection.The fruits are having green skin colour.Tolerant to powdery mildew.Fruits are green in colour and straight. Higher yielder than Poona Khira and Himangi. Released by MPKV, Rahuri (Maharashtra).

***Phule Priyanka :*** High yielding and released by MPKV, Rahuri, Maharashtra.

***K 75 :*** High yielding under temperate condition. Released by Dr. YSPUOHF, Solan, Himachal Pradesh.

***K 90 :*** High yielding under temperate condition. Released by Dr. YSPUOHF, Solan, Himachal Pradesh.

***Solan Hybrid :*** This is $F_1$ hybrid. Released by Dr. YSPUOHF, Solan, Himachal Pradesh, suited to the temperate region. It is high yielding.

***AAUC 2 :*** Fruits long (26 cm), cylindrical, medium size (450 g) and whitish green. Days to first harvest 60-65 days. Yield 45-50 t/ha. Released by Assam Agril. University, Jorhat.

***Swarna Poorna :*** Fruits long, cylindrical, medium size (300 g) and light green without placental hollowness. Tolerant to powdery mildew. Days to first harvest 55-60 days. Yield 30-35 t/ha in 120 days. Released by HARP, Ranchi.

***Swarna Sheetal :*** Fruits long, cylindrical, medium size (250 g) and whitish green without placental hollowness. Days to first harvest 60-65 days. Yield 25-30 t/ha. Released by HARP, Ranchi.

***Swarna Agethi :*** Developed through mutation breeding at HARP, Ranchi, Fruits long, cylindrical, medium size (200 g) and green without placental hollowness. Days to first harvest 45-50 days. Yield 30-32 t/ha in 120 days.

***CO1*** **:** A selection from Kanyakumari district (TN). Fruits are long with a characteristic narrow neck at early stage, light green colour at salad stage (7-8 days), with dark green stripes at vegetable stage (16 days) and bright yellow with greenish yellow strips at ripening stage. Yield is 14 t/ha and released by T.N.A.U., Coimbatore.

***Hybrid No. 1*** **:** A highly productive slicing cucumber with predominantly female flowering. The rind colour is green and certainly lighter than poinsette. Fruit diameter is 4.0-4.5 cm, with average length of 17-20 cm. The fruits have extremely small seed cavity, fewer small seeds.

***Balam Khira*** **:** It is released from Saharanpur (UP). Early to medium, very popular variety in that region. Fruits are small, tender and oval in shape, high yielding.

***Poona Khira*** **:** Early to medium, very popular variety in Saharanpur region. Fruits are small, pale green, 12-14 cm long.

***Priya (VK1)*** **:** Early, fruits 25 cm long, dark green, bitterness free, resistant to downy and powdery mildew, angular leaf spot, mosaic, anthracnose and scab.

***Rajdhani*** **:** Extra early maturing, prolific bearing hybrid, with predominantly female flowers. First harvest start 35 to 40 days after sowing. Fruits cylindrical, smooth 14 to 16 x 4 cm, white with crispy flesh. Fruits are formed at every axil after the 5$^{th}$ axil. Yield 12 to 15 mt/ha under optimal conditions. Concentrated fruit set and prolific yielder. Excellent for fresh market.

***Uphar*** **:** Early maturing vigorous hybrid with more number of female flowers. First harvest start 45-50 days after sowing. Fruits cylindrical, straight, light green with light yellow lining at blossom end 22 to 25 x 6-7 cm size. Flesh white and crispy and free from bitterness. Yield 15-20 mt/ha under optimal conditions. It has wide adaptability.

***NS 601 (F1)*** **:** Vigorous plant growth with medium sized leaves. Fruits measure around 25 cm long with white spines. They are attractive cream, white in colour with light yellow lining at blossom end. It is a prolific yielding hybrid.

***Himangi (OP)*** **:** It is developed by Namdhari seeds. The medium vigorous plants bear fruits of length 18-20 cm with white spines, fruits are uniformly cream in colour. It has crisp texture with very good keeping quality. It is a very good consistent yielder.

***Mohini*** **:** Plants has very strong growth and foliage density. This variety produce quality fruits of fleshy compact with green colour and weight 150 g approximately. Flowering starts within 47-52 days. Long duration reduces deformities in fruits under good management practices.

***Supriya*** **:** Fruits are green, 9-10 cm long. Average weight 100 g, high yielder.This O.P. variety grown in winter season (November-December).

***Nirmalya*** **:** It is a summer sowing, early maturing variety. First fruit harvested within 35-45 days from the date of sowing. Fruits are green, weighing 100 g each and heavy yielder.

***Rainy special*** **:** This variety is grown in rainy season. Fruits are bigger, light green and mature in 60 days from the day of sowing.

***L-04 (F1)*** **:** Plants are vigorous with many branches, medium slicing type. Fruits 4 x 18-20 cm size, medium green and white powder on skin make the fruits look nice. Fruits are cylindrical, thick flesh, few seed have long storage life, average weight 200-250 g per fruit. Very early harvest, about 35-38 days after sowing and give continuous harvest for very long period.

***L-05 (F1)*** **:** This hybrid produce mostly female flowers, set the laterals as well as main stem. It produce uniform, medium green fruits of size 4 x 16-18 cm, cylindrical with fruit weight 170-190 gram per fruit, matures in 35-38 days after sowing and slicing type. It is resistant to anthracnose, CMV, DMR,PM Oidium and has excellent keeping ability.

***L-06 (F1)*** **:** Plants gynoecious, vigorous, with many branches, can grow all the year round. Fruits uniform dark green, cylindrical, size 6 x 20-22 cm. Thick flesh with few seeds, average fruit weight 225 g. Fruits mature in 60 days after sowing and give continuous harvest. Good fruit setting and very long harvest period.

***L-07 (F1)*** **:** Plants gynoecious, vigorous with many branches. Fruits, medium green, cylindrical, size 6 x 20 cm, mature in 55-60 days after sowing. Fruits yield 35-45 t/ha with continuous harvesting for longer period.

***Mekong (L-08; F1)* :** Fruits medium, green with light green end, slender cylindrical, slicing type, size 3.6 x 14 cm with average weight 105 g. Fruits mature in 30 days after sowing. Good for fresh market, highly resistant to disease and heat.

***L-09 (F1)* :** Plants very strong, develop more branches and produce good plant canopy. Each plant gives many fruits (15 fruits/plant). Fruits light green to white, long and cylindrical, average weight 295 g, good quality, thick flesh with little seeds. First harvest starts 35-38 days after sowing. Harvest time is long and continuous. Long storage life and good for remote transportation. Tolerant to CMV, Scab, DM, PM, anthracnose, angular leaf spot and resistant to rain and high temperature.

***L-999 (F1)* :** Plants grow and develop very strongly, branch development is very good and each plant gives many fruits (15 fruits/plant). Fruits light green to white, long cylindrical, average weight 280 g, good quality with little seeds. Good for remote transportation. First harvest start 32-36 days after sowing, with longer and continuous harvesting. Tolerant to CMV, Scab, DM,PM anthracnose, angular leaf spot and resistant to rain and high temperature.

***Eclipse (F1)* :** It is pickling type, medium green fruits with light green short and cylindrical size 6 x 15-16cm. First harvest start 55 days after sowing. Resistant to anthracnose, CMV, DM, PM.

***Panak (F1)* :** It is slicing type, fruit dark green, cylindrical, size 4 x 19-25. First havest starts 60-65 days after sowing. Average fruit weight 175 g, resistant to anthracnose, CMV, DM, PM, medium.

***Storm (F1)* :** This hybrid is very early, widely adapted, gynoecious, especially suitable for tropical and subtropical areas. Vigorous dark green plant that sets fruit abundantly. The plant begins production approximately 50 days after emergence. Fruits dark green, uniform, smooth firm, cylindrical with slight tapering at blossom end. Fruits size 22 x 6 cm, excellent holding and shipping ability. The flesh is creamy, white and flavourful. Tolerant to CMV, Scab, DM, PM, anthracnose and angular leaf spot.

***MCTH-101* :** It is a high yielder with long (40-45cm), Light green fruits which are dull in colour, straight and easy in harvesting.

***MCUH-401* :** Fruits are white at tender stage and change to brown at both end when fully ripen. Fruits are of small size with dark spine and 15 to 20 cm long. It is high yielder.

***Pusa Sanyog*** : This is an F1 hybrid between a Japanese gynoecious line and Green Long Naples. It matures in 50 days. Fruits are 28-30 cm long. Cylindrical and have dark green skin with yellow spines. It is suitable for temperate region and hills of subtropical region. It has been released by IARI Regional station Katrain.

***DCH-1* and *DCH-2*** : These are newly developed tropical gynoecious hybrids named as DCH-1 and DCH-2 at IARI have shown great promise for cultivation in plastic green house recording 220 and 192.6 tonnes fruit yield/ha respectively (6-9 fold increase in productivity compared to the open field conditions).

***Asterix*** : It matures in 60 days from sowing. It gives blocky and cylindrical fruits. It is tolerant to downey mildew and is very high yielding. It can be sown in all three seasons. The optimum size for harvesting is at 7 gm/ fruit. It is used for pickling industry for exports.

***Ajax*** : This pickling cucumber is predominantly female flowering hybrid, has vigorous plant growth and gives stable total yield of straight fruits. Medium to dark green coloured fruits borne in cluster have length/diameter ratio of 3 : 1. Its earlyness, total high yield, uniform cylendrical fruit size, excellent internal quality, good toerance to hot and humid weather conditions and excellent processing quality makes this hybrid the best choice for processing industry.

***Stimora*** : This pickling cucumber hybrid has vigorous plant growth and gives stable and high production till the end of cropping period. The fruits are uniform, cylindrical in shape. Smooth and dark green in colour. This hybrid has excellent processing quality. Resistant to powdery mildew, scab and tolerant to CMV.

***Tripti (19)*** : This newly developed hybrid has vigorous plant growth and uniform fruit size. Cylenderical fruits are mottled green in colour, crispy, solid flesh with small seed cavity, 22-25cm in length and 6-8cm in diameter. Bitter free and has a long shelf life.

***Rani (50)*** : A high yielding and vigorously growing hybrid produces large uniform bitter free fruits (30 cm) with spines on fruit surface. Fruits are crispy, green in colour and has excellent shelf life.

***Summer green*** : This hybrid is early and prolific bearer with good diseases tolerance. Fruits are shinning dark green in colour and spindle in shape. Excellent crispy and tender fruits makes it ideal for both fresh vegetables and frying.

***Chaman (98013) :*** This hybrid is early in maturally (60days), vigorous in plant growth and prolific bearer. Fruits are attractive green in colour with spines.large fruit size (15x16cm) and 75-80 grams in weight.

***Green gold :*** Vigorous plants with short internode and highly productive. Fruits are long (50cm), green in colour, uniform in size with attractive strips. fruits can be harvested in about 50days from planting.

***Raveena :*** Plants vigorous in growth,highly productive. Attractive long fruits (50-55cm), light green in colour and firm. Suitable for long transportation.

***Namrata (95003) :*** Vigorous vine growth.Short internode and highly productive. Fruits are (50-55cm) long, light green in colour, 8cm in diameter with medium fermness and bitter free.

***Lata (95006) :*** Newly introduced hybrid is high yielding, vigorous in growth. Fruits are uniform bright green, firm with attractive stripes.First harvest possible in 45-48 days after planting, Suitable for long transportation.

***Sun 230 :*** A full season hybrid which produces large vigorous determinate vines combines with high yield of heart shaped bright red firm fruits, resistant to verticillium F1 & F2 and tolerance to early blight during wet season.

***Sun 145 :*** An early maturing hybrid with medium sized determinate vines produces round, firm and attractive red colour fruit with uniform ripening. It has verticillium. F1 & F2 resistance.

***Sun 176 :*** A full season hybrid which produces medium large vigorous vines, fruits are firm, heart shaped, smooth and bright and resistant to verticillium. F1 & F2. An ideal hybrid for growing during wet season, if TLCV is not a problem.

***Green star :*** An early maturing poinsett type hybrid. Predominantly female plant produces dark green long cylindrical fruits of length 20–22 cm and 18–19 cm girth. Crisp fleshed fruits have good tolerance to CMV, PMR and DMR.

***Sneha :*** A very high yielding early maturity gynoecious hybrid. Vigorous vines produce medium green blocky fruits with LD ratio of 3.0–3.1.

***Shruti :*** A predominantly female hybrid. Produces medium green cylindrical fruits with LD ratio of 3.2 –3.4, very vigorous vines are highly productive.

***Champion :*** Highly productive slicing cucumber variety. Predominantly female flowering plants produce poinsett type dark green fruits of length of 20 cm.

***Lenyadri :*** New hybrid variety plants with vigorous vines. Predominantly female flowering variety produces medium long fruits of light green colour.

***Zeenat :*** A very excellent gynoecious Bait alpha cucumber consumed extensively in the Middle East. Firm, fruits are round, cylindrical, slightly tapered glossy green. Fruits of size 16 X 5 cm mature in 55 days. Tolerant to CMV and MMV 1. Suitable for open field cultivation.

***Calypso :*** Most popular choice for Gherkin pickles. Very productive and widely adaptable hybrid produces dark green colour, well shaped pickling cucumber. Calypso is gynoecious and has exteremly vigorous and productive vine growth.

***Octopus :*** Medium vigorous,sturdy vine with dark green leaves. Fruits are fairly uniform,short,cylindrical,medium green,white spined and firm. 1-2 fruits borne per axil. Prolific bearing.

## KAKROL (SWEET GOURD)

Kakrol *(Momordica cochinchinensis* Spreng); 2n = 14; belongs to the family Cucurbitaceae, also known as sweet gourd is the indigenous, dioecious, perennial vine crop of high nutritional, medicinal and economic value. It is grown for their immature tender green, nutritious fruits which are cooked as a vegetable. The roots are tuberous and used for propagation as a planting materials. The vine remain dormant in winter. The flowers are white or pale yellow. The fruits are green when unripe turning to an orange yellow colour when ripe. The fruit is ovate, 10-20 cm long, pointed, densely covered with conical spines. Seeds are numerous in fruit embeded in orange red pulp. Seed ovoid, dark grey brown to black in colour and sculptured on both sides.

Fruit contains moisture (84.09 %), protein (2.61 %), fat (0.66 %), carbohydrate (5.69 %), crude fibre (5.93 %) and mineral matter (1.20%), calcium (21 mg), phosphorus (148 mg),iron (2.5 mg) per 100 g edible portion. Kakrol has medicinal uses, the seeds are used as aperients and in the treatment of ulcers, sores and obstructions of liver and spleen. Kakrol fruits and leaves are used for external application in lumbago, ulceration and fracture of bones. Its seed contains an oil which is used as an illuminant in Indo-China. It maybe used as admixture with drying oils in the formulations of paints and varnishes. The root frothes in water and may be used as a soap for washing.

## Varieties and Hybrids

There is no any improved varieties and hybrids of this under-utilized vegetable crop. In all parts of the country only local varieties are grown. Recently CHES, Bhubaneswar has released one variety which is given below :

***Neelachal Arakta (CHMC-2) :*** Rich in carotenoides and lycopene. Its fruits are large, globular weighing 500-900g, 10-15cm long, and 8-9 cm across. Fruits fleshy and reddish.

## MUSK MELON

Muskmelon (*Cucumis melo* L.); 2n=24; syn; Kharbuza, Kharbooj, Khurmuj, Kharbuja, Sweet melon, Kalinga, Chira, Kasturi tarabuja, Velapandu, Velapalam, Thaikumbalom, Kharbuj, Bachang, Melonegurke, Chiral, Sakkartoti, Kekkarike; belongs to the family cucurbitaceae; is an important dessert fruit, valued for its cooling effect during summer. It is an annual climbing or creeping herb having large hairy leaves. Many sex forms like monoecious, gynomonoecious, gynoandromonoecious and perfect are present in muskmelon and it leads to high amount of cross-pollination. Its fruits are generally spherical, oval or elliptical, of varying size and colour.

There are different forms and growing regions of muskmelon :

(i) *Cucumis melo* var. *callosus* - bitter.

(ii) *C. melo* var. *acidulus* - sour.

(iii) *C. melo* var. *cantalupensis* - bland.

(iv) *C. melo* var. *reticulatus* - sweet.

*Cucumis melo* ssp. *rigidus :* Melons growing in hot dry regions (Central Asia and Iran)

*C. melo* ssp. *orientale :* Melons growing in hot and humid regions (Asia minor).

C. *melo* ssp. *europees* : Melons growing in cool, dry regions (Asia minor).

C. *melo* ssp. *flexuosus* : Semi feral melons - snake melons.

C. *melo* ssp. *chinensis* : Chinese melons (East Asia).

C. *melo* ssp. *spontaneum* : Wild melons.

Good muskmelon should be thick skin, thick flesh with good consistency and flavour, attractive outer colour, small or negligible hollowness of fruit and resistance to diseases like powdery mildew and downy mildew.

## Varieties and Hybrids

Large number of cultivars differing in colour, shape, size and sweetness of fruits and resistance to disease and pests are grown in different parts of the country. Some important botanical varieties are given below :

(i) Var. *reticulates* (netted melons) - fruits small with ribbed and netted surface.

(ii) Var. *Cantalupensis* (Cantaloupe melons)- fruits warty, scaly, and rough, with hard rinds, surface often warted. It differ from the other having twice the number of chromosomes (2n = 48).

(iii) Var. *inodorous*, winter melon, casaba melon-fruit with little of musky odour, ripening late and keeping into the winter; surface usually smooth.

(iv) Var. *flexuosus* - Snake or serpent melon-fruit long and slender 2 to 8 cm diameter and 45 to 90 cm long, curved and crooked. Use to some extent for preserves.

(v) Var. *Dudain* - Fruit small, about the size of an ordinary orange surface marbled with rich brown, very fragrant, grown mostly for ornamental and strong scent.

(vi) Var. *Chito* - Mango melon or lemon cucumber; fruit small, the size of a melon, used in making preserves called mango preserves. The fruits are known as orange melon, melon apple and vegetable orange.

There are large number of traditional varieties of musk melon with different size, shape, and colours grown in different parts of the country are given below :

***Uttar Pradesh* :** Lucknow safeda, Bati, Kanpuria, Jogia, Mathuria, Batti, Kajra, Allahabad Kajra, Jaunpuri, Netted (Jametha), Mau, Bhagpat, Mahaban.

***Guajrat* :** Sankheda, Galteswar, Balteshwar (Panam) Kharri, Jalgaon Type, Kabri, Gurbeli, Kavita.

***Rajasthan* :** Sanganer, Hara Gola, Motia, Tonk, Harela.

***Andhra Pradesh* :** Sharbat-e-Anar, Bathesa, Papay (papaya), Chiranji, Ladoo.

***Maharashtra* :** Batti, Goose, Jam, Neel.

***Karnataka* :** Kadapa.

***Harayana* :** Kutana, Bhagpat.

***Punjab* :** Haridhari, Musa.

***Non* - Dessert types**

***Andhra Pradesh* :** Nakka Dosa Kaya, budam Kaya.

***Tamil Nadu* :** Vellarikkai.

***Kerala* :** Chavathakkai, Mudikode.

***Maharashtra* :** Chibur.

***Uttar Pradesh, Bihar, Madhya Pradesh* :** Kakri, Tar, Phoot.

***Rajasthan* :** Kachri.

A systematic research on the improvement of muskmelon has resulted in the development of several improved varieties in the country.

Some improved varieties are, Arka Jeet, Arka Rajhans, Pusa Sharbati, Pusa Madhuras, Durgapura Madhu, Punjab Sunehri, Punjab Rasila, Sarda Melon, Selection 1, Sharbat-e-Anar, Hingan, NDM1,NDM2, MH5 and MH13 (Hisar Madhur), RM43 (Rajasthan), Gujarat Muskmelon 1, Gujarat Muskmelon 2.

**Varieties resistant to Downy mildew-** MR-12 (Punjab Rasila).

**Varieties resistant to cucumber green mottle mosaic virus-** DVRM-1, DVRM-2.

**Varieties resistant to fusarium wilt -** Golden Gopher, Iriquois, Minnesota, Midget, Harvest Queen, Delicious 51.

**Variety Resistant to alternaria** - Pordue 44.

**Varieties tolerant to High soil pH** - Jobner 96-2 and60-3.

## Hybrids

*P.A.U.* : Punjab hybrid, MH-10.

*IIHR* : Arka jyoti.

*IARI* : Pusa Rasraj.

*ARS* : MHY3.

*MAHYCO* : MHC-2, MHC-5,MHC-9,MHC-10.

*Century seeds* : Punjab hybrid, Madhumati, Madhubala, Madhurmilan.

*Bejo Sheetal Seeds* : BSS 361, BSS 362.

*Unicorn Seeds (Choice)* : Gayatri, Deepti, MMT –033, Sabita, Royal Gold.

*Namdhari Seeds* : Abhijeet, NS 7455, NS 9, NS 89.

*Nunhems Seeds* : 6281, Castelle, Drake.

*Tropica Seeds (Green CO.)* : Alien, Indo, Supra, Rock Star, Yellow Rocket.

*Golden seeds* : Madhu Raj

*Sungro Seeds* : Punjab Sunheri (OP), Madhuras (OP).

**Detail characteristics of varieties and Hybrids are given below:**

*Arka Jeet* : A selection from the Batti strain of Uttar Pradesh. An improvement over a local collection (IIHR-103) from Lucknow. Released by the IIHR, Hessaraghatta. Fruits small, round with flat ends, each weighing 300-500g. Skin smooth and orange- coloured. Flesh white with medium soft texture, excellent flavour, very sweet (15-17 percent TSS), juicy and rich in vitamin C. Early maturing. Yields 15 tonnes/ha in 90 days.

*Arka Rajhans* : Selection from a local collection (IIHR-107) from Rajasthan. Released by IIHR, Hessaraghatta. Recommended for cultivation in southern regions of India. Fruits round to slightly oval with netted creamy white rind. Fruit weight 1.25-2.0 kg. Flesh thick, firm white and sweet (11-14 per cent TSS). Good keeping and transport qualities. Resistant to powdery-mildew, early maturing and yields 28.5 tonnes/ha.

***Pusa Sharbati*** **:** Developed from a cross between Kutana and PMR 6 of the USA. Pusa Sharbati was released by the IARI, New Delhi. Suitable for riverbed cultivation and northern India. Fruits round with netted rind. Salmon-orange flesh firm and thick with small seed cavity, moderately sweet (11-12 per cent TSS). Keeping quality is good. Yields 15 tonnes/ha.

***Pusa Madhuras (No.445)*** **:** It is mid-season selection from a Rajasthan collection with roundish-flat fruits weighing a kilogram. Released by the IARI, New Delhi. The skin is pale green, sparsely netted with dark green sutures and salmon orange flesh, juicy and sweet (12-14 per cent TSS). Keeping quality is poor. Yield 15 tonnes/ha in 90 days.

***Durgapura Madhu*** **:** A selection from the local material of Rajasthan. Recommended by the Department of Agriculture, Jaipur, and Rajasthan. Very early cultivar confined to Jaipur region of Rajasthan. Fruits are oblong, weighing 500-600 g, pale green rind, light green flesh with dry texture, seed cavity big, very sweet (13-14) per cent TSS.

***Hara Madhu*** **:** This variety is an inbred selection from a local collection of Uttar Pradesh (Kutana type). Released, by the Punjab Agricultural University, Ludhiana. Its vines are long and vigorous. Leaves sparsely lobed; light green; fruits round, tapering towards the stalk end with 10 green sutures; weight 1 kg; does not slip from the stalk; rind is thin, smooth and pale yellow at maturity; flesh green, thick, very juicy, melting and sweet (13 per cent TSS); matures late; its post-harvest life is short, transportability poor. Yield 12.5 tonnes/ha.

***Punjab Sunehri*** **:** A selection from a cross of Hara Madhu X Edisto. Released by the Punjab Agricultural University, Ludhiana. Vines are medium- long. Leaves are lobed and dark green and petioles are short; fruits are round to elliptical, devoid of distinct sutures; fruit weight 800g, slips from the stalk; rind thick, netted and dull yellow at maturity; flesh thick, orange, fairly juicy and sweet (11 per cent TSS). Early maturing, has long post-harvest life and excellent for distant transport. Yield 16 tonnes/ha.

***Hisar Madhur* : (Mh 13)** : It has been developed for earliness by hybridization of Pusa Sharbati (5445) and 75-34. Pedigree selection was followed. The prevalence of downy mildew is low due to earliness. The plant is indeterminate with spreading habit; foliage green, andromonoecious, fruit, round with about 10 ribs, yellow brick-red skin with green strikes on ripening, orange flesh, small cavity, musky flavour, juicy, TSS, 8.5 percent, high yielding (Average 99.2 tonnes per hectare). It takes 75-80 days for first picking. The muskmelon can be picked up in 78 days.

*RM-43* : It has been selected from a heterozygous local collection of Rajasthan. Individual plant selection has been followed to obtain the pure line. The vine of "RM 43 grows vigorously. It takes 49 days for flowering and 83 days for first picking. Average fruit weight is 550g. the fruit is oblong round with a hard green sutured rind and green flesh enclosing a medium sized seed cavity. The fruit yield is 14.51 tonnes/ha. The TSS content is 12.6 per cent. Mature fruits slip out easily from the vine, leaving a circular depression. This is known as the full slip stage. The fruits are also harvested on half- slip stage. However "RM 43 takes 105 days for final picking.

***Punjab Rasila* :** The muskmelon Punjab Rasila is a cross of WMR 29 X Hara Madhu. It provides round fruits with green thick, juicy and flavoursome flesh. It takes 73 days from transplanting to first picking, 10 days earlier Hara Madhu. It is long vines with fruits setting close to the base of the vine. The rind is light yellow and green sutured with medium netting. The flesh is green thick, juicy and flavoursome. Fruit develops full-slip stage at maturity. The TSS is 10 per cent and above. Average fruit weight is 600 g. The average yield is 60 tonnes/ha.

*MH-10* : This hybrid has been developed by cross of WI-998 x Punjab Sunehri at PAU, Ludhiana, plant has medium vine growth, early maturity and monoecious sex expression; fruit setting takes place close to the base of the vine; fruits are nearly round or oval, netted rind, non-sutured, thick, brown on maturity, average weight of 9.15 g; flesh is thick, orange, medium juicy, flavoursome with 9.10% of T.S.S.; fruit develops full-slip stage at maturity; good shelf- life and suitable for distant transportation; gives an average yield of 24 t/ha.

*MHY 5* : A cross of Durgapura Madhu x Hara Madhu, MHY 5 has well developed long vines (2.0-2.5 m), vigorous and moderately spreading. It takes 45 days for flowering, the fruits are ready for picking 95 days after sowing and it takes 110 days for final picking. The fruits are round, flat, tapering at stem-end, green yellow, smooth rind, light green, soft flesh with medium seed cavity and sweetness (13-16% TSS). Average fruit weight is 700-800 g with fruit yield of 150-200 q/ha. The variety is moderately resistant to root rot, powdery mildew, resistant to virus and free from downy mildew under field conditions.

***Punjab Hybrid*** : This is an F1 hybrid between a male sterile line (MS 1) and Hara Madhu. Released by the Punjab Agricultural, University, Ludhiana. An early-maturing hybrid with orange flesh and netted rind, stable sweet (12 percent TSS), flesh thick, orange, juicy, melting and flavoursome. Yield 16 tonnes/ha. Fruit develops "full slip" stage and rind turns light yellow at maturity; shalf life of 2 days; poor transport quality.

***Pusa Rasraj*** : (M 3 Hybrid)- A F1 hybrid between Monoecious 3 and Durgapura Madhu. Released by the IARI, New Delhi. Early-maturing; fruits are oval with greenish yellow, smooth and suture less rind, green-fleshed and oval in shape. Very sweet, TSS 11-12 per cent. Yield 17.3 tonnes/ha. To a vine 3-4 fruits are produced and each weighs about a kg. It takes 75-80 days for the first harvest from the date of sowing. The hybrid is suitable for garden lands as well as river bed cultivation. It is very productive yielding almost 250 q/ha.

***MHC -2*** : Luxuriant vine growth and high yielder. Fruits round and middle shaped. Weighs 0.8 to 1.0 kg. TSS 12-14 brix. Yellowish flesh and sweet. Early hybrid harvested after 75-80 days when the skin turns yellow. It is susceptible to wilt.

***MHC-5*** : Creepers grow luxuriantly. Fruits middle shaped round (oval). Weighs 0.8-1.0 kg. TSS 14 brix. Netted skin, harvested when turns yellow. Early bearer and high yielder, susceptible to wilt.

***MHC-9*** : Vines have luxuriant growth and high yielder; fruits are oval shaped and big, weighs 1.0 to 1.5 kgs. TSS 10-11 brix. Netted skin with developed lines. Flesh orange colour. It is medium duration hybrid. Fruits mature in 80-85 days, good flavour and aroma.

***MHC-10*** : It grows very well and very high yielder, fruits big and round. Weighs 1-1.5 kgs. TSS 12 brix. Netted skin but no developed lines on the fruits surface. Orange colour flesh. Medium duration.

***BSS 361*** : Fruit weight is one kg and shape is round with netted skin and ridges with green flesh, ready in 75-80 days after sowing.

***BSS 362 :*** Fruit weight 800 g to 1 kg. Shape is round with netted skin. Ready in 80 days after sowing. Flesh is green in colour. Very sweet in taste. Has thick skin & hence good for transport. Very high yielding.

***Drake :*** Medium vigorous plants with good leaf colour produces netted,ribbed round fruits weighing1.2-1.4 kg. Tolerant to powdery mildew and Aphids (VAT gene). Excellent shipping quality.

***Castella :*** Medium early maturing hybrid bears oval round fruits having green skin before maturity which turns grey green at full maturity. Semi-netted fruits have sutures with high sugar content and weighs approximately 1200-1400 gms. Deep orange flesh has good aroma, suitable for long distance transportation. This hybrid is resistant to F.O.M. races 0, 2 and partial resistant to powdery mildew.

***Madhurmilan :*** An early hybrid maturing in 75 days. The rind is deeply netted with slight sutures. Fruits large (av. 2.0 kg) with average polar diameter of 18 cm and equatorial diameter of 15 cm. It has thick, sweet and delicious orange flesh. It is strongly recommended for short distance shipment and early local markets.

***Madhubala :*** A round very early (70 days) hybrid (50 % harvest by 85 days) bearing medium sized (600-700 g), firm fruits turning to coppery-brown colour with deep green strips on maturity. Flesh thick, salmon coloured and highly flavoured and sweet (12-14 % TSS). This hybrid is quite cold tolerant and can be successfully sown for early market.

***Madhu Raj (F1) :*** Early maturing vigorous hybrid with prolific fruit setting habit, released by Golden seeds. Maturity 80-85 days after sowing. Fruit large sized, round, well netted without sutures. Flesh is thick crisp, deep salmon very sweet (TSS 14-15 %) with good aroma and flavour. Seed cavity is small and seeds are tightly held. Av. Fruit weight 1.4 to 1.6 kg. Yield 12 to 15 mt/ha under optimal conditions. Excellent keeping quality and ideal for long distance transportation. Tolerant to downy mildew disease.

***Abhijeet (F1) :*** This is a medium maturity hybrid. The fruits are round to oval with good netting without sutures. Deep salmon thick flesh, very sweet (TSS 13 %), very good texture, flavour and aroma. Fruits weigh 1.5-2.0 kg. Very good transport and keeping quality with tolerance to downy mildew and fusarium wilt in North India.

***NS 7455 (F1) :*** An early hybrid with oval fruits, one of the best selling, full slip fruits with fine netting without any sutures. Flesh colour is deep salmon, medium thick with small cavity, very sweet (TSS 13-15 %). Very good melting texture. Fruit weight is 1.5-1.8 kg. Mature in 80 days after transplanting. Very good transport and keeping quality.

***NS 9 (F1) :*** An early maturing hybrid has an attractive golden yellow rind colour upon maturity. Fruits are round to oval in shape with excellent netting and without sutures.Each fruit weighs 1.5 to 2.0 kg, has very good flesh colour and texture with small tight cavity. TSS is 13-15 %. The hybrid matures in 60-65 days after transplanting and is a prolific yielder.

***NS 89 (F1) :*** Early to medium maturing, eastern shipper type hybrid. Fruits attractive oval shape with sutures and netting. Excellent deep salmon flesh colour with good firm texture and 12-14 %. TSS. Each fruit weigh about 1.5-1.2 kgs and has a small cavity. Matures in 70 days after transplanting and is an excellent yielder.

***Alien (F1) :*** This hybrid variety with vigorous growth. Prolific, medium early maturity (90 days from sowing), it can be harvested within 45-50 days after flowering. Round fruits, on average 1.2 kg. Light yellowish to green rind, densely netted, green juicy flesh. Very strong flavour. Tolerant to powdery mildew and downy mildew.

***Indo (F1) :*** Vigorous hybrid variety, tolerant to powdery and downy mildew, prolific, medium early, can be harvested within 45-50 days after flowering. Round fruits with green skin and netting, weighing around 1.0 kg per fruit with green flesh. Very juicy and sweet.

***Rock Star (F1) :*** Excellent fruit setting and fine netting-deep globe fruits of 1.3-1.5 kg, maturity around 55-60 days after flowering. Resistant to fusarium and powdery mildew.

***Supra (F1) :*** Very vigorous and disease tolerant (Fusarium, DM, PM) hybrid variety, medium early maturity (90 days from sowing). Fruits sets easily and is very juicy, tender and has a strong aroma. Very good shipper, easy to grow.

***Yellow Rocket (F1) :*** Early maturing hybrid (80 days after sowing) and can be harvested 40 days after flowering. The rind is tough, light yellow at full maturity. Fruits are oblong and weigh around 1.5 to 3.0 kg. White to light orange flesh. Very high sugar content (14-16%), the flesh is tender and very sweet. Resistant to fusarium and DM. Good shipper.

*6281* : Late maturing hybrid bears fruits weighing 800-900 gms. Fruits have dense netting, very good taste and high sugar content. Suitable for long transportation. Tolerant to Powdery mildew.

*Gayatri* : Very promising hybrid plants produces round fruits weighing around 0.6-1.3 kg, fruits with yellow netted rind, have sweet and aromatic light green flesh.

*Deepti* : Very pleasing hybrid plant produces round fruits. Orange, sweet and aromatic flesh. Fruits have yellow netted rind and each fruit weighs around 0.75-1.5 kg.

*MMT-033* : Excellent hybrid Ananas type variety. Fruits of size 1-2 kg are oval and rind is netted with yellow orange colour. Flesh with yellow colour is sweet and aromatic.

*Savita* : An early hybrid variety with concentrated fruiting. Yellow netted round fruits are sweet and flesh with orange colour is very aromatic. Rind changes colour from green to yellow on maturity. Fruits weigh about 0.65-1.4 kg. Good tolerance to powdery mildew race 1 and Fusarium race 0.1.

*Royal Gold* : A new introduction of hybrid Muskmelon. Fruits are oval to oblong with rough yellow skin. Flesh is creamy white and very aromatic. Fruits can be harvested in 90 days.

## POINTED GOURD

Pointed gourd; *Trichosanthes dioica*; 2n = 24 (syn. Patol, Adav patola, Peyupadal, Parmal,Parwal, Kabi padavala,Kadapaddoola, Patola, Wild snake gourd (*Trichosanthes contourne, Schlangen fruchtigahaar* blume), commonly known as Parwal a perennial minor dioecious cucurbit vegetable grown for its tender fruits, which considered a delicacy in Punjab and fetches premium price. It also find a place in the list of exported vegetables. Mostly grown in Bihar Bengal, Orissa, M.P. and Eastern U.P. This is a major crop of river bed cultivation in Diara land.

### Varieties and Hybrids

Varieties are based on morphological variations (shapes, size and striation) of parwal fruit. The plants can be grouped into four categories

(a) Plant bearing 10-13 cm large,dark green fruits with stripes.

(b) Plants bearing 10-16 cm long, thick, dark-green fruits with very faint pale-green stripes.

(c) Plants with small (5-8 cm long) roundish, dark-green striped fruits.

(d) Plants with small fruits tapering towards the ends, green and striped.

A large number of local varieties are popular with the growers. They are "Dandali", "Kalyani", "Guli", "Bihar Sarif", "Sopari safed","Niria", and "Santokhiya".

From a large number of germplasms collected at Narendra Deva University of Agriculture and Technology, Faziabad, three varieties, "FP-1", (round green), "FP-3" (Spindle-shaped, green-striped) and "FP-4" (Spindle-shaped, light green)- have been developed for commercial cultivation. Besides these, Damodar, Kazli Bombay, Swarna Alaukik and Swarna Rekha also developed for cultivation.

**Detail characteristics of varieties and hybrids are given below:**

**FP 1 :** Fruits are round and dark green in colour.

**FP 3 :** Fruits are pointed and light green in colour.

**FP 4 :** Fruits are spindle shaped and light green in colour.

***Patar*** : It is commonly grown in eastern UP. Its fruits are small, roundish and dark green striped.

***Local un-named variety*** : Its fruits are about 10 cm long with pointed ends and swollen in the middle and are pale green when young, changing to deep orange colour when ripe.

***VRP 101*** : Fruits are medium in size and light green in colour. Average yield is about 250-300 q/ha.

***VRP 102*** : Fruits are long, having minimum seed and light green in colour.

***VRP 103*** : Fruits are round, dark green in colour with well developed stripes.

***VRP 104*** : Fruits are long (10-25 cm) with well developed stripes.

***HP-1*** : Fruit round, green with white stripes, fruit weight 20-30 gm.

***HP-3*** : Fruits 6-8 cm long, thick, green with white stripes, flesh slightly yellowish, fruit weight 25-30 gm.

***HP-4*** : Fruit 8-10 cm long, thick, glossy with light green in colour with white stripes. Fruit weight 20-25 gm.

***HP-5*** : Fruits 6-8 cm long, thick, glossy, with light green in colour with white stripes. Fruits weight 20-25 gm.

***Swarna Rekha (CHES-Sel-1)*** : This variety has been developed by the Central Horticultural Experiment Station., Ranchi. Fruits white greenish striped long (8-10 cm) cylindrical and solid with soft seeds. Fruiting season starts in March and completes by October. Suitable for rainy season, resistance to bacterial blight and fruit fly infestation. Shelf life 8-10 days, yield 250-275 q/ha.

***Swarna Alaukik (CHES-Sel-2)*** : This variety has also been developed by the CHES, Ranchi. Fruits light green and medium (5-8 cm) in size. Fruiting starts from March and completes by October. Less seeded fruits with long keeping quality. Tolerant to moisture stress and to bacterial blight and field resistant to fruit fly infestation. Yield 270-300 q/ha.

***Neelachal Kirti (CHPG-15)*** : High yield, solid core, soft seed, high market preference. CHPG-15 has been identified as most promising line of pointed gourd. It has been identified and collected from Balasore region of Orissa. The genotype has dark green fruit colour with 3-4 fragmented stripes of cream colour, which has high market preference. The fruits are long, elongated spindle shaped with solid core and soft seed. The fruit weight is 45-50 g at appropriate harvesting stage and one plant produces more than 28 kg fruits in one season of 10-11 months. With better management practices, 50-60 t/ha fruit yield (2500 plants/ha) can be obtained from CHPG-15 under upland conditions. CHPG-15 is a dioceous line which requires sufficient (10%) male plants for fruit set and proper fruit growth. The plants are vigorous with dark green foliage. Plants start flowering after 60 days of planting and fruits are generally harvested at weekly interval. Plant shows moderate susceptibility to red mite on foliage particularly during high humidit period. The variety has been found tolerant to powdery mildew and Fusarium wilt under field condition.

Other promising genotypes of pointed gourds are CHPG 11, CHPG 16, NP 702, Rajendra Parwal-7.

***Rajendra Parwal 1 :*** Fruits green with stripes, long and tapering at both ends. Average fruit weight 40 g. Yield 17.5t/ha. Developed at BAC, RAU, Sabour, Bihar.

***Rajendra Parwal 2 :*** Fruits drum shaped, whitish green with light stripes. Average fruit weight 30 g. Yield 16.0 t/ha. Developed at BAC, RAU, Sabour, Bihar.

***Konkan Harita*** : Fruits dark green, 30-45 cm long, tapering at both ends, 10-12 fruits/plant. Developed at KVK, Dapoli.

## Pumpkins and Squashes

Pumpkin and squashes (syn. Vegetable marrow, musky gourd) have been associated with man in the Americas for at least 10,000 years. The five cultivated species such as *Cucurbita moschata*. Duch ex poir. (Pumpkin), *C. pepo* L. (summer squash), *C. maxima* Duch (Winter squash), *C. mixta* Pang. (Winter squash) and *C. ficifolia* Boucha. (Fig leaf gourd / malabar gourd), were selected by American Indians long before 1492 and they were staple items of their diet. The fruits of many wild species of genus *cucurbita* are of same size and colour to attract the attention of primitive man. The fruits had hard and tough rinds with excessively bitter flesh. The seeds, however, are non-bitter, tasty and nutritious.

The genus *Cucurbita* consists of 13 species or species groups. There are only 5 domesticated species of *Cucurbita* mentioned above and they are separated by sterility barriers as well as a convincing number of morphological features of the trichomes, leaves, calyxes, corollas, stamens, pedicels and seeds. The key characters of genus *cucurbita* are fruits fleshy, many seeded pepo, flowers - solitary or fasciculated, lemon yellow to deep orange, leaves deeply or shallowly lobed not pinnatifid and corolla companulate, gamopetalous, lobed, half way down. The plant is monoecious, highly cross-pollinated and entomophilous.

The morphological characters of important cultivated species of *cucurbita* have been given in following tables :

| Cultivated species | Common name | Skin texture | Fruit flesh, colour and texture | Stage of edible fruit |
|---|---|---|---|---|
| *C. moschata* | Pumpkin (Sitaphal) | Hard rind, smooth surface | Fine grained, gelatinous substance sticking to seed, yellow flesh. | Ripe, strong flavour, long keeping quality |
| *C. pepo* | Summer squash (Chappan kadu) | Hard, angular | Coarse grained, deep yellow flesh | Immature, poor keeping quality |
| *C. mixta* | Winter squash | Hard | Coarse grained, deep yellow flesh | Immature |
| *C. maxima* | Winter squash (Vilyati kadu) | Hard | Fine grained, without gelatinous substance sticky to seed | Immature |

The other species having importance in the genus are :

1. *C. lundelliana* (Buffalo gourd) - resistant to powdery mildew
2. *C. foetidissima* - resistant to several viruses
3. *C. ecuadorensis* - resistant to several viruses

All cultivated species are secondary polyploides with 2n=40. *C. moschata* has a key role as a connecting bridge between other cultivated and wild species.

## PUMPKIN

Pumpkin (*Cucurbita moschata* Poir.); 2n = 40; syn. Sitaphal, Kaddu, Kadua, Kashiphal, Mithakaddu, Saphuriikumara, Pilam Kohalum, Lal Phopla, Gummadi Kayi, Poashani, Gumbala, Kannada, Duddini, Prala Kalu, Red Gourd, Lal Bhopla, Lal Kumhara, Meetha Kaddu; belongs to the family Cucurbitaceae; is an important and most extensively grown as vegetable in tropics. Low cost of production, long keeping quality and adapatability to a wide range of climatic conditions, favours popularization of the crop. Crop is mainly grown for its mature as well as immature fruits which are considered good source of carotene. Fruit is a pepo with hard rind and smooth surface. Flesh is yellow to orange in colour and is slightly sweet.

## Varieties and Hybrids

Pumpkin (*Cucurbita moschata* Poir) has large numbers of varieties in India. They vary in shape, size and colour of flesh. The common varieties are- large red, Large round, Yellow flesh, Red flesh, Summer King Kong, Katki King let, Banaras Local (Summer bush type; Rainy season), Small Sugar, King of Mammoth, Punjab Large, Thudiylur, Pollachi (madras), Co1, Co2, Baidyabati, Guamal, CM-14, Sel-124, Solan Badami, Large Cheese, Hubbard, Delicious, Buttercup. Pusa Vishwas, Pusa Vikas, Kentucky Field.

***IARI, New Delhi*** : Pusa Vishwas, Pusa Vikas, Pusa Hybrid 1.

***Kerala Agril. University, Trissur*** : Ambili, Suvarna, Saras, Sooraj.

***TNAU, Coimbatore*** : CO1, CO 2.

***IIHR, Bangalore*** : Arka Chandan.

***UHAF, Solan*** : Solan Badami.

***Tropica Seeds (Green CO.)*** : Clown Giant, Negra, Angel, Dream, Dakara, Martinica, Stone.

**Detail characteristics of varieties and hybrids are given below:**

***C0-1*** : A late variety, matures in 175 days. Fruits with broad proximal end and tapering to distal end, dark green at immature stage and turning light brown on maturity. Flesh is yellow, fruits are attractive, globular, medium large (7 to 8 kg); each vine produces 7 to 9 fruits; yield 288 q/ha.

***CO-2*** : Fruits small flat round, weighing 2 kg with orange-coloured flesh. Yield 226 q/ha.

***CM-14*** : It was developed from a local collection at KAU, Vellanikkara and identified in 1987 for Zone VI, where it gave 21% increased yield over control. It has spreading plant habit, fruit flat round in shape and weigh about 6 kg each, green in colour with shallow furrowed surface. Flesh thickness is about 4.30 cm. Average fruit yield/ plant is 15 kg.

***Pusa Vishwas (Sel 107) :*** It is a selection from local germplasm collection of Delhi area. The vine growth is vigorous; average vine length 4.5m. Leaves are lobed their surface non-hirsute, almost soft in feel, spotted white due to the slightly raised epidermal cells. Fruits are spherical, of medium build. The average fruit weight is 5.0 kg and flesh, a deep golden-yellow. In pusa vishwas at least 400 quintals of fruit may be harvested from one hectare.

***Pusa Vikas :*** Vine is semi-dwarf to dwarf (2.0-2.5 m). Leaves are lobed and surface of leaves is non-hirsute and almost soft in texture with light green or yellow spots due to raised epidermal cells. The fruit is small and round flat, weighing on an average 2.0 kg each. Fruit flesh is yellow, a precursor of vitamin A. Pusa Vikas is suitable for growing in spring-summer season (Feb-March to June). It takes an average 115 days from sowing to harvest its mature fruits. On an average yield is 30 tonnes/ ha, 27% higher than "Solan Badami'.

***Arka Chandan :*** It is a high yielding, short duration variety, producing on an average yield 300 q/ha in 120 days. The fruits are medium in size, weighing 2-3 kg, round with slightly pressed blossom-end and light brown rind. The flesh is thick, firm, bright orange and rich in carotene content (3331IU/ 100 g flesh). It has good cooking and keeping qualities.

***Pusa Hybrid 1 :*** It is first hybrid of Pumpkin available for commercial cultivation. Its vine growth is medium with an average length of 4.0 cm. The leaves are lobed and surface of leaves is non-hirsute and soft in texture with white spots due to raised epidermal cells. The fruits are rounded-flat, and medium in size. On an average a fruit weight 4.75 kg each. The flesh is deep golden-yellow in colour. It gives 30% higher yield than the hitherto popularly-grown pumpkin 'Pusa Vishwas'. Its fruits may be picked up 100 days after sowing, the average yield being 520 q/ha.

***Baidyabati :*** It is cultivated in winter season. It has long, round, big fruits, average weight 12.0 kg each. Fruits turn to yellowish gray colour when matured.

***Saras :*** Fruits elongated, orange fleshed and medium sized. Av. Fruit weight 2.7 kg; yield 39.0 t/ha.

***Sooraj :*** Fruits round, orange fleshed and medium sized, average fruit weight 3.32 kg, yield 35.0 t/ha.

***Solan Badami :*** Fruits small, yield 22.5 t/ha, developed by Univ. of Hort. and Forestry, Solan.

***S-101 :*** Smooth to shallow ribs, brown skin with cream patches and green lacerations. Oval to round with pressed poles fruits. Flesh colour with thickness 2.46 cm and fruit size 20.8 x 17.8 cm.

***S-07 :*** Smooth to shallow ribs, light brown rind with cream patches and green yellow pencil lining. Fruit shape oval to round with pressed poles. Flesh colour yellow with thickness 2.31 cm. Fruit size 22.68 x 16.82 cm.

***Guamala Improved :*** Suitable for summer and rainy season. Round green even after mature. Average fruit weight is 6.0 kg.

***Vishal :*** It is developed by Ankur Seeds. Vines are vigorous in growth. Fruits are dark green coloured, oval, oblong with longitudinal sutures with ribs. 50 % female flowering in 51-55 days. It has good yield potential.

***Vaibhav :*** It is developed by Bejo Sheetal Seeds. The hybrid mature in 100-110 days. Fruits are round to flattish round with yellowish green at maturity. The flesh is orange yellowish. Fruit weight vary from 6-7 kg. Its standing ability is 80-90 days after maturity.

***MPH-1 :*** This hybrid is developed by Mahyco, suitable for all pumpkin growing areas. Fruits oblong to long, turns orange yellowish with regular mosaic at maturity. Fruits mature in 100-110 days with average weight 7-8 kg each and can be stored upto 3 months. Yield potential is 80-85 t/ha.

***MHPK-2 :*** This hybrid is released by Mahyco and produce first marketable fruits in 85-90 days. Fruits, round and flat, turns orange yellow with regular mosaic at maturity. Average fruit weight is 3.0-3.5 kg each. Mature fruits can be stored upto 3 months. Yield potential is 70-75 t/ha. Suitable for all pumpkin growing areas of the country.

***MHPK-4 :*** This hybrid is evolved by Mahyco and produce first marketable fruits in 105-110 days. Fruits are oblong to long with yellowish green skin at maturity, orange yellow flesh. Average fruit weight is 4.5-5.0 kg each. Yield potential is 80-85 t/ha. Mature fruits can be stored upto three months and suitable for all pumpkin growing areas of the country.

***Ambili*** **:** Fruits flat round with shallow furrows, light green at mature stage and turning light brown at maturity, yellow flesh, average weight 6.0 kg. Yield 34 t/ha.

***Suvarna*** **:** Fruits flat round, green with white patches and spots at immature stage, flesh orange in colour. Av. Fruit length 13.5 cm; L : B ratio 0.6; Av. fruit weight 3.5 kg; yield 37.0 t/ha.

***Negra (F1)*** **:** Fruits medium, flat, dark green skin, flesh bright yellow, sweet and delicate flavour. Fruits mature in 80-85 days from sowing. Fruit weight 3-5 kg and heat tolerant.

***Angel (F1)*** **:** Fruits have ribbed silver-gray skin, golden flesh, elongated, shape, appetizing flavour. Mature in 80-85 days after sowing, fruit weight 2-3 kg and very good heat tolerant.

***Clown Giant (F1)*** **:** Fruits globe shaped with bright orange skin, flesh orange. Fruits mature in 75-80 days after sowing and resistant to powdery mildew. Average fruit weight is 15.0 kg.

***Dakara (F1)*** **:** Fruits elongated globe shape with bright orange skin, mature in 75-80 days after sowing, weight 7-8 kg, flesh orange and resistant to powdery mildew.

***Dream (F1)*** **:** It is a medium sized, flat shape hybrid variety, with dark green skin and bright yellow flesh, sweet flavour. Fruit weights around 2.0-3.0 kg each.

***Martinica (F1)*** **:** Early maturing, pear shaped hybrid with grey green skin, golden yellow flesh, resistant to viruses.

***Stone (F1)*** **:** Fruits medium sized, flat shape, bright yellow flesh hybrid, dark green skin, light yellow at the bottom, flavour is very sweet and delicate. Fruits mature in 82-85 days after sowing, fruit weight 3.0-3.2 kg. This hybrid is tolerant to heat.

## RIDGE GOURD

The ridge gourd *(Luffa acutangula* L.) Roxb. (2n = 26) is also known as angled loofah, ribbed gourd; is an annual monoecious climber. It is also seen growing as an escape in south and east Asia. Young fruits are used as vegetable. The fruits are 15-40 cm long and club-shaped, the surface distinctly ribbed.

## Varieties and Hybrids

The promising varieties are Pusa Nasdar, Pusa Sadabahar, Co-1, Co-2, PKM-1, Arka Sujat, Arka Sumeet, Arka Harit, IIHR-8, Satputia, Nasdar, 10 Ridges Gourd, One frooti special malabari (Sadaṣhiv), Pocha's Gheea Torai Smooth, Satputia (7 fruits at one point, ridge gourd), Konkan Harita, Punjab Sadabahar, Pant Torai-1, Swarna Manjari, Swarna Uphar, Utkal Manjushree, Utkal Trupti.

## Hybrids

*Mahyco* : Surekha (MRGH-1), MRGH-2, MRGH-3, MHRG-6.

*Nath sluis* : N.R.G.S.-7.

*Sandoze India* : Rohini.

*Modi Agritech* : M.A.R.G.H-21, MA.R.G.H-289.

*Sungro Seeds* : Gaurav.

*Century Seeds* : Harsha.

*Bejo Sheetal Seeds* : Radhika.

*Novartis Seeds* : Rohini.

*Unicorn Seeds (Choice)* : Vineeta, RGT-101.

*Sreemaa Seeds* : (All O.P. varieties) Jaipuri Long, Dhara, Kanya Kumari, Barapatri.

*Namdhari Seeds* : NS 3, NS 401, NS 403, Sanjeevini.

*Golden Seeds* : Golden 50.

*Ankur Seeds* : Revati, Ruchika, Latika.

*Nunhem Seeds* : Green gold, Namrata, Lata.

*Tropica Seeds (Green CO.)* : Nagar

**Detail characteristics of varieties and hybrids are given below :**

*Pusa Nasdar* : A selection from Madhya Pradesh collection. An early maturing variety, producing club shaped fruits. It is a medium early variety. It flowers in about 60 days in summer. Fruits are ridged and light green. Flesh yellow with firm texture, good flavour and high yield. 15-20 fruits per vine. More suited for rainy season.

***Pusa Sadabahar*** : Fruits are ridged and light green good flavour and high yielder.

***Co-1*** : Moderately vigorous, 10-12 fruits per vine weighing 3 to 4 kg per vine. The fruits are long 60-75 cm and 30 cm in girth, first harvest taken in 55 days. Fruits are long green with prominent ridged and light greenish yellow coloured flesh. The duration is 125 days. Average yield is 150 q /ha.

***Co-2*** : It is pure line selection. Long fruits are about 120 cm. Light green with shallow grooves numbering 11 with mean weight of 350 to 380 g each fruit. The fruit length is about 120 cm. Duration 110 days and it gives 27.8 tonnes / ha yield.

***PKM-1*** : It is an induced mutant, suitable for cultivation in June-July and December-January season. Mean fruit weight 300 g. crop duration of 160 days and yield is about 280 qt/ha.

***Arka Sumeet*** : It has been selected from a cross between an early fruiting 'IIHR54' and a long fruited' IIHR24'. The advanced generation selectin 7 was released as 'Arka Sumeet'. Vines are medium vigorous, growing upto 4.5 m length. Fruits are 55 cm long with 25 cm girth. It bears 1st female flower on 15th node onwards and takes 52 days for first picking. Each vine produces 13-15 fruits of 380 g each in 120 days. Fruits are light green, cylindrical with slow seed maturity. On an average fruit yield is 53 tonnes /ha.

***Arka Sujat*** : It is a selection from a cross between an early-maturing, medium-fruited 'IIHR54' and a long fruited 'IIHR18'. It bears first female flower from 13-15 node in a medium vigorous vine 5.5. m long 40 days after sowing. Fruits are light green, clyindrical measuring 50-55 cm with an average fruit weight of 350g. Its yield potential is 52.5 tonnes/ha. It is moderately resistant to downy mildew.

***Pant Torai-1*** : Fruits 5 cm long and club shaped, yield 10 t/ha, more suitable for rainy season. Released by GBPUAT, Pantanagar.

***Swarna Manjari*** : Fruits elongated, medium sized, highly ridged, green with soft flesh. First harvest 65-70 days after sowing. Yield 18-20 t/ha in 140-150 days. Released by HARP, Ranchi. It is tolerant to powdery mildew.

***Swarna Uphar*** : Fruits elongated, medium sized, ridged with soft flesh, first harvest 65-70 days after sowing. Fruits green, cylindrical, 55 cm long, 2.5 cm girth, delicate aroma. Released by HARP, Ranchi.

***Surekha*** : Fruits are attractive tender, very thin and 40-45 cm long, weighs 125-140 g, first fruit comes after 40-45 days and continues upto 120 days. Use fungicides against the diseases for long fruiting of the hybrid. It is a high yielder hybrid.

***IIHR.8*** : It produce attractive, round fruits. A vine bear 40-45 fruits beside high yield and moderately resistant to downy mildew. The fruits developed on 15-16th node on medium vigorous vines. The variety is medium - late in maturity. It takes 55 days to first-harvest. Each plant produces 40-45 fruits weighing 100g on an average in 120 days, the yield potential being 43 tonnes/ha. The fruits are attractive light-green with good cooking qualities especially for stuffing purposes. 'IIHR 8' has been developed through a pure line selection from a native collection of Karnataka.

***Punjab Sadabahar*** : Fruits long, 3-5 cm thick and slightly curved, yield 10 t/ha.Released by PAU, Ludhiana.

***Konkan Harita*** : Fruits dark green, 30-45 cm long, tapering at both ends, 10-12 fruits/plant. Released by KKV, Dapoli.

***Utkal Manjushree*** : It is a good creeper, flowers are hermaphrodite in nature i.e., bisexual, an exception in ridge gourd. Fruits are borne in clusters of 4-8 in numbers and mature in 55-60 days after sowing in kharif season. Fruits are medium small in size and ridges are less prominent. Average yield is 110.89 q/ha.

***Utkal Trupti*** : Average creeper length is 2.96 m. Fruits are long, straight, ridges are prominent green in clour. Fruits mature in 63 days after sowing. Average yield is 80.5 q/ha. Both varieties developed by Orissa University of Agriculture and Technology, Bhubaneswar.

***Surekha (MRGH-1)*** : Fruits long slender, shining green, length 40-45 cm, weight 125-150 g each. Fruit harvest starts in 40-45 days. Yield potential 45-50 t/ha.

***MRGH-3*** : Fruits long, slender, light green with prominent ridges, length 32-35 cm. Average fruit weight 100-120 g. Yield potential 45-50 t/ha.

***MHRG-6*** : Fruits long, slender, light green with prominent ridges, fruit length 30-35 cm, average fruit weight 100-120 g. Yield potential 50-55 t/ha. Fruit harvest start at 45-48 days after sowing.

***Radhika*** : It is very vigorous growing and high yielding hybrid with shining light green colour. Fruit straight growing and long with approximate weight 125 g, length 40-50 cm. First harvest start at 55-60 days after sowing.

***Harsha*** : Fruits of this early hybrid are borne over a considerable long period. The fruits are deeply ridged, slender, straight with dark green colour, with an average length of 40 cm. Fruit flesh is tender and devoid of bitterness. Vines keep good vigour in high temperatures. Maturity 50-55 days from sowing.

***Golden 50*** : An early maturing vigorous hybrid, yielding first harvest 50-55 days after sowing. Fruits are long and straight, 50-55 cm, bright green with well marked strong ridges. Average fruit weight 250-300 g. Inside flesh pure white with tender seeds. Keeping quality very good. Yields 115-120 mt/ha.

***Green Gold*** : Vigorous plants, short internode and highly productive. Fruits are long (50 cm), green in colour, uniform in size with alternative stripes. Fruits can be harvested in about 50 days from planting.

***Namrata*** : Vigorous vine growth, short internode and highly productive, light green fruits are long (50-55 cm), medium firm and bitter free.

***Lata*** : High yielding hybrid, vigorous in growth, fruits are uniform, bright green, firm with attractive stripes. First harvest possible in 45-48 days after planting. Suitable for long transportation.

***NS 3 (F1)*** : An early medium, prolific and continuous bearing hybrid. Medium long (30-40 cm), straight, attractive green fruits, each weighing 200-250 g. Flesh is very tender and white in colour with very good cooking qualities. Slow seed maturity, less seeded and has very good keeping quality.

***NS 401 (F1)*** : Early hybrid, vigorous plants with prolific and continuous bearing. Fruits are long straight (50-55 cm) attractive uniform green, each weighing 250-300 g. Flesh is white and very tender with excellent cooking qualities. Less number of seeds per fruit, seeds development is slow and very good keeping quality.

***NS 403 (F1)*** : An early medium vigorous prolific continuous bearing hybrid. Medium long straight (35-40 cm), attractive green fruits each fruit weighing 200-250 g. Flesh is white, tender with good keeping and cooking quality.

***Sanjeevini (F1)*** **:** Early hybrid, medium vigorous plants, attractive medium to long green fruits. Mature in 50 days after sowing. Fruit length 30-35 cm, weight 250-300 g. Rind colour green, slow seed maturity.

***Revati (F1)*** **:** Early bearing hybrid, fruits are small, green coloured weighing about 120-130 g, picking starts in 50-55 days. Good yielding potential and best for packing.

***Ruchika (F1)*** **:** Uniform vigorous growth, fruits are medium long, slender, attractive green coloured weighing 130-150 g. Picking starts in 55-60 days. High yielding potential.

***Latika (F1)*** **:** Very strong plant vigour. Fruits are long, slender, green coloured weighing about 170-200 g. Picking starts in 57-62 days, very high yield.

***Nagar (F1)*** **:** Light green fruits are 30 cm long. Maturity around 40 days after planting. High disease tolerant and high yielder.

***Jaipuri Long*** **:** Attractive colour, long and slender fruits of 40-45 cm, average weight 300 g, mature in 55 days.

***Dhara*** **:** Small fruits, ridge less, 10-11 cm long, 50-60 g weight, bears fruits in cluster. Heavy yielder variety.

**Barapatri :** Early variety, first fruit matures between 45-50 days. Fruits are 6-7″ long, 125 to 150 g weight.

***Satputria*** **:** It is a hermaphrodite cultivar, produces pale green small fruits in clusters. In each cluster 5-7 fruits are borne. Yield is 20-25 t/ha. Commonly grown in North India.

***Kanya Kumari*** **:** Medium fruits 22-25 cm, weight 100-120 g, bear in cluster, first harvest 50-55 days after sowing. Heavy yielder.

***Jagruti*** **:** It is suitable for both season. It bear fruit solitary and medium in size with ridges.

***Vineeta*** **:** Plants are strong with vigorous vines. First picking starts between 42–45 days. Dark green fruits are 30–35cm long with deep ridges. Fruits weigh about 250 –270gm.

***RGT 101*** **:** Vigorous hybrid plant produces long light green slender fruits of length 30–45 cm with very high yields. Harvesting begins in 45–50days.

*Rohini (F1) :* Vigorous growth with medium sized (2.5-3m) vine with dark green foliage. Fruit is long (40-45 cm) with green ridges having very few seeds. Quality fruit weighing 140-160gm, gets ready in50-55 days after sowing.

*Gaurav* : It is very early hybrid, give first picking in 45 days after sowing, fruit shape long and slender, colour dark green, length 45 cm and average weight 350 g.

## ROUND MELON/ROUND GOURD

### (Syn. Indian Squash)

Round melon/Round gourd/syn. Indian Squash (*Praecitrullus fistulosus* Pang (2n = 24); also known as Tinda, Dilpasand, Tindu, Tendsi, Tindas, Tendu, Dhemsa, Indian squash, squash melon; is one of the popular summer vegetable in north India, especially in Punjab and Uttar Pradesh. Fruits harvested at tender stage before hardening of seeds and used as cooked vegetable. Fruit has a cooling effect and contains good amount of vitamin A. Fruits are used by patients suffering from dry cough and for increasing blood circulation.

### Varieties and Hybrids

There are two types of gourds commonly grown in the country. One type has light green colour and other is with dark green colour. Some high yielding varieties are Annamalai selection, Improve Shipper, Arka Tinda, Punjab Tinda.

### Hybrids

*Mahyco* : Tinda.

*Ankur Seeds* : Amol.

**Detail characteristics of varieties and hybrids are given below:**

*Annamalai Selection :* It has been developed at TNAU, Coimbatore, good fruit type and higher yielder. Yield about 110 qt/ ha. Crop duration is 100 days.

*Arka Tinda :* The fruits are more or less round, light green, with few seeds. This is a early summer and rainy seasons variety and tolerant to fruit fly. It produces up to 100q fruits/ha.

***Punjab Tinda :*** It has been developed at Punjab Agricultural University Ludhiana. It is high yielder with medium fruit size.

***Whitish Green :*** Fruits are medium in size, pale or light green in colour and are ready for harvest after 50 to 55 days of sowing. The leaves are light green in colour.

***Dark Green :*** Fruits are medium in size, dark green or slightly blakish in colour, and are ready for first harvest after 55 to 60 days of sowing. The leaves are dark green in colour.

***Tinda-S-48 :*** It produces medium sized fruits, weighing on an average 50 g each. The fruit colour is light green and they are ready for first harvest after 65 to 70 days of sowing.

***Improved shipper :*** It is high yielding and medium maturity variety.

***Mahyco tinda :*** It gives flowering 30-35 days after sowing. profuse bearer, sizes 100 to 120 cm long and leaves and stems are tender, fruit size 80 to 100g and fruit shape is elliptical, depressed at both end with green colour and tender skin, fruits are very tender and tasty. Fruit reach to marketable maturity within 45 t0 55 days. Fruits should be harvested at one to two days interval.

***Amol (F1) :*** Plant strong and vigourous; fruits are flat round attractive light green, coloured weighing 85-100 g per fruit. Picking starts in 49-55 days. Good yielding potential.

***Ludhiana special :*** This improved variety mature in 70-80 days after sowing. Fruits flattish round, green, medium size 50-60 g each, gives 8-10 pickings.

## SMALL GOURD

Little gourd *(Coccinia grandis* (L.) Voigt, 2*n*=24) popularly known as *'kundru'* is so far a neglected and underutilized species (NUS) of family cucurbitaceae. Kundru is mainly consumed as fried vegetable in eastern and southern parts of the country where it is grown. It can also be used for various preparations of mixed vegetable and salad. Little gourd is indigenous to India, particularly the Eastern India where a rich gene pool is available in wild as well as in homestead gardens. It is a hardy vegetable and has a vast scope of growing in marginal lands where it gives fruits almost round the year. In tribal areas of Orissa, Jharkhand, Chattisgarh and Andra Pradesh presently, the fruits of kundru are being gathered from natural forests or are being grown in homestead garden and small pockets. However, it is gaining popularity among the consumers and producers due to its nutritive/medicinal value and round the year production potential. Owing to non availability of standard variety of kundru, the farmers are cultivating only land races which yield variable fruits.

Small gourd *(Coccinia grandis* (Voigt); syn. *C. indica*) ; 2n = 24, 36, belongs to the family cucurbitaceae, also known as little gourd, ivy gourd (Kundru or Tondli) is a semi-perennial dioecious plant with small fruits widely grown in eastern, western and southern states in India. It is a semi-perennial, dioecious creeper, consisting of distinct male and female plants and highly cross-pollinated. Plants climbing or prostrate perennial herb, soft stemmed, evergreen creeper.

The small gourd is propagated by its tuberous roots, or by stem cuttings. It is one of the rare cucurbit which is planted by multiplication of stem cuttings. Fruits ovoid or elliptical botanically "peppo", small and round or thick and long, green with or without white stripes when immature, becoming bright scarlet when ripe.

## Varieties and Hybrids

Being a vegetative propagated crop, variability is rather limited in small gourd. Plants differing in size, shape and weight of fruits, pattern of strips on fruit skin, presence of mucilage in fruits, leaf lobbing etc. are noticed. Fruit shape varies from ovoid, oblong to long cylindrical.

In Coccinia there are two distinct types. They can be broadly classified as "bitter" and "sweet" based on the edible quality of the tender fruits. It was also reported that sweet variety is under cultivation in different part of the country.

In the *Coccinia cordifolia* there are two varieties namely, 'Allahabad' and 'Aligarh'. There are so many good genotypes available in the locality i.e., in small gourd are of Orissa. Bengal and Bihar and used for cultivation in those regions.

**Detail characteristics of varieties are given below :**

***Sulabha :*** This variety has been developed by Kerala Agricultural University, a high yielding (60 t/ha) variety having a fruit length and weight of 9.25 cm and 18.48 g respectively.

***CHIG-15- (Neelachal Kunkhi) :*** A promising genotype CHIG 15 has been identified for high culinary quality and yield. This is an early fruiting variety which starts fruit production 40 days after planting (DAP). The variety produces light green, long slender, cylindrical fruits of 7.5-8.5 cm length with soft texture and has attractive stripes in skin. The variety CHIG-15 has sequential fruiting habit and fruit develops by means of vegetative parthenocarpy hence, there is no need of male plants for pollination and more number of female plants can be accommodated per unit area. It has an excellent organoleptic quality as fried vegetable as well as vegetable curries. A single plant of CHIG-15 produces a yield of 20 kg fruits in one growing season. Being a high yielding genotype with better quality, it would give higher economic returns to the farmers and would also meet the local demand. The fruits fetch good market price with consumer satisfaction as they have good physical appearance and soft texture.

***CHIG-33 (Neelachal Sabuja) :*** High yield, hardy plant type, tolerant to bruishing. A selection from Khordha district of Orissa. It is similar to desi types in its appearance. Fruits are dark green in colour with fractured stripe and conical shape. The fruit has excellent cooking quality and taste. Planting is generally done in June-July/October-November. CHIG-33 starts flowering at 10-12 nodes after 30-35 days after planting in first year. Subsequent years good sprouts take place from stubs which produce fruit profusly. Individual fruit weighs around 25-30g. Plant is highly vigorous reaches up to 10m.and has 4-6 branches. The genotype produces fruits by means of vegetative parthenocarpy and it does not require pollinizer for fruit set. It gives nearly 70-80 harvests per season (10-11 month). Yield per plant in this genotype is 40-50 kg/year.

## SNAKE GOURD

Snake gourd, *Trichosanthes anguina* L. (2n = 22, 24) is an annual creeping cucurbits, cultivated for tender fruits, which are used as vegetable. It is known as Chichinda in Hindi, Chinese name of the snake gourd is 'Manka' means cucumber of the southern barbarians, indicated India or the Indian Archipelago as a possible centre of origin. A good variability of the crop exists in India. The fruits are harvested and cooked green and have a fairly good nutritive value. The wild form *T. cucumerina* L. is an ancient cultivar and occurs wild in India. Its fruits are small in size, highly bitter and have medicinal value.

## Varieties and Hybrids

Cultivars/varieties/cultures/hybrids differe in length, thickness, colour, presence or absence of white stripes or striations on fruits. Many varieties were evolved through selection.

The commercial varietis, cultures and strains developed in snake gourd are : Col, Co2, MDU 1 ; PKM 1, H 8, H 3171 and H 375, IHR 16-A, TA 19, Maharashtra panni pudal, No.20-5-5 (Short), No.29-1-9 (Medium) and No.30-2-2 (Long) and Aduthurai for Madras. Some green long and white long are best according to the colour of the fruits. Nepali (Sadashiv & Co.); Verma's Long green, Long white and Pocha's Extra Long are some of the varieties available in different states.

**Detail characteristics of varieties and hybrids are given below:**

***Konkan Sweta :*** Fruits 90-100 cm long, having whitish surface with greenish stripes. This variety is recommended for growing in Konkan region of Maharashtra. The yield is about 25-27 t/ha in 125-130 days. It is developed by KKVP, Dapoli.

*Co 1* : It is a selection, bearing long fruit 160 to 180 cm with white stripes, weighing 500 to 750 g. It gives a yield of 18 tonne /ha with a crip duration of 135 days.

*Co 2* : It is a selection, bearing short fruits (30 cm length), light greenish white. It yields 35 tonne per ha, with a crop duration ranging from 105 to 120 days. It is suitable for high density planting.

***MDU 1*** : It is a hybrid between a long-fruited selection 1 and a short fruited pannipudal. The plant is less spreading. The fruits are green with white stripes and medium in length (70 cm). It yields 32 tonnes per ha with a crop duration of 143 days.

***PKM 1*** : It is a mutant with a crop duration of 145 days. The mean fruit weight is 500-800 g with a mean length of 155 to 170 cm. It gives a mean yield of 24 tonnes /ha.

***APAU Swetha*** : Fruits long having whitish background with green stripes. Yield 28 t/ha in 125-130 days. It is developed by APAU, Hyderabad.

*TA 19* : It is new collection recommended by the Kerala Agricultural University. Fruits are about 60 cm long. At immature stage they are light-green with white-stripes at the styler end. The average fruit weight is 600 g and the fruits are ready in 65-70 days from sowing.

***Kaumudhi*** : Uniform white large fruits with acute tip, average fruit length 100 cm, av. fruit girth 29.4 cm, av. fruit weight 1.34 kg, productivity 50 t/ha.

***Baby*** : Small attractive white fruits suitable for easy handling and marketing, av. fruit length 36.5 cm av. fruit girth 22 cm, av. fruit weight 474 g, productivity 40.0 t/ha.

***Manusree*** : Uniformly white and medium long (67.72 cm), cylindrical fruits with green markings at the pedicel. Av. fruit weight 775.0 g; Yield 32.5 t/ha in 120-140 days. Above three varieties are developed by Kerala Agril. University.

## SNAP MELON

Snap melon or phoot (*Cucumis melo* var. *momordica*), 2n = 24, is a popularly grown as vegetable in a arid and semi arid regions of Rajasthan. It matures but unripe fruits are used to prepare rayta. Ripe fruits are used as dessert or as salad. They are also dehydrated for off-season use. These are locally called Khelera. Ripe fruits which bursts spontaneously on ripening is used as a substitute for muskmelon due to its cooling effect. Due to its bursting tendency on ripening, the fruits is known as snap melon. Seeds kernels are extracted and are extensively used in sweet and bakery products.

### Varieties and Hybrids

Genotypes were characterized for earliness, fruit size, number of fruits/plant, flesh quality and taste. Of the promising genotypes AHS6, AHS 10, AHS 50, AHS 54, AHS 64 and AHS 82, have been recommended for cultivation. These have been developed by Central Institute for Arid Horticulture, Bikaner, Rajasthan. These genotypes have now been widely accepted for commercial cultivation in arid regions.

**Detail characteristics of varieties and hybrids are given below:**

***AHS10*** : It is an early, high yielding selection from local land races of arid region. Fruit become ready for harvesting 67-69 days after sowing. Its fruits are oblong and medium sized, 850-950 g in weight, 17-20 cm in length and 9.7-10.5 cm in diameter. Edible flesh thickness is 2.1-2.6 cm and the fruit cavity is 5.5-6.1 cm wide. The whitish pink flesh is sweet in taste having 4.5-5.0, brix, male and female flowers appear 26 to 38 days after sowing. Ripe fruits can be harvested 67-120 days after sowing. Its vines are medium in vigour with 2.5m length and 8.5 branches. It is highly productive, giving 4.0-4.5 fruits/plant (202-222 q/ha).

*AHS 82* : This selection yields uniform, medium sized and long fruits. Sowing can be done twice in a year, i.e., summer (Feb-March) and rainy season (June-July). Male and female flowers appear 28-35 days after sowing, respectively. First harvesting can be done 68-70 days after sowing which continues upto 110-115 days. Each vine bears 4.5-5.0 fruits giving an yield of 225-248 q/ha. Fruits are 900-950 g in weight, 22.5-24.5 cm in length and 9.0-9.2 cm in diameter. Edible flesh thickness is 2.1-2.41 cm with 5.4-6.0 cm wide fruit cavity. The flesh is sweet (4.3-4.9° brix), tasty and light pink in colour.

***Pusa Shandar*** : The fruits are green in immature stage and yellowish white in full mature stage. They are cylindrical; immature fruits are pubescent and average fruit weight is 700 g. The smooth surface of fruits cracks or bursts when mature. The fruit flesh is light pink and 1.8 cm thick and T.S.S. being 6.5 ° Brix. Seeds are creamy white in colour. It is suitable for packaging and long transportation. It is duel purpose variety and yields 370 q/ha in kharif and 402 q/ha in summer.

Recently 30 genotypes were tested at IARI, New Delhi during spring-summer season. Out of which some genotypes were found suitable in different yield attributing characters. These are :

*SM 4* : It is early maturing than other genotypes.

*SM 7(O)* : It is vigorous and showed maximum yield. Profuse vine length, fruit diameter and fruit weight.

*Sm 15-1* : It is very sweet and has highest TSS (9%) than other culture.

*SM 18* : It has very long fruits and good yielder.

*SM 25* : This genotype has very low fruit cavity and give more edible portion and has maximum ascorbic acid.

*DBSM* 3 : The fruits of this culture has maximum flesh thickness and total carotenoids.

*AP 6* : The fruits of this culture have good taste and it has maximum reducing and non-reducing sugar.

## SPINE GOURD

Spine gourd (*Momordica dioica* Roxb. 2n=28) belongs to the family Cucurbitaceae; is popularly known as 'Desi kankad' (Orissa), 'Kankodan' (Bombay presidency), 'Kartoli' (Marathi), 'Kuntola', 'Kantolan', (Gujarathi), 'dhar karela' (Punjabi), 'pugakara' (Telugu) and 'pazhu pagal' (Tamil). Presently, it is found in wild and semi-wild form. However, its cultivation is picking up gradually. It possesses several medicinal and curative properties. Its leaf decoction reduces fever, tuberous roots help in relieving headache, sweating and stone formation, whereas fruits are quite useful in controlling diabetes. Its fruits are found to oval with soft spines, green when unripe and bright yellow to orange when ripe. The ripe fruits open in three irregular pieces. Seeds 10-20, compressed, smooth surfaced with bright red arid.

It is an important dioecious and minor cucurbit. Plants are semi perennial creeper. Stems angular, long tuberous root. The fruits are small, 2.5 to 5.0 cm long, ovoid, ellipsoidal, densely echienate with soft spines. Seeds are slightly compressed 3.0 to 7.0 mm long.

The immature tender green fruits are used for culinary purpose, as sole vegetable. Some time mix with other vegetables and prepares wholesome curry. Fruits are also used in ulcers, piles, sores and obstructions of liver and spleen. They possess several medicinal properties and are good for those suffering from cough and other digestible problems. The unripe fruits act as appetizer and astringent. The seeds are used for chest problems and stimulate urinary discharge. The roots contain a triterpenoid saponin, which give water and can be used as a substitute for soup. Spine, gourd is quite useful in removing piles, migraine, excess sweating, cough and stone etc.

### Varieties and Hybrids

Spine gourd shows a wide genetic diversity. A large numbers of local strains / varieties are popular with the growers. They are having the fruits of different shapes and size and used for cultivation. There is no named variety and a great degree of variation is observed among seed population.

Recently Central Horticultural Experiment Station, Bhubaneswar, has started improvement work on spinegourd. Forty genotypes have been collected and are being maintained in field gene bank. Out of that some promising genotypes have been identified.

**The detail characteristics of varieties/genotypes are given below:**

***CHSG-28 (Neelachal Shree)*** : An elegant spine gourd variety Neelachal Shree has good appearance, high yield (4-5 kg/plant) and high market preference. The vine of this variety is thin and spreading which grows very well on three line wire trellis system. The variety is moderately resistant to anthracnose and downy mildew under field condition, whereas it is moderately susceptible to angular leaf blight and *Diaphania indica*.

## SPONGE GOURD

Sponge gourd, *Luffa cylindrica*, M.J. Roem. (2n = 26) syn. *L. aegyptiiaca* Mill; also known as towel gourd, smooth loofah, vegetable sponge, dish cloth gourd, is an annual, monoecious climber. The tender fruits are eaten fresh or more commonly cooked as vegetable.

### Varieties and Hybrids

There are not many recognised varieties of the sponge gourd. The promising varieties are Pusa Chikni, Kalyanpur Chikni, Pusa Supriya, Phule Prajakta.

### Hybrids

***Mahyco*** : Harita, MSGH –1, MSGH-3, MHSP-8, MHSP-10.

***Century Seeds*** : Utsav.

***Nath Sluis*** : NSGH-10

***Sungro Seeds*** : Chetak.

***Bejo Sheetal Seeds*** : Gita, Sita.

***Unicorn Seeds (Choice)*** : Vasudha.

***Namdhari Seeds*** : NS 441, NS 442.

***Golden Seeds*** : Priya.

**The detail characteristics of varieties and hybrid are given below:**

***Pusa Chikni*** : A selection from Bihar collection, early fruiting variety, flowering in about 45 days. The fruits are smooth and dark green colour, more or less cylindrical, 15-20 fruits per vine, suitable for both spring-summer and rainy seasons. Released by I.A.R.I., New Delhi.

***Pusa Supriya*** : The vines are 2.5-2.8m long, with 4-5 branches. Leaves are dark green, leaf size smaller with 6 lobed margin, non-hairy, smooth, long petiole, tip pointed, simple. Inflorescence recemose, unisexual flowers monoecious, companulate; staminate flowers large in number in cluster on long pedicels, large- sized, dark yellow. Pistillate flowers, solitary, large in size with long peduncles. Fruit distinct, pale-green, medium-long, 15-20 cm long, straight and slightly curved at the stem end, pointed distal-end smooth, non-hairy, slim, narrow green linings and spots, tender end,ocarp, skin very thin, on an average 10-12 fruits weighing a kg. Fruits become ready for picking 50-55 days after sowing in spring-summer and 44-48 days after in kharif seasons. It yields 83.20-96.6 q/ha during spring summer and 94.70 – 108.30 q/ ha during kharif season.

***Pusa Sneha*** : Pusa Sneha is an improved variety over hitherto very popular Pusa Supriya. The vines are 2.5-3.0 m long with 4-5 branches. Leaves are dark green, small in size with 6-lobed margin, non-hairy, smooth, long petioles. Inflorescence recemose, unisexual flowers, monoecious, companulate; staminate flowers large in numbers, in cluster on long pedicels, large-sized, dark yellow, pistillate flowers solitary, large in size with long peduncles. Fruits are attractive, dark green, 20-25 cm long, straight, peduncle long, smooth, non-hairy, flesh tender, skin hard suitable for long distance transportation. It is suitable for growing in both spring-summer and kharif seasons. Fruits become ready for first picking 50-55 days after sowing in spring summer and 40-45 days in kharif season. On an average Pusa Sneha provides 107.5 q /ha fruits in spring-summer and 121.17 q /ha in kharif season.

***Phule Prajakta*** : It is a selection from local grermplasm. It is released by M. P.K. V. Rahuri (MS). It is suitable for summer and kharif season. Fruits are medium green, flesh white, sutures are dark green. Yield 15 t/ha.

***Rajendra Nenua 1*** : Fruit, long, greenish white, smooth and thick, resistant to fruit fly and fruit rot. Yield 25 t/ha.

***SEL 99*** : Short vine growth, fruits pale green in colour, smooth, non-hairy, tender flesh, number of fruits per vine varies from 12-16. Suitable for sowing in both spring- summer and kharif seasons. First picking takes 50-53 days in summer and 45 days in kharif season. Average yield is 100-140 q/ha.

***MSGH-1*** : Fruits tender, with slender shape and 20-25 cm long, weighs 100-125 g. shining skin light green and smooth. Matures after 40-45 days and fruiting continues till 120 days, yield potential is 35-40 t/ha.

***Harita*** : Fruits are tender, slender, round and 20-25 cm long, weighs 100-125 g, skin rough but shining and deep green colour, matures after 40-45 days and fruiting continues upto 100-120 days, yield potential is 35-40 t/ha when staked.

***Gita*** : It is a mid long to long hybrid and very productive. Fruits long green smooth, with faint stripes. Fruit length 25 cm. 1st harvest starts after 60 day of sowing.

***Sita*** : It is a very high yielding hybrid, which can be grown in summer, rainy season and mild winter. Fruits dark green, smooth and soft with faint strips, fruit length 20-25 cm. First harvest starts 60 days after sowing.

***Chetak*** : This hybrid give first picking in 40 days after sowing, fruit shape long and thin, with average length of 30 cm and tender, fruit colour dark green, smooth, average weight 250g.

***NS 441 (F1)*** : Early, vigorous, prolific and continuous bearing hybrid, long (20-30 cm) and attractive dark green cylindrical fruits, each weighing 200-250 g. Flesh is white and tender with good cooking quality and slow seed development.

***NS 442 (F1)*** : Medium early, prolific and continuous bearing hybrid, medium long (20-30 cm) attractive light green to whitish green cylindrical fruits each weighing 200-250 g. Flesh is white with slow fibre maturity.

***Utsav*** : Fruits are cylindrical, tender, attractive, glossy, Algae-Green colour with average diameter of 3.0 cm and length of over 30 cm. The fruit pulp has good flavour with small seeds devoid of bitter principles. Average weight is 135 g. The vines remain vigorous and bear over a considerably long period.

***Priya*** : Early maturing hybrid with vigorous plants and prolific fruit setting habit. Fruits are uniformly cylindrical, smooth, tender and dark green in colour, 25-30 cm long and weigh 200-250 g. Performs well under dry and rainy conditions. Prolific fruiting habit allows continuous harvest for 5-6 months.

***Vasudha*** : An excellent hybrid with strong vine, produce light green cylindrical long fruits of size 25-45 cm and weighs 90-140 g.

## SUMMER SQUASH

Summer Squash (*Cucurbita pepo* L.), 2n = 40; is known as Chappan Kaddo, early maturing and have bushy or trailing habit,. Most cultivars are bushy in nature and can be planted at close spacing. Varieties differ in size, shape and colour of fruits.

### Varieties and Hybrids

Some important varieties / cultivars are : Early Yellow Prolific, Australian Green, Butter nut, Sugar, Green Hubard Golden Hubbard, Connecticut Field, Yellow Crook neck, English Marrow, Punjab Chappan Kaddo, Patty Pan, Pusa Alankar etc.

***Century Seeds*** : Parikarima.

***Golden Seeds*** : Chandrika.

***Indo-American Export*** : Ambassador, President Early butter-nut.

***Unicorn seeds*** : Himanshu, Laila, Abtab

***Early Yellow Prolific*** : An early type bushy plant, fruit medium sized, warted, tapering towards stem end, skin light yellow turning orange-yellow on maturity. Flesh tender at the stage it is consumed as vegetable. Released by IARI, Regional Station, Katrain.

***Australian Green*** : An introduction, very early, bush type, fruit dark green with longitudinal white stripes, 25-30 cm long fruits per plant 15-20, yields 150-200 q/ha. Very tender at edible stage. Released by IARI, Regional station, Katrain.

***Punjab Chappan Kaddo 1*** : It is an inbred selection from the segregating local material of Punjab and recommended for cultivation in 1982. Plants are bush type, foliage thick and erect, leaves non-lobed and green without white specks, petiole and leaves hairy, fruits green, disc shaped, mildly ribbed with flat stem-end and attractive, average fruit weight about 800g and a plant bears about ten fruits, early maturing and is ready for first harvest in about 60 days from sowing. It has a predominant female tendency, field resistance to downy mildew and tolerance to CMV, powdery mildew and red pumpkin beetle. It has high yield (about 200-225 q/ha). It has been released by Punjab Agricultural University, Ludhiana.

***Patty Pan*** : It is an introduction from United States and recommended for cultivation during 1972. Fruits disc shaped, chalky white, tender and very attractive at edible stage. It is a short duration variety with yield of 200 q/ha in 85 to 90 days. Released by IIHR, Hessarghatta.

***Pusa Alankar :*** It is an F1 hybrid cross between EC 207050 and Sel-1P1-8 (a derivative from cross between chappan Kaddo and Early Yellow Prolific), early maturing, uniform dark green fruits with light coloured stripes, slightly tapering towards the stem end, tender delicious and matures in 45 to 50 days. It has been bred at IARI regional station, Katrain.

***Himanshu :*** It is a newly developed summer squash hybrid. Strong and open plant, lobed and silver spotted leaf. It is a great producer of cylindrical and uniform fruits. showing dark green minutely white speckled colour.

***Laila :*** Summer Squash hybrid is Lebanese Marrow type, specially, selected for the Himalayan region and Middle East market. Fruits are cylindrical slightly bulbous with light green exterior and creamy interior. Plants are extremely prolific, producing fruits during a long period of time. Maturity cycle is approximately 50 days under desirable conditions.

***Aftab :*** This Summer Squash hybrid variety is Lebanese Marrow type. Very popular in Middle East and Mediterranean area. Plants are medium size, semi open bush, single stem. Uniform bearing fruits are pale green, bulbous in shape and 18x5 cm in size. Maturity cycle ranges from 40-50 days under desirable conditions.

***Chandrika :*** It is extremely early maturity bushy hybrid, developed by Golden seeds.First harvest commences after 35-38 days of sowing. Fruits are round, slightly flattened at the blossom end and pale green in colour. Prolific yielder and tolerant to squash mosaic virus.

***Parikarima :*** The plants are dwarf, vigorous with short internodes, bush type hybrid suitable for out door as well as poly-culture. Fruits almost globe shaped, creamish, apple green with very good flavour. Excellent quality produce can be harvested in about 8-12 days of flower fertilization. Spraying ethrel (Ethephon) at 250 ppm induces the formation of flowers and increases the fruit yield.

***Ambassador :*** One of the most attractive zucchinis available in the market. High yields over a long season combined with earliness have made this hybrid widely popular. The tender fruits are waxy, medium green outside with white crisp interior flesh. The open plant habit offers easy picking for commercial growers and home gardeners. Plant compact bush open habit easy picking; fruit, mature in 51 days after sowing. Cylindrical smooth fruits with medium dark green beautiful appearance and size 18-20 cm with high fruit yield over long season.

***President*** : This hybrid is ideal for plastic green house and home gardens. Fruits are dark green with light green flecks. Plant upright, open habit.Fruit maturity 49 days after sowing, fruit size 18-20 cm, cylindrical slight taper long and smooth fruits.

***Early butternut*** : This is an early hybrid with semi bush compact plant and produces large yields of high quality fruits; ideal for market gardens. The medium sized fruits have a fine texture and sweet flavour and mature in 82 days. Fruits mediumsize, fairly blocky, even tan colour, with 25-30 cm size at marketable stage.

## TEASEL GOURD

Teasel gourd *(Momordica subangulatata* sp. *renigera*) is an important cucurbitaceous crop of eastern and north-eastern India. It is commonly known as *vilayati kakad* or *Assam kakad*. Teasel gourd is a hardy species and produces fruits of high nutritive value which is consumed as vegetable at immature stage. A wide range of diversity in teasel gourd exists in the natural system owing to continuous regeneration through seed. Presently, the fruits are collected from natural system by the local people and a wide variety of fruits is sold in the market. These teasel gourds can be seen in the vegetable market during July-September. Variation in fruits is mainly due to colour of fruit, shape, size and mellowness of seeds. The general preference of the consumers is for green type of fruits with soft seed and high pulp content. However, growers prefer a type which produces green fruits with fairly long fruiting period (July-October).

### Varieties and Hybrids

On the basis of consumer survey and visualizing the need of the cultivators, a large number of teasel gourd lines have been collected and evaluated at Central Horticultural Experiment Station (IIHR), Bhubaneswar.

Out of the 25 variable germplasms collected from different ecosystem, most promising culture has been selected and named as Neelachal Gaurav (CHT G-2), which detail characteristics are given below.

***Neelachal Gaurav (CHTG-2) :*** The fruits of Neelachal Gaurav is dark green and oval having small spines. The fruits of CHTG-2 are 6.0 cm long and 3.8 cm thick (diameter at centre) with an average 50 g fruit weight. The plant produces 230-250 fruits in full cropping season with sufficient pollinators. The variety needs hand pollination for assured yield which varies between 12-15 kg per plant per year. The selected variety Neelachal Guarav is vigorous with dark green foliage, strong vine and fairly long growing period (15th June-15th October). The variety produces high number of female flowers which shows its potential of high yield if proper pollination is ensured. The variety shows moderately resistant reaction to *Diaphania indica* infestation, and moderately susceptible reaction to anthracnose and downy mildew diseases under field condition.

## WATER MELON

Watermelon (*Citrullus lanatus* (Thumb.) Mansf.; syn. *Cucurbita citrullus, Colocynthes citrullus* (L.) Kuntze, *Citrullus vulgaris* Schrad); 2n=22; is also known as Tarbooj, Kalinda, Jamaika, Matira, Paniphal, Kariju, Turmuj, Kalingad, Tarubhuj, Purcha, Palam Panna, Thannir Mathan, Kallangadi, Chayapula Kuttoowombi, Tandur, Belikh, Zichi, Pilchagnadi, Komardu, Pha-rai, Melond can pasteque, Wasser malone; is a popular dessert fruit in tropics and has been grown widely on both sides of equator. It is a slender hairy, monoecious or andromonoecious annual, often sprawling over the ground. Fruit is a many seeded pepo with red, green, yellow or whitish flesh and white, black or reddish yellow flat seeds.

Shimotsuma (1965) classified watermelon as *Citrullus lanatus* (Thumb.) Mansf., *C. ecirrohsus, C. colocynthes* (L.) Shrad - possible ancestor of watermelon - *C. naudinaus, C. fistulosus* (Round melon), and *C. vulgaris* var. *citroides* (Bailey) Mansf. (Citron melon).

Watermelon is classified into four major categories based on rind colour, fruit shape and seed colour.

(a) Rind ivory white with conspicuous green stripes.
   (i) Fruit cylindrical, seed white, seed dark tanned;
   (ii) Fruits cylindrical, brownish tinge or black seeds.

(b) Rind dark green or bluish - fruit size variable. Seeds are brown with markings.

(c) Rind is grey, seeds are white, fruit size varying.

(d) Rind white or pale-green with dark green stripes, seeds white.

The fruit contain 95 per cent water, 0.2 per cent protein, 0.3 per cent minerals and 3.3 per cent carbohydrates per 100 g fresh weight (Aykroyd, 1963). Among all members of cucurbitaceous crops, Watermelon is the rich in iron content. Most of the cultivars have deep pink or pale pink flesh colour with a slightly reddish tinge, containing largely lycopene and anthocyanin pigments. It is a common man's fruit relished by millions rich and poor, young and old. The juice is an alternative to drinking water in dessert areas. It has cooling effect. It is not usually advised to take watermelon fruit immediately before or after food.

## Varieties and Hybrids

There are large number of varieties, grown all over the country. This being a cross pollinated crop and as hardly any care is taken to maintain its purity. Some important varieties are Jaunpuri local, Improved Shipper, Asahi Yamato, Sugar Baby, Delhi local, Pusa Rasal, Katagola, Pusa Bedana, New Hampshire Midget, Special no.1, Arka Manik, Durgapura Meetha, Durgapura Kesar, Fuken (HAU). Jobner-21, and Jobner 18-1. Bush Baby, Yellow Baby. Red Glory etc.

There are several cultivars, locally grown which are named after the region in which they are available, such as Farrukhabadi, Moradabadi, Faizabadi of U.P. Most of them have fruits with dark green colour or pale green with black stripes, oblong to round shape weighing 8 to 10 kg with thick rind. The flesh colour would be variable from pale pink to pink with big flat black to brown seeds. In Rajsthan, Mateera cultivar is grown in rainy season around Bikaner. Varieties differs in colour, shape, flesh thickness of rind, keeping quality, flavour, seed size etc.

***Varieties suitable for long transportation :*** Peacock, Crimson Sweet, Arka Manik, Durgapura Meetha, have tough rind for long transportation, (Indian and American varieties).

***Varieties with thin rind :*** Developed from Taiwan and Japan - Yellow Baby, Carnival and Red Glory.

***Varieties with bright red colour :*** Giza or deep crimson, Arka Manik and Crimson Sweet are preferred in Indian market.

***Varieties with yellow flesh :*** Yellow Baby, and Durgapura Kesar (do not gain with the Indian costumers).

***Varieties with high quality (high TSS, granular texture and fine aroma) :*** "Arka Manik, Paradise, Sweet Marvel, and Sugar Delicata.

***Varieties resistant to fusarium wilt :*** Calhoun Gray, Summit, Smokylee, Dixilee, Sugarlee, and Texas W-5.

***Varieties resistant to anthracnose :*** Congo, Fairfax, Charlston Gray, Crimson Sweet and Arka Manik.

## Hybrids

***Indo American Hybrid Seeds :*** Madhu, Milan, Mohini, I.A.-9305, I.A. 9307.

***Mahyco :*** Amrit (MHW-5), Santripti (MHW-4), MHW-6, MHW-11, MHW-16, MHW-17, MHW-23, MHW-24, Suchitra (MHWM-285), Shrinidhi (MHWM-505).

***Century Seeds :*** Hybrids no. 1, Millionare, CWH-1 CWH-5, CWH-7, CWH-8.

***Charlis India :*** Sugar green, Sweet heart.

***Nath Sluise :*** Nath Apurva, Nath Madhuri, NWMH-1

***Nuzividu Seeds :*** N.W.M.H-4,N.W.M.H-5 and N.W.M.H-6.

***Sandoz India :*** M.H.W-6 Crimson Tide.

***Clause International :*** Paradise, Rosia, Fawida, Sultan.

***Annapurna Seeds :*** Sugar Boy.

***Tokita Seeds :*** Tokita 786.

***Tropica (Green CO.) :*** Dark Dragon, Dark Belle, Missile, Apolo, Black Rocket, Sugar Belle, Long Dragon, Big Boy, Golden Tiger, Baby Dragon, New Hope, Sugar Dragon, Uranus, Dragon Giant, Sweet Baby, Golden Dragon, King 999, Dark Sweet, Yellow Moon, Yangtse, Red Glory (Sunlight).

***Nunhems Seeds :*** Madhuri (64), Madhubala (80), Khushboo (10), Red Honey, Maya, Surabhi, Mumtaz (99031).

***Namdhari Seeds :*** NS 295, NS 246, NS 1004, NS 2332, NS 294, NS 296, Mayur (NS 261), NS 271, NS 281, NS 321, Tejas (NS 251), NS 2726, NS 4088, NS 54134.

***Golden Seeds :*** GC-282, GC-285, GC-286, Disco (GC-287).

***Novartis Seeds :*** Augusta, Crimson Tide, DR 27-2, Sugar King, Super Dragon.

***Unicorn Seeds (Choice) :*** Sultan, Emerald, Maharaja, Vandana, Black Baby, Red Sweet, Sweet Ruby, 113.

***Bejo Sheetal Seeds :*** BSS 358, BSS 360, BSS 359, Swapnil, BSS 408, Sweet Baby.

***Modi Agritech :*** M.A.W.H-243 (Arka Manikpari), M.A,W.H-308 (Sugar baby pari).

***Indo American Export :*** Weetdiv, Madhu, krimpia, Alempia, Pataregra, A.I.A. 9305, A.I.A. 9306.

***IIHR, Bangalore :*** Arka Jyoti.

***Sungro Seeds :*** Rasila, Raj Ras, Rangeela, Rasmadhu.

***Annapurna Beej :*** Hunter.

**Detail characteristics of varieties and hybrids are given below:**

***Special No-1 :*** Fruits are round and small with red flesh and red seeds. It is early in maturity. The average T.SS is slightly lower than shipper. Released by Punjab Agricultural university Ludhiana.

***Sugar Baby :*** This is a variety originally introduced by M. Hardin, Geary, Oklahoma, USA in the year 1956. It is early and resistant to drought. From USA, it was introduced to India by IARI. The vines are medium long. Fruits are smaller (3-5 kg), round in shape. The rind is bluish black. Flesh is sweet (11-13% TSS) with deep pink colour. Seeds are brown and small. Yield potential is 150 q/ha.

***Asahi Yamato :*** It is a mid-season Japanese introduction by IARI, New Delhi. Fruits are medium weighing 6-8 kg. The fruits are round oblong, non-striped with light green rind. Flesh is deep pink with 11-13% TSS. Fruits ripen in about 95 days. Yield is 225 q/ha.

***Pusa Bedana :*** It is a seedless triploid variety of watermelon developed at IARI, from a cross of Tetra –2 (4x) X Pusa Rasal (2X). The fruits have dark green skin with faint stripes. Fruits are somewhat triangular in shape with tough rind, red flesh and white remnants of false seed. TSS is 12-13% Average fruit weight is 5-6 kg. The number of fruits per vine varies from 3 to 6. It takes 115-120 days for first fruit harvest. It could not become popular due to irregular fruit shape and high cost of seed.

***Arka Manik :*** This has been bred at IIHR, Bangalore from a cross of IIHR 21 X Crimson Sweet. Fruits are round to oval with light green to dark green stripes and weight is about 6 kg. The flesh is deep crimson, with excellent granular texture. It is very sweet with 12-15% TSS. It has multiple resistance to powdery mildew, downy mildew and anthracnose. Yield potential is 600 q/ha. It has good keeping and transport quality.

***Bush Baby :*** It is a dwarf, mutant, help in restricting vine growth to 0.5-1 m and in controlling multiple branching from the crown of the plant. It helps in accommodating larger population (15,000-20,000 plants/ha) in a unit area so yield is higher.

***PKM 1 :*** Fruit dark green, flesh pinkish red. Yield 36-38 t/ha in 120-125 days. Released from TNAU.

***Fuken :*** Resistant to anthracnose disease. Released from HAU, Hisar. It is high yielder and produce good quality fruits.

***IIHR-20 :*** Fruit are oval to round, green round with deep blue angular strips, red flesh with 11-12% TSS, crop duration is 90-100 days.

***Crimson Sweet :*** Fruits are oval, light green ground with green to dark green strips, flesh is crimson; crop duration is 110-115 days.

***New Hampshire Midget :*** It is an early variety introduced from the USA. Fruit is small weighing 1.5 to 2 kg. Shapes is oval, skin colour bright green with dark green lacerations and red flesh inside. It take 28-30 days from the date of pollination to ripening. It is an early home-garden variety.

***Improved Shipper :*** An introduction from the USA, a big sized melon weighing 20-35 kg. The fruit is dark green with moderate sweetness (8 to 9% TSS).

***Durgapura Meetha :*** A late cultivar maturing in 125 days, fruits round with light green. Rind is thick with good keeping quality, flash sweet, TSS around 11% with dark red colour, average fruit weight 6 to 8 kg, seed with black tip and margin, released by Agricultural Research Station, Durgapura Rajasthan.

***Durgapura Kesar :*** Also a late cultivar, fruit weight 4-5 kg. Skin green with stripes, flesh yellow in colour, moderately sweet, seeds are large. Released by Agricultural Research Station, Durgapura Rajasthan.

***Arka Joyti*** **:** It is a mid-season F1 hybrid cultivar evolved at IIHR, Bangalore by crossing a local watermelon of Rajasthan (IIHR 20) with Crimson Sweet. Fruits are oval with deep blue angular stripes. Average fruit weight is 5-6 kg and each vine bearing 2-3 fruits. Flesh is bright crimson, sweet with 12-14% TSS. Yield potential is 800 q/ha.

***Santrupti*** **:** Fruits oblong, colour blackish green, reddish flesh, thick and hard skin. Average fruit wt. 4-6 kg, T.S.S 9.5 to 10 brix. Fruits mature in 95-100 days after sowing. High yielder. Resistant to wilt.

***Amrit*** **:** Fruits oblong, long, deep greenish skin with lines. Hard and thick skin. Deep red flesh. Average fruit wt. 6-8 kg, T.S.S. 10-10.5 brix. Fruits mature in 95-100 days after sowing., high yielder. Resistant to wilt.

***M.H.W. 6*** **:** Grow luxuriantly and high yielder. Fruits oblong, weigh 6-8 kgs,T.S.S,10.5 brix. Light green colour with blackish lines. Flesh colour blood red and sweet. Fruits mature in 100-110 days, high yielder. Resistant to wilt.

***M.H.W. 11*** **:** Fruits medium shaped, round, weigh 4-5 kgs, flesh red and tasty, T.S.S 11.0 brix.. Blackish green and hard skin. Good keeping quality. Medium duration. Better yield if planted within Nov to Jan. High yielder.

***MHW-16*** **:** Luxuriant growth and high yielder. Fruits round, weigh 4-5 kg. T.S.S 11.0 brix. Blackish skin with deep black lines. Reddish flesh and good yielder if planted during Nov.-Jan. It is a early hybrid.

***Suchitra (MHWM-285)*** **:** Fruits oblong, light green with dark green striped, rind flesh is scarlet red and firm, mature in 95-103 days after sowing. Average fruit weight 7-8 kg, TSS (brix)10.This bybrid is suitable for early sowing, export potential, good transportation quality. Yield potential is 110-135 t/ha.

***Shrinidhi (MHWM-505)*** **:** This hybrid is suitable for early sowing, export potential, good transportation quality. Fruits oblong, whitish green with dark green striped rind, flesh is scarlet red and firm. Fruits weighing 5-7 kg each. TSS (brix) 8 and mature in 95-105 days. Yield potential is 100-110 t/ha.

***BSS 358*** **:** Fruit round in shape, weight 4 to 6 kg. fruit colour light green with dark green strips. Ready in 85 to 90 days after sowing.

***BSS 359*** : This hybrid has vigorous growth which gives high production. Fruit is round, colour light green with darker green strips. Fruit weight is 4-6 kg and mature in 80-85 days after sowing. Flesh is bright red, sweet delicious and crispy. Tough rind stands handling and long transportation.

***BSS 360*** : Plants vigorous with good yielding ability.Fruits oblong, medium green with dark green strips; mature in 85-90 days after sowing. Fruit weight is 6 to 8 kg.

***Swapnil*** : Fruits oblong, dark black, 8-10 kg in weight. Fruits mature in 85-90 days after sowing. This hybrid is good for long transport and has very good and bright red coloured sweet flesh. It is high yielding.

***BSS 408*** : Fruits round, green to black with 4-6 kg weight each and mature in 85-90 days after sowing. This hybrid has good and sweet flesh and is good for long transport.

***Sweet Baby*** : Fruits oblong, netted, slightly green with 4 to 6 kg weight each and mature in 90 days after sowing. It is very sweet hybrid and good for transport.

***CWH-1*** : A Sugar Baby type hybrid with heavier yield, maturing in about 115 days. Vigorously growing vines hold longer in the field, enabling fruits to thicken and achieve heavier yields. Rind colour similar to sugar baby but thicker. Fruits round, each weighing 7.0-8.5 kg. Flesh colour deep red, quite firm and TSS 9.0-9.7. Recommended for long distance shipping.

***CWH-5*** : Medium size, middle maturity, vigorous growing and resistant to different diseases. Oval fruits, thin rind with dark-green strips. Red flesh, crisp, high sugar content. Good shipper.

***CWH-7*** : Large size, middle maturity, vigorous growing; highly resistance; oval fruits, thin rind with dark green stripes. Red flesh, crisp. High sugar content. Good shipper.

***CWH-8*** : Large size fruits (7-9 kg) of light green stripes. This hybrid is widely adapted. Exhibits robust growth and excellent fruit set. High yield potential. Short oblong fruit feature crisp, red flesh. The strong outer rind permits long-distance shipping.

***GC-282*** : This is early maturing hybrid. Fruits can harvested in about 70-75 days after sowing. Fruits are round, blackish green and weigh-5-6 kg. The pulp is deep red, very sweet and crisp. Resistant to fruit cracking and good for long distance transportation.

*GC-285* : Early maturing hybrid with vigorous and prolific vines. Fruits can be harvested in about 85-90 days after sowing. Fruits are medium to large, weigh 7-9 kg, uniformly oblong, light green contrasted with dark green stripes. Flesh uniformly red, firm and very sweet. The hybrid has the ability to adopt to different soil types and good for long distance transportation.

*GC 286* : This hybrid has vigorous and prolific vines. Fruits can be harvested in about 75-80 days after sowing. Fruits round, nearly black in colour and weigh 6-7 kg. Flesh bright red in colour, crisp and very sweet.

***Disco (GC 287)*** : Prolific growing vines with excellent fruit setting habit. Fruits canbe harvested in about 70-80 days after sowing. Fruits are small, oblong, light green with dark green stripes and weigh 2.0-3.5 kg. Flesh uniformly red, sweet and crisp. Excellent for long distance transportation and ideal for consumers who prefer small size fruits.

***Maya*** : Medium early maturity hybrid with vigorous vine which produces high yield and protects fruits from sunburn. An excellent choice for early market. Flesh is sweet, bright red in colour. Fruit weighing between 10-15 kgs. Its blocky shape and dark green colour, makes it an ideal candidate for supermarket display. Resistant to Fusarium wilt race1.

***Madhuri (64)*** : This early maturing hybrid is vigorous in growth and produces oval to oblong fruits weighing upto 12kg. Flesh is red in colour, juicy and granular in texture with small internal seeds and sweet (11-12Brix).

***Khushboo (10)*** : This noval variety has very vigorous vine growth, produces round to oblong fruits weighing upto 13kg.Fruit colour is very attractive shiny and dark tiger stripe skin. Flesh is juicy and dark red with granular texture and possesses high TSS (13%). Excellent quality for storage and shipment.

***Madhubala (80)*** : Medium early, vigorous and prolific bearer, large oblong fruits have shiny light green rind with dark green stripes. The flesh is red and firm, juicy with about 12% Sugar context. This flesh is red and firm, juicy with about 12% sugar content. This variety features excellent keeping quality and strong outer rind for long distance shipping. Avoid stress conditions.

***Surabhi :*** This early maturing diploid hybrid with a vigorous growth, produces more fruits and protect fruits from sun burn. Globe shaped fruits are dark green in colour and good for long transportation and storage. Flesh is firm, crispy, sweet and juicy. Partial resistant to high temperature and fusarium wilt race 1.

***Mumtaz (99031) :*** This hybrid is highly productive with excellent fruit setting and well adopted to wide range of growing conditions and climates. Large oblong fruits are light green with dark green stripes weighing about 12-13 kg. The bright red flesh is sweet, crisp and granular. Good self life and excellent for shipping.

***Red Honey :*** This very early maturing hybrid has round fruit shape. Medium size fruits weighing 4-5 kg. and have attractive tiger skin. Bright red flesh and granular texture has high TSS (11-12%). A good shipper and very productive.

***Sultan :*** A proven dragon type hybrid produces oval –oblong shaped fruits with maturity period of 36–38 days from flowering. Thickness of rind is about 1.0cm. Flesh is crisp and juicy and fruits weight about 7–14 kg. Variety tolerant to fusarium, anthracnose and also to dry and moist conditions.

***Emerald :*** A dragon type oval–oblong hybrid. fruits with light green rind and dark green stripes weigh about 8–14 kg. flesh is crimson red, crisp and juicy. Fruit matures in about 90 days from sowing. Extremely good for shipping.

***Maharaja :*** An early maturity dragon type hybrid produces big oval–oblong shaped fruits weighing 10–12 kg., flesh is dark red and very sweet with high brix.

***Vandana :*** This crimson sweet type hybrid is vigorous and highly productive. Oval shaped fruits are striped, flesh is deep red, sweet and fibreless. Fruits weigh 8–12 kg and are very uniform in size throughout harvesting period. Tolerant to fusarium and anthracnose.

***Black Baby :*** An early maturing sugar baby type hybrid produces round very dark green fruits of about 4–8 kg, flesh is crispy red with delicious flavour. Very good shipping quality.

***Red Sweet :*** Early maturity sugar baby type hybrid produces round fruits having green rind and very sweet crispy red flesh of delicious flavour, fruit weighs 6-9 kg and keeping quality is excellent. Variety tolerant to fusarium 0&1 and Anthracnose.

***Sweet Ruby :*** An early maturity high yielding Charleston Grey type hybrid. plants produce oval-oblong shaped light green fruits of size 7-12kg., flesh is sweet and attains dark red colour on maturity. Good for shipping due to tough rind. Tolerant to Fusarium wilt and Anthracnose.

***Hybrid 113 :*** This dragon type medium early hybrid plants produce oval-oblong shape fruits with crisp sweet flesh, light green rind with dark stripes, fruit weighs 8-12kg.

***Augusta :*** Augusta is a large fruited sugar baby type hybrid. Fruit size ranges from 6-10 kg. fruits are globe shaped and have solid dark green exterior rind. Maturity is about 75–80 days after sowing. Flesh colour is bright red and very sweet in taste. Seeds are relatively small and light brown in colour.

***Crimson Tide :*** Vine grows vigorously and matures in 80-85 days. It is one of the earliest hybrid. Fruit is light green, blocky round with dark green broken stripes. Flesh is bright red sweet and tasty. Average weight of fruit is 8-10 kgs. Seed are small, dark brown in colour. Strong,sturdy fruit being ideal for long distance shipper. Yield 100-110 mt/ha in ideal conditions.

***DR27-2 :*** A seeded, red flesh hybrid in the striped Oriental-type. High round fruits of approx. 6-8 kg have clear, medium–thin stripes. Plants are of medium early maturity with strong vigour. Big attractively – striped fruit with juicy tender flesh. No hollow hearts. Yield 100-110 mt/ha in ideal conditions.

***Sugar King :*** This vigorous, good fruit setting hybrid is resistant to fusarium wilt and produces round, dark black skinned fruits. The rind is hard and can stand long distance shipping. Each fruit weighs about 8 to 10 kg and has good taste. Yield 100-110 mt/ha.

***Super Dragon :*** Strong resistance to Fusarium wilt and good shipping quality. Medium late maturing hybrid with plant vigour and excellent fruits setting. Oblong shaped. Rind is light green in colour with good stripes. Flesh is bright red and firmly textured. Consistently stable sweetness. Highly resistant to Fusarium wilt. Better shipping and storage quality.

***Rasila :*** Crop mature in 80 days after transplanting; fruit very sweet (TSS 10%), fruit shape round, dark green in colour, flesh deep red juicy, fruit weight 2.0-4.0 kg.

***Raj Ras*** **:** Fruit mature in 75-80 days after transplanting, sweet with high TSS, shape round with green and dark stripes, flesh red, crispy, less seediness, fruit weight 2.5-4.0 kg.

***Rangeela*** **:** Fruit mature in 75-80 days after transplanting, fruits very sweet (TSS 11%), shape elliptical (oblong), rind colour light green, flesh deep red, juicy and crispy, fruit weight 2.5-4.0 kg.

***Rasmadhu*** **:** Fruit maturity 80 days after transplanting, fruit very sweet (TSS 12 – 13%), shape oblong, rind colour green with dark green stripes, flesh crimson red, juicy and crispy, fruit weight 3.0-6.0 kg.

*NS 295 (F1) :* This is a jubilee type, medium to early (90 days) hybrid with good fruit setting ability. Fruits are blocky oblong in shape with an average fruit weight of 7-8 kg. Large fruits weigh upto 14.0 kg. The rind is light green with dark green stripes. The flesh is deep crimson with good granular texture and high TSS (12-13 %). Good transport and keeping qualities.

*NS 294 (F1) :* Medium to early jubilee type, hybrid with dark green stripes on green rind pattern. Fruits are oblong blocky in shape with an average fruit weight of 5-7 kg. Has good flesh colour and granular texture with a TSS of 11-12 %.This hybrid is an excellent yielder and matures in 85 days.

*NS 296 (F1) :* This hybrid produces blocky oblong fruits with jubilee pattern. It has deep crimson coloured flesh with good crisp texture and high TSS (11-12 %). It is a medium maturity hybrid and a prolific yielder with an average fruit weight of 6-7 kgs.

*NS 246 (F1) :* Medium to early (90 days) hybrid with large sweet fruits; oval blocky fruits weigh 7-8 kg. The rind is dark green with blue stripes. The flesh is deep crimson with fine texture and good TSS (12-13 %). This jubilee type hybrid is resistant to fruit cracking and has good transport quality.

*NS 1004 (F1) :* An early (80-85 day), jubilee type hybrid with excellent fruit setting ability, very attractive round fruits, shiny light green rind with dark green stripes. Flesh colour is deep crimson with very good texture and good TSS (12-13 %). Fruits are medium in size (5-7 kg) and well suited for long distance transport with good keeping quality.

*NS 2332 (F1) :* An early, jubilee type hybrid with very good fruit setting ability. Very attractive round to oval fruits with shiny green and dark green stripes on rind. The flesh is very attractive crimson red in colour. Each fruit weigh 6-8 kgs. The texture is very good and tastes sweet (TSS 11-12 %).

***Mayur (NS 261; F1) :*** Fruits are large blocky oblong, green rind with cal sweet type, dark green stripes and an early hybrid. The vine is strong which gives excellent protection from sun scald. The fruits weigh 8-12 kgs with brilliant red flesh and high TSS (11-12 %). It is suitable for long shipping. Performs well in a wide range of condition under good management.

***NS 271 (F1) :*** This is medium to early crimson sweet type hybrid with good fruit setting ability. Very attractive round to oval fruits which are medium to big in size (7-8 kg), flesh is crimson in colour with very good texture and TSS (11-12 %). It is well suited for long distance transport being of good keeping quality.

***NS 281 (F1) :*** Medium sized round fruits weighing 6-7 kgs. This is a Charleston grey round type (rind pattern) early hybrid has very good setting ability and wide adaptability. Rind is shiny light green with green veins. Flesh is bright red and juicy with TSS 10-11 %. It is good for long distance shipping and a heavy yielder.

***NS 321 (F1) :*** This hybrid has good vigorous vines. It bears fruits which are round to oval in shape. The rind is dark green in colour with an excellent fruit appearance (sugar baby type). Each fruit weigh 6-8 kg. Internal quality is superb with excellent flesh colour, crisp texture and sweetness (TSS 2 %). It is an early maturing hybrid with an excellent yield potential.

***Tejas (NS251) F1 :*** This is an early maturing sugar baby type hybrid produces round fruits of 6-7 kg with deep green rind and attractive red flesh of high TSS and granular texture. This superior yielding hybrid gives high quality fruits with excellent uniformity in size and shape.

***NS 2726 (F1) :*** This hybrid is ice box type with each fruit weighing 2-3 kg has jubilee pattern on the rind. Fruit is of oblong shape and has attractive deep crimson coloured flesh with excellent texture. Fruit will attain TSS of 12-13 % upon maturity (80 days). The rind is tough making it excellent for long distance transport. It is a heavy yielder.

***NS 4088 (F1) :*** Excellent ice box type hybrid with fruits of 2-3 kg and jubilee pattern on the exterior. Fruits are oval blocky in shape with excellent flesh colour and texture. Excellent tasting hybrid with good TSS of 13-14%. This hybrid matures in 80 days and is a high yielder.

***NS 54134 (F1) :*** Green Charleston type ice box hybrid. The fruit weighs 3 kgs and has an excellent oval shape with attractive red colour and crisp texture. It has good sweetness (11-12 %) and mature in about 80 days. It is an excellent yielder.

***Tokita 786 (F1) :*** Vigorous vine, green to dark green foliage, blocky oblong, rind is light green with dark stripes. Average fruit weight 10-12 kg. Flesh is very sweet crispy, deep red with granular texture. Thick and hard rind, good for long transportation and storage. TSS 12-12.5 %.

***Dark Dragon (F1) :*** Very early and vigorous, sets fruits easily. Very high yield. Fruits are oblong with smooth rind. Maturity 65-70 days after sowing. Dark green rind and bright red flesh. Tender and sweet. Fruits are uniform in size, around 2.0-2.5 kg. Resistant to fusarium, anthracnose and phytophthora.

***Dark Belle (F1) :*** Very uniform fruit set, high yield. Fruits are oblong with dark green stripes. Smooth and thick rind good for transportation, red flesh and very sweet (sugar content about 14 % brix). Early maturity, about 60-70 days after sowing. Fruit weight 3.0-3.5 kg. Resistant to fusarium, anthracnose and phytophthora.

***Missile (F1) :*** This widely adapted hybrid exhibits robust growth and excellent fruit set. The short oblong shaped fruits are light green with dark green stripe. The fruit size is 2.0-4.0 kg and is excellent for shipping and has a good shelf-life. The flesh is red and crisp. This mid-early hybrid mature in 70-75 days.

***Apolo (F1) :*** This very early baby Charleston type hybrid features small easily marketed fruit. The flesh colour is deep red and sweet.This hybrid is an excellent shipper.

***Black Rocket (F1) :*** This hybrid exhibits excellent growth, has a strong vine and the ability of tolerating to disease especially fusarium and highly yield potential. The oblong shape fruit has high brix with dark green rind and vague stripe, red flesh. Fruits weigh from 4-5 kg, ship very well and have excellent shelf life. This hybrid matures in 55-60 days.

***Sugar Belle (F1) :*** Strong growth and superior to other hybrid. Fruits are oblong shape, green rind with dark green stripes, uniform fruit.Average fruit weight is about 3.0 kg, red flesh, high brix, few seeds good for transportation. Maturity is 55-60 days after transplanting.

***Big Boy (F1)*** **:** This robust growing hybrid bear compact fruit with few seeds which is excellent for shipping. The fruit is oblong with light green rind and stripe. Fruit weighs from 4-5 kg, red flesh, and high brix. It is highly tolerant to disease and harsh weather.

***Long Dragon (F1)*** **:** This hybrid performs excellent growth and superior to other. The fruit is oblong shape with red flesh and sweet, light green rind and dark green stripe. This thick rind is an excellent shipper and perform compact fruit in the field, fruit weighs from 7-8 kg. This is superior to other in tolerating to disease and high yield potential.

***Golden Tiger (F1)*** **:** It is highly productive hybrid, bear sweet and very compact fruits. Maturity 90 days after sowing. Fruits oblong, light green rind with dark green stripes, weighing 5-6 kg, yellow flesh. Resistant to fusarium 1.

***Baby Dragon (F1)*** **:** This hybrid is highly productive bear sweet and very compact fruit, mature in 90 days after sowing. Fruit oblong, light green rind with dark green stripes. Fruits weighs 5-6 kg and flesh yellow, resistant to fusarium 1.

## WINTER SQUASH

Winter squash *(Cucurbita maxima* Dutch) 2n = 40, is vining type annualor rarely bush, only slightly harsh to touch, stem soft round, androecium short, thick, columnar, peduncle soft basically round, but enlarged by soft cork, immature fruits are edible, flesh fine grained, without gelatinous substance sticking to seed, funicular attachment of seed acute asymmetrical; seed margin smooth, white or pale brown, not separating easily from pulp.

### Varieties and Hybrids

The important varieties and hybrids of winter squash are :

Mammoth, Hubbard, Delicious, Buttercup, Arka Suryamukhi, Monitor,

***Nunhems seeds*** **:** Monitor (8506).

***Arka Suryamukhi*** **:** It is developed at IIHR, Bangalore. It is a selection from foreign introduction. The fruits are attractive, small, each weighing 1-1.5 kg, round and deep orange, with bright orange flesh. Fruits have good cooking, keeping and transport qualities. The variety is resistant to fruit fly and yields up to 350 q/ha in 100 days.

***Monitor (8506) :*** This hybrid is suited for open field cultivation and has semi open leaves with semi upright plant habit, long petioles facilitate easy harvesting.This is mid season maturity hybrid with long harvest season. Fruits are long cylindrical in shape, very glossy and suitable for long distance shipping.This hybrid is resistant to CMV + ZYMV and also has partial resistance to WMV2.

***Gold Keeper :*** This is a hybrid of winter squash with compact vine habit. Fruits mature in 90-100 days. Fruits are oblate shape with orange red colour of size 20-23 cm at marketable stage. High TSS content and suitable for processing.

***Table Ace :*** It is a superior acorn winter squash. This hybrid has solid, tasty flesh with almost no fibre or stringiness. A compact semi-bush plant allows more plants per acre. Fruits mature 5-10 days earlier than open pollinated acorn varieties. Fruit mature in 70 days after sowing. Full across rounded ribs, blunted blossom end, colour dark (black) green with 13-15 cm size at marketable stage.

□□□

# 8

# Leafy Vegetables

## ALTERNANTHERA

Alternanthera (*Alternanthera sessilis*); syn. Ponnanganni Greens; belongs to the family Amaranthaceae; is a small annual or perennial prostrate herb grown in hotter parts of India up to an altitude of 1200 m. Leaves are variegated green and yellow or bronze and green or red and pinkish brown. Leaves are fleshy when grown in moist conditions. Leaves and tender shoots are clipped and used after cooking. Due to ornamental value it is grown as edge plant in garden. Alternanthera is propagated through cuttings or by division.

### Varieties and Hybrids

There is no any improved varieties in this crop. Only local materials/genotypes available with growers are used for cultivation.

## AMARANTH

Amaranth (*Amaranthus* spp.); 2n = 32, 34, 64; syn. Chaulai, Khada, Sag, Arakeerai, Marsa, Banopatanate, Tandalji, Elam, Chemcheera, Math, Tandukeerai, Norpa, Serukeerai, Pungikeerai; is one of the oldest and most common leaf vegetable grown in India. Leaves and succulent stems are used as vegetable. They are very good source of iron and vitamin A.

It belongs to the family Amaranthaceae, genus *Amaranthus*, *A.* spp. Amaranthus is one of the main genera of the large and taxonomically diverse group of leaf vegetables. The genus *Amaranthus* includes 50-60 species, cultivated for leaf (green) and grains and a few are wild species.

Majority of leaf cultivars grown in India belongs to *Amaranthus tricolor*. Major species found in india are :

*A. tricolor* (syn. *A. gangeticus*),

*A. mangostanus*, *A. polygonoides* – cultivated-leaf type.

*A. Dubius* – cultivated leaf type.

*A. blitum* (syn. A. lividus) - cultivated leaf type.

*A. tristis* – cultivated leaf type.

*A. viridis* – wild leaf type.

*A. spinosus* – wild leaf type.

*A. hypochondriacus* – cultivated grain type.

*A. cruentus* – cultivated grain type.

*A. caudatus* – cultivated grain type

## Varieties and Hybrids

Cultivated leaf amaranthus varieties and cultivars differ in size,shape and colour of leaves and stems, position of inflorescence etc. and belong to different species. A list of varieties and hybrids developed by different institutes is given below.

***T.N.A.U., Coimbatore :*** CO1, Co-2, Co-3, Co-4, Co-5, Sirukeerai.

***I.C.A.R.I., New Delhi :*** Pusa Chotti Chaulai, Pusa Badi Chaulai, Pusa Kirti, Pusa Kiran, Pusa Lal Chulai.

***IIHR, Bangalore :*** Arka Suguna, Arka Arunima.

***OUAT, Bhubaneswar :*** Utkal Mayuri, Utkal Cheka Leutia.

***Annapurna beej :*** Champal Nata, Cheka Leutia.

***Tropica Green Co :*** Green, red green, red.

**Detail characteristics of varieties and hybrids are mentioned below:**

***CO.1 (A. dubius) :*** It is a pure line selection and is suitable for tender greens as well as matured stems. It yields 2.150 kg of greens per 3 sq.m. area (7165 kg/ha) at 25$^{th}$ day after sowing. The leaf stem ratio is 2.0; lacks initial vigour but makes rapid growth after 30$^{th}$ day, thus making it specially suitable for late harvest. Leaves dark green with ridged appearance; stem dark green, round and succulent; takes 50 days for flowering and 90 days for seed maturity; plants tall 150 cm; seed yield of 1.5 t/ha. Seeds are small and black, tolerant to pests and diseases.

***CO-2 (A tricolor) :*** It is a selection from local type and suitable for early harvest. It yields 3.230 kg of greens per 3 sq.m. area (10780 kg/ha) at 25$^{th}$ day after sowing; the leaf/stem ratio is 1.8; early germination and vigorous growth make it more suitable for early harvest as tender greens; leaves green, lanceolate, slightly elongated; stem green, succulent, take 42 days for flowering and 80 days for seed maturity; plants medium tall with 125 cm; seed yield 2.80 t/ha; seeds bolder and black; tolerant to pests and diseases.

***CO-3 (A. tristis) :*** This variety is a clipping type; clipping can commence 20 days after sowing and can provide continuous supply for three months. It gives a yield of 30 t/ha. This cultivar lends itself for 10 clippings commencing from 20 days after sowing. The leaf-stem ratio is high and this enhances the palatability of the cooked greens to a great extent.The leaves of this cultivar is nutritious with 35.9 mg of ascorbic acid per 100 g of fresh matter. It flowers in 35-40 days after sowing and matures at 85-90 days.

***CO-4 (A. hypochondriacus) :*** This grain type makes rapid vegetative growth within a period of 20-25 days. Plants are dwarf; grain yield 2.0-2.5 t/ha in 80-90 days.

***CO-5 (A. tricolor) :*** Leaves double coloured with green and pink and is free from fibre. It gives a rosette growth in early stages and first harvest is possible in 25 days; yield 40 t/ha in 55 days.

***Sirukeerai (A. polygonoides) :*** It is a traditional cultivar in Tamil Nadu, suited for uprooting at 25 days after sowing; leaves are small, ovate with blunt bifurcated tip and have long petiole; collar region is dark pink and at leaf axil a miniature branch initiates.

***Pusa Chhoti Chaulai (A. blitum)*:** This cultivar has been developed through selection and suitable for leafy shoots. Plants erect with thin stem, slightly dwarf; leaves small, green in colour,; responds well to cutting; flowers are borne in clusters in leaf axils. It is best suited for early summer and can be grown in rainy season also.

***Pusa Badi Chaulai (A. tricolor)* :** It is also developed through selection. The economic product of this cultivar is leafy shoots. Stem thick, green; leaves large and green; responds to cutting and distinguishable by its much larger growing period. It is best suited for sowing in summer and rainy seasons.

***Pusa lal Chaulai (A. tricolor)* :** Amaranth Pusa Lal Chaulai is a high yielding and red pigmented variety. Its stems and leaves are bright red (magenta), it is ideal for ornamental purpose also. It provides stem and leaf a 1 : 5 ratio; provides first flush of leaves 33 and 24 days after sowing in spring summer and kharif seasons. On an average its yield is 49 and 45 t/ha in spring and kharif seasons.

***Pusa Kiran (A. tricolor)* :** It is a promising selection No.S-44-6 was developed from a natural cross between *A. tricolor* and *A. tristis*. The selection is a combination of good character of both species. Leaves are glossy green with broad ovate lamina. The lamina is 7-9 cm long and 6-7 cm wide.The petiole is 5.0-6.5 cm long.The thickness of the clipped stem is 0.45-0.55 cm with internode length of 1.3-1.8 cm. The stem : leaf ratio is 1 : 4.6.It is ready for harvest 21-25 days after sowing, the duration of harvest being 70-75 days. It take 95-100 days to flower. It supplies 35 t/ha and suitable for kharif cultivation.

***Pusa Kirti (A. blitum)* :** The leaves of Pusa kirti are green with broad oval lamina, 6-8 cm long and 4-5 cm wide, petiole is 3-4 cm long.The stem is green tender with an internode distance of 1.3 to 1.5 cm; the thickness of the clipped stem is 0.45-0.50 cm. Stem : leaf ratio is 1 : 3.4. It produces an average yield of 55 t/ha and has the largest duration (70-84 days) of harvest and it takes 100-120 days to flower. First flush, it leaves is ready for harvest 30-35 days after sowing. It is highly suitable for summer cultivation and it can be grown in kharif also, the yield is distinctly higher in summer.

***Arka Suguna (A Tricolor)* :** It is a pure line selection from Taiwanese introduction (IIHR 13560), provides highly succulent stems along with attractive broad and green leaves which are rich source of proteins, minerals and vitamins A and C. It is ideal for multicut without becoming fibrous and become ready for first picking 25 days after sowing. The leaves may be cut at 10-12 days intervals (upto 90 days duration). It regenerates at the faster rate, the total yield being 27 tonnes/ha of fresh greens. Moderately resistant to white rust, its leaves and stems are rich in calcium and iron.

***Arka Arunima (A tricolor) :*** It is a multicut, purple amaranth variety. Its leaves are broad and dark purple in colour. It becomes ready for first picking 30 days after sowing. Its two subsequent cuttings at 10-12 days interval canbe taken with a total yield of 27.4 t/ha. It has been developed through pureline selection from 'IIHR 49'. Its leaves are rich in calcium and iron and low in antinutrient factors like Oxalate and nitrates. It is resistant to white rust.

***Utkal Mayuri :*** Plant medium in stature and highly branched, leaves are multicolour (pink with green border). Both stem and leaves are tasty. Regular leaf plucking is possible and it gives number of leaf harvests and better yield. Suitable for both the seasons of summer and kharif. Average plant weight is 570 g to 925 g. Mature plant weigh upto 3.50 kg.

***Utkal Cheka leutia*** : Plant dwarf and spreading type. Average height is 18.4 cm. First harvest in 37 days after sowing. Leaf green and small type. Prunning gives better yield. Perform very well in summer and kharif season. Average weight of 10 plant is 65 g.

***Lal Sag*** : It is grown quite popularly in many states of the country. It belongs perhaps to *A. mangostanus* and is a high yielding Indian amaranth variety. It produces seed early but has small flowers.

***Amaranthus cruentus*** : It is a edible leafy amaranth, it complete germination in 5 days, light green coleoptiles of sprouting seedlings. Seedlings are slow growing, leaves are dark green, shiny, petioles are short, roots are white, free from diseases and pests.

***Amaranthus hypochondriacus*** : It is another edible sp. of amaranth and it complete germination in 4 days. Anthocyanin coleoptile of sprouting seedlings, which are fast growing. Leaves are light green, shiny. Petioles are long, roots are anthocyanin in colour. Free from diseases and pests.

***Cheka Leutia*** : It is green from February to May, leaves are ready for first cutting, 35-40 days after seed sowing. Stems are red and leaves green and round. After first cutting it produces more branching.

*Champa Nata* : It becomes ready for harvesting, 30-45 days after seed sowing. Plants are tall and leaves are round. It is suitable for one time harvest.

*Green* : Green leaved variety. The leaves are green and elongated in shape. Mature in 25-30 days and produce yield around 12 t/ha. A very fragrant tropical herb, good for frying or making soup. It is main season variety.

*Red Green* : Red green is a leaves variety. Sharp and somewhat narrow leaves mature in 25-30 days., yield, around 12 t/ha. A very fragrant tropical herb.

*Red* : It is a red leaved variety. Broad dented leaves, mature in 25-30 days. Produce yield around 12 t/ha. A very fragrant tropical herb.

## BASELLA

Basella (*Basella alba, B. rubra*); 2n = 44, 48 belongs to the family Basellaceae, also known as Indian Spinach, Malabar Spinach, Malabar Night Shade, Vine spinach, Climbing Spinach, Gambian Spinach; is an annual or biennial twining or much branched herb distributed in tropics. It also behave as a perennial leafy vegetable grown against support in home garden and as an unstaked crop in a few areas in tropics. It is popular in Assam, West Bengal and parts of South India. Both types are used as wholesome vegetables and the tender stem, leaf stalks and leaf lamina are chopped and cooked. A colouring dye is extracted from red berries for colouring eatables, jellies, pastries and preserves in certain countries. Chinese sell the bright red juice in the form of a powder known as Gintjoo. The crop also has ornamental value. Many medicinal uses are reported in basella. It has a cooling effect and is used in digestive disorders and in treatment fistula, pustules and inflammatory tumours. It possesses antiviral properties and is applied to eyes for curing conjunctivitis.

The plant is a short-day and sensitive to extreme temperatures and come up under partial shade. Flowering is precluded at a day length of more than 13 hours but water stress promotes early flowering. Flowers are white or pink, small sessile in cluster on elongated thickened peduncles in an open branched inflorescence. Fruits enclosed in fleshy perianth and globose, dark purple, juicy and shining. This C4 plant is mainly propagated through seeds.

## Varieties and Hybrids

*There are mainly two distinct types* : (a)One type (*B. rubra*) is with reddish, petioles, and stems, consumed as vegetables and common in W.B, Assam, South India.(b) Second type is green leaves, petioles, and stems. It is commercially cultivated and belong to var. *alba*. It is more common in Orissa, U.P., Punjab and also in west Bengal, Assam, and south India. (c) Some others are *Basella lucida, Basella cardifolia, Basella nigra and Basella saponica*. These are not common as vegetable.

## BATHUA

Bathua (*Chenopodium album* L.) syn. Chakkravarthi greens, Goose foot or common pig weed, wild spinach; 2n = 18, 32, 36, 54; belongs to the family Chenopodiaceae, is nutritious leaf vegetable and tender shoots are cooked as vegetable or eaten as raw in salads. The crop is known as 'Paripu keerai' in Tamil. It is a polymorphic, mealy white erect herb. Though the plant is slow growing, under cultivation it grows tall and leafy. The stem is slender, angled and green, red or purple coloured. Leaves are waxy coated and dull green in colour with pale green centre and waxy margin. Flowers are seen in clusters forming a compact or loosely panicled spike in axils. Seeds are shiving black in colour. Grain yielding forms are also available.

Bathua is propagated through seeds and is either direct sown or transplanted. Crop can be used as indicator of magnesium status of soils as its content in plant varies with contents in soil.

## Varieties and Hybrids

Two leaf varieties namely 'Pusa Bathua' from IARI, New Delhi and 'Ooty 1' from TNAU, Coimbatore are developed in bathua. Detail characteristics are given below :

***Pusa Bathua No. 1 :*** It is developed through a pure line selection from local collection. The plants are grown to a height of 2.25 m with purplish green leaves 10.5 cm x 3.0 cm in size. The thickness of clipped stem is 0.45 cm and length of internode 2.5 cm. The purplish-green stem is tender in texture. Its leaves in tender stage may be picked up 45 days after planting. They may be harvested for 150 days. The plants flowers 175-180 days after planting. It may be grown during October-March. On an average yield is 30 tonnes/ha.

***Ooty 1 :*** It has been developed from TNAU, Coimbatore. It grows to a height of 38-40 cm and yields 28.9 t/ha in 55 days in hills and 17.0 tonnes in 50 days in plains. Leaves are dark green and unfolded. Leaves are pinkish in colour. It is suitable to grow for its grain (seeds) production (1.2 t/ha) in 145 days crop duration. It shows resistance to Cercospora leaf spot, Colletotrichum leaf spot and Macrophomina foot rot.

Some other genotypes are Sel-2, Local-1 and Local-2.

## BEET LEAF (PALAK)

Beet leaf (syn. Spinach Beet), *Beta vulgaris*, var. *bengalensis* Hort., 2n = 2x = 18 belongs to the family chenopodiaceae; also known as Indian Spinach, Palang sag, Palang, Meetha palang, Teegabatchali, Dumpabucchale, Vusavyeley, Pasalai, Spinak soppu, Spinach, Keerai and Palak, in Hindi, is the most common leaf vegetable in India. Palak belong to same species in which Spinach beet, garden beet, beet root, sugarbeet and Swiss Chard are included. Since Palak was grown earlier in Bengal, it is known as *Beta vulgaris* var. *bengalensis*. It is mainly grown during winter in the plains and spring summer in hills.

### Varieties and Hybrids

Varieties/cultivars/cultures, based on pigmentation of mid-rib and leaf veins, are divided into two groups. One is with reddish mid-rib and leaf veins and the other with green mid-rib and leaf veins.

All Green, Pusa Palak, Pusa Jyoti, Pusa Harit, Jobner Green, Palak No.51-16, Banerjees Giant, Pusa Bharati, Arka Anupama, Ooty-1, HS-23, Pant Composite are important commercial cultivars.

***Sungro seeds :*** Sungro All Green.

***Annapurna Beej*** : Red Kanka, White Kosla, Light red Kosla,

***Tokita Seeds*** : ($F_1$) upright

**Detail characteristics of varieties and hybrids are given below:**

***All Green*** **:** It is a pure cultivar and produces uniformly green tender leaves and gives about 6 to 7 cuttings at 15-20 days interval. It is a heavy yielder and yields about 125 q/ha of green leaves. During winter season it produces seed stalks in 75 days.

***Pusa Palak*** : It is a selection from a cross between Swiss Chard and local palak. It produces uniform green leaves without any purple pigmentation. It is a late bolting cultivar.

***Pusa Jyoti*** : This cultivar is evolved as a result of treating the seeds of cv. All Green with 2 % colchicine solution for 24 hours and selection made from subsequent generations upto 10 years. It is a giant-leaved strain and produces large, green, thick, tender, succulent and crisp leaves of good taste even when eaten raw as salad. When cooked they get easily minced and develop nice flavour and colour. Plants are very vigorous, quick growing and regenerate quickly after each cutting. It gives 6 to 8 cuttings and yields over 49 t/ha. It is rich in potash, calcium, sodium, iron and ascorbic acid and can be grown throughout the year.

***Silver*** : The improved variety 'Silver' supplied 10-15 t/ha of leaves in two picking in July and September-October. Two additional picking may be taken if the root of the existing crops are sown in poly house.

***Pusa Bharati*** : It is similar to Pusa Jyoti, the hitherto popular variety. Its cordate shaped leaves are 25 cm long and 14 cm in breadth, the petiole length being 15 cm smooth and tender, the leaves of Pusa Bharati are pure green without any red pigmentation. They may be picked up 30-40 days after sowing for 150 days duration. Its plants flower 175-180 days after sowing. On an average it gives 50 tonnes/ ha leaves in winter and early-spring seasons. It is ideally suited for both kharif and summer seasons.

***Arka Anupama*** : It provides medium-large, dark-green, succulent, crinkled and attractive leaves. It is a late bolter with vigorous initial growth and regenerated at a faster rate. Its leaves become ready for first picking 35 days after sowing. After that they can be cut at early 10-12 days interval (up to 75 days). The total yield comes 41.08 tonnes/ ha of fresh green from 4 picking. It has been developed by crossing IIHR 10 and IIHR-8 followed by pedigree method of selection. It is resistant to cercospora leaf spot under field condition. Leaves are rich in chlorophyll, carotene, vitamin E and iron contents.

***Pusa Harit*** : It has been developed by hybridization between sugar beet and local palak and suitable for cultivation in the hills throughout the year mainly because of its chilling requirement, a character inherited from its sugar beet parent. Plants are upright, vigorous growing with uniformly thick, green, slightly crinkled giant sized leaves. It is a heavy yielder with remarkable ability for rejuvenation. It has late bolting habit and wide range of adaptability to varying climates. It can also tolerate alkaline soils.

***Jobner Green*** : It was evolved by spontaneous mutation (point mutation) detected from the population of a local collection sl. no.5. It produces uniform green, large, thick, succulent, tender leaves with strong flavour having entire margin. Leaves are easily cooked and taste is comparable to cv. All Green. It is a prolific yielder and gives an average yield of about 296 q/ha.

*HS-23* : This cultivar is released by HAU, Hisar. It produces dark green, large thick, tender and juicy leaves. It is a quick growing cultivar, gives first cutting about 30 days after sowing and a total of 6-8 cuttings at 15 days interval.

***Palak No. 51-16*** : This cultivar has been developed by the Department of Agriculture, Maharashtra State. It produces green leaves, gives several cuttings and yields about 160 q/ha.It is a late bolting cultivar.

***Banerjee's Giant*** : It is a very popular cultivar evolved by S.P. Banerjee in West Bengal. This cultivar has been developed from a cross between local palak and beet root. It produces large thick leaves with very succulent stem and fleshy root.

***Ooty 1*** : A variety developed at the Horticultural Research Station, Udhagamandalam of TNAU by selection from a Himachal collection. First harvest is possible in 45 days followed by three more cuttings at 15 days intervals.

***Pant composite*** : A cercospora leaf-spot resistant variety developed at GBPUA&T, Pantnagar. It produces tender, juicy uniform green colour leaves.

***Red Kanka*** : It has red blood colour, grow very fast, soft and succulent leaves, generally cultivated in winter season. It is ready for harvest 25 days after seed sowing. One time harvest is common with this variety.

***White Kosala and Light red Kosala*** : Its cultivation is done in rainy and winter season. Both types are preferred by different locality and grown as a common leafy vegetable in states like Orissa, Bengal, Bihar, Jharkhand and Parts of Chattisgarh.

***$F_1$ Upright*** : It has shiny, very dark green leaves and 18-20 leaves per plant, very tasty, easy for harvesting. Resistant to downy mildew. Rich in minerals.

***Sungro All Green*** : It has uniform green, smooth, tender leaves. Start first cutting 25-30 days after sowing. Give 5-6 cuttings at 15-20 days interval. A late bolter, has very good market preference.

## CELERY

Celery (*Apium graveolens* L.); 2n = 2x=22, belong to the family Apiaceae, (Umbelliferae); known as Ajmud, Karas, Salari in Hindi, rank second in salad crops. It is a biennial plant, although grown as annual crop. There are three distinct cultivated forms or taxonomic varieties, grown for different economic purposes are classified as follows:

i) *Apium graveolens*, var. *dulce*, common name celery. The plants of this type develop succulent, solid petioles which are consumed mostly raw in salads or as celery 'sticks' and are often cooked in soups and stews. It is widely grown in Europe and in the Americas.

ii) *Apium graveolens* var. *rapaceum*, common name celeriac or root celery. It is characterized by the development of enlarged hypocotyl and root tissue. The resulting globe like structure, which has a strong flavour, is consumed mostly cooked in stews and soups or grated on salads. This crop is grown mostly in northern and eastern Europe.

iii) *Apium graveolens* var. *secalinum*, common name smallage or leaf celery. The plants of this group have slender, leafy and often hollow petioles and are used mostly for their leaves as a condiment garnish and for medicinal purposes. Smallage is consumed mostly in Asia and in the Mediterranean countries. Annual forms of this varieties are also grown for seed, which is used as a condiment and for beverage.

### Varieties and Hybrids

Vilmorin (1950) reported that two varieties of solid stem Italian celery were available in France, 'Solid celery' and Striped rose'. In 1875, Chemin selected a line known as 'Solid Golden White Celery' which became the variety 'Paris Golden' Yellow Self-blanching. In 1884 Pascal selected a green celery line from 'Paris Goden Yellow self blanching' which was known in the seed trade as pascal celery. In 1887, both varieties were introduced in North America under the names of 'White Plume' for the self-blanching type and 'Giant Pascal' for the green type.

Celery cultivars are two types, viz., self-blanched and green cultivars.

***Self blanched type*** : Self blanched or golden self blanch is tends to blanch easily and has less demand due to low vitamin 'A' content. They are Michigan Improved Golden, Cornell 19 (thick petioles susceptible to brown spot), Cornell 619 (thick petioles, resistant to fusarium yellow), Supreme Golden, Golden plume or Wonderful, Golden self-blanching, Golden No 15 and Florida Golden.

***Green cultivars*** : Green cultivars are gaining importance now a days. They are : Utah (dark green), Utah 52-70 (resistant to brown spot), Utah 10-B (have boron and magnesium deficiency), Giant Pascal, Emerson Pascal, Summer Pascal (resistant to brown spot) and Ford hook. All green cultivars are resistant to *Fusarium* yellows. Emerson Pascal is resistant to early and late blights also, but has a tendency to bolt.

***Cultivars from IARI*** : Some of the celery cultivars acclimatized at IARI, sub station, Kulu valley, are given below:

***Standard Bearer*** : It is an early medium tall plant, produces medium pink stems with longitudinal streaks, stalk solid and of good size and flavour.

***Wright's Grove Giant*** : This cultivar, is medium late, tall growing, high yielding produces large white stalks of fine quality.

***Fordhook Emperor*** : It is late cultivar, plant dwarf and stocky, produces solid, white, thick, broad and tender stalks.

***Triumen*** : It matures in 85 to 90 days, has dark green coloured attractive foliage and medium long petioles. It has good standing ability.

***Celery Paris*** : It is a resistant to heat and rain. The foliage is tall and dark colours. Stalks are thick smooth and fleshy. Seeds are very small (more than 2500 seeds per gram). Plants can be harvested 2-3 months after transplanting.

***Celery France leaf*** : It has vigorous dark green and sturdy foliage with long, thick and smooth stalks.

Some other cultivars are Utah - 52-70, Golden Self Blanch, White Gem, Grant Red, Belder, White Solid.

## COLEUS

Coleus (*Solenostemon rotundifolius*; syn. *Coleus parviflorus* Benth); 2n=32 belongs to the family Laminaceae, is commonly known as chinese potato or koorkan, a minor and seasonal tuber crop which produces tubers only during December-February. There is no seed production in this crop and is propagated only through vegetative means. Coleus is a unique plant of genus *Coleus* and has under ground tubers of economic importance. The another species *C. forskohlii* (Poir.) Briq. (syn. *C. barbatus* Benth.) is grown of which roots are used as spice and condiment and preparation of medicines.

### Improved Varieties

There are usually two types of coleus. One is having small sizes tubers and the other one large-sized. The varieties having small tubers are good in flavour. Whereas large tuber type produces higher yield.

***Shree Dhara :*** It is an improved variety. It yields about 20-25 t/ha over a 6 month period. It is released by the CTCRI, Thiruvananthapuram in Kerala. In other places local types are grown.

## CORIANDER

Coriander (*Coriandrum sativum* L.); 2n = 22, syn; Dhania, Dhaniya, Dhaniyalu, Kothamalli, Kottambri, Dhane, Kastumbri, Dhanyaka, Coriander, Kusbara; belongs to the family Apiaceae (Umbelliferae) is a tropical plant and is capable of tolerating drought and heat. Its name is derived from the Greek words 'Koris', meaning bed bug because of the unpleasant, fetid, bug-like odour of the green unripened fruits. The leaves are used for flavouring soups and other foods. The fruits are an important ingredient of curry powder, usually contributing the greatest quantity of all the spices followed by turmeric. Coriander is grown throughout the year for leaf purposes.

### Varieties and Hybrids

For essential oil production small seeded varieties are preferred. Whereas for spice use, large seeded varieties are mostly grown.

**Detail characteristics of varieties and hybrids are given below:**

***Sadhna (CS-4) :*** Plant semi-erect, tall, good bearing habit. It is suitable for rainfed condition. Early maturing, seed yield 10-12 q/ha in 115 days. Oil is 0.2 per cent. It is resistant to aphids and mites. Grown in Andhra Pradesh.

***Swathi (Lam CS-6) :*** Plants grow to medium height. It is suitable for rainfed areas of Andhra Pradesh. It is early maturing which helps in escaping powdery mildew incidence. The seed yield is about 9-10 q/ha in 95-105 days. Essential oil content is about 0.3 %. It is olerant to powdery mildew.

***DH-5 :*** It is developed by CCSHAU, Hisar through selection from a local collection and identified in 1993 for irrigated conditions. The plants are medium-tall and bushy. It matures in 120-130 days and produces an average yield of 18 to 20 q/ha under good management. Seeds are round, attractive and medium-sized (13.8 g/1000).

***Katwantype :*** It is recommended for Madhya Pradesh. It is very popular among the farmers due to maximum green leaf cutting, late maturity and taller growth habit. Crop duration is 150-170 days. The average yield is 10-12 q seed/ha.

***Lam CS-2 :*** It is a medium tall (80 cm) variety with more branching and bushy in nature. Mericarp is straw coloured. Crop duration is about 105 days. The average yield is 1300-1400 kg/ha. It is reasonably tolerant to pest and diseases as compared to local varieties. It is suitable for cultivation in Andhra Pradesh, Rajasthan and Gujarat.

***Lam CS-4 :*** A fairly tall rariety with good branching habit and more number of basal leaves. Crop duration is about 115 days. The variety is reasonably tolerant to pest and diseases as compared to the local. The average yield is 14-15 kg/ha.

***Lam CS-7 :*** It is a selection from north Indian acclimatized to condition of Andhra Pradesh. It is dwarf variety with more leaves of good flavour. Average yield is 9 q/ha. The essential oil content of seed is also high (0.3-0.48 per cent).

***RCR-435 :*** The cultivar is developed by RAU, Rajasthan through half-sib recurrent selection from a local collection of Jalore district and identified for release in 1995 for irrigated conditions. The plants are bushy with quick early growth and medium-sized grains (14.0 g/1000). It matures in 110-130 days and produces average yield of 10.05 q/ha.

*RCR-436* : It is developed by RAU, Rajasthan through half-sib recurrent selection from a local collection of Kota and identified for release in 1995 for limited moisture conditions. Plants are bushy with quick early growth and bold grains (16.0 g/1000). It matures in 90-100 days and produces average yield of 11.0 q/ha under limited conditions.

*RCR-446* : It is also developed by RAU, Rajasthan through half-sib selection from a local collection of Jaipur district and identified for release in 1997 for irrigated conditions. The plants are leafy and erect with higher number of grains per umbel. Seeds of this cultivar are medium in size (13.5 g/1000). It matures in 110-130 days and produces an average yield of 12 q/ha.

*Sindhu* : Plant growth is medium, suitable for unirrigated condition of Andhra Pradesh. High yielding (10 q/ha seed yield), tolerant to disease and pests.

*Gujarat coriander-1* : Plant growth medium, suitable for unirrigated conditions. The variety is bold seeded having higher yield potential and seed yield is about 11 q/ha in 110 days. Doing well in Gujarat.

*Gujarat Coriander-2* : Plant growth medium, suitable for unirrigated areas. It is bold seeded having higher yield potential of about 14 q/ha seed yield in 120 days duration.

*Naraul selection* : Plant growth medium, water requirement is less, bold seeded and high yielding (15-20 q/ha). Mostly grown in Haryana.

*Gwalior No. 5365* : This variety is resistant to fusarium wilt, medium seed size, produces 8 to 10 q/ha seed yield in 100-130 days. Generally grown in Madhya Pradesh.

*Chimpa S-33* : Pant grows tall, late maturing, high seed yield,18-20 q/ha in 180-210 days. it is an introduction from Bulgaria, Suitable for cultivation in Madhya Pradesh.

*Merrocan* : Seed yield 10-12 q/ha in 140-160 days. Oil content is high. Suitable growing in Madhya Pradesh.

*Rajendra Swati* : Plant grow medium in height, resistant to stem gall, disease and fruitfly. Variety is an early maturing. Seed yield is about 12-14 q/ha in 100 days. Seeds are rich in oil content. It is suitable for intercropping. It is released from Rajendra Agricultural University, Dholi Campus, Bihar.

***RCR-41*** : Tall erect plants, round and small seeds. Suitable for irrigated tracts, oil content 0.25 %, resistant to stem gall and tolerant to wilt, late type with 130-140 days duration, seed yield 12 q/ha.

***RCR-47*** : Plants grow in medium height, resistant to stem gall disease, seed yield is about 12-14 q/ha in 120-140 days. This variety is suitable for irrigated areas.

***Kiran (UD 14)*** : This variety is suitable for irrigated areas, resistant to stem gall, seed yield 12 q/ha in 130-140 days, oil content is 0.25 per cent.

***UD-20*** : Plant growth is medium, bold grain, suitable for growing in unirrigated areas, seed yield is 12 q/ha in 120 days.

***CS-287*** : Earliest type with 78 days duration, small bold grain suitable for rainfed crop. Tolerant to wilt and grain mould. Yield 5.10 q/ha.

***CO-1*** : Tall plant, suitable for both, leaf and seed production. Water requirement is not so high and cultivated as irrigated as well as rainfed. Crop duration is 110 days. Seed yield is 5.0 q/ha.

***CO-2*** : This is dual purpose variety. The yield potential for seed production in 90-100 days is 6-8 q/ha and the green leaf production is 10 q/ha in 40 days after sowing. Tolerant to drought, oil content 0.3 %.

***CO-3*** : This variety is superior to both CO1 and CO 2 in adaptability and seed production (8-10 q/ha). Suitable both for kharif and rabi seasons, lower incidence of fusarium wilt. Seed colour is attractive brownish yellow. All these three varieties (Co1, CO2 and CO3) are released by Tamil Nadu Agril. University.

***Pant Haritima*** : This variety is developed at G.B.Pant University of Agril. and Technology, Pantnagar. It yields better as green leaf as well as seed. Plant growth is heavy to medium, fine seeded, late maturing, high potential for seed production (15-20 q/ha). It is tolerant to stem gall disease (*Protomyces mecroporus*).

***NP (D) 92*** : It is a selection from local material grown around Delhi area. Seeds are small, light green in colour having less fibre.

***Small seeded*** : Seeds are small with less fibre and is most suited for growing in temperate regions. Volatile oil content is higher than bold seeded ones. The seed colour is light green.

***Bold seeded*** : Seeds are bold having more fibre and is most suited for growing in tropical and sub-tropical regions. Volatile oil (Coriandrol) is lesser than small seeded ones and seed colour is light brown.

***Caribe*** : It is developed by Bejo Sheetal Seeds, Jalna, Maharashtra. This hybrid has got broad leaves, nice aroma and is tolerant to bolting. Mature in 50 days after planting. It can be planted in kharif, rabi and in mild summer. It gives multiple cuttings.

## CURRY LEAF

Curry leaf (*Murraya Koenigii* (L.) Swreng. (syn. *Bergera koenigii* Koen; Chaleas koenigii Curz ex. Swingle); 2n = 18; is also known as mitha-neem, Kadhi limb, Burusunga and Basango, belong to the family Rutaceae. The genus *murraya* has been given after Professor John Adam Murray of Gottingen. It is an unarmed semideciduous aromatic pubescent shrub or small tree 3-5 m high with slender but strong woody stem and branches covered with dark grey bark and closely crowded shady crown. Used as a flavourant and leafy vegetable by the early inhabitants (Dravidians) of this tract. The crop has been introduced to South India when the Dravidians moved to south with the arrival of Aryans from Persia, the present middle East.

Curry leaf is a perennial, self-pollinated handsome aromatic and the tree is a true diploid with 2n = 18 (x=9). The seeded nature, viability of seed and fertility of seedlings indicate the true diploid nature of the plant.

There are two species *Murraya paniculata* (L.) Jack indigenous to Burma and *Murraya exotica* L. (Syn. *Chaleas. Paniculata* Mant, *Camminum sinense* Rumph, *Marsana buxifolia* Sonnerat) an ornamental shrub indigenous to India, Ceylon and China are related to *Murraya Keonigii*. The leaves are gland dotted and strongly aromatic. Flowers are terminal corymbose cymes. The berries are subglobose or elliposoid purplish black when ripe and two seeded. Curry leaf is propagated through seeds.

## Varieties and Hybrids

There are two types of curry leaf, viz., broad leaved and small leaved types. The small leaved types are more fragrant and hence used for extraction of essential oil.

In addition to these two types there are few eco-types under cultivation in south India. In Tamil Nadu, 'Senakambu' (Red petioled) is a local type and the leaves have better aroma and flavour due to higher oil content. Recently some new varieties have been developed. They are DWD 1 (Suwasini) and DW 2.

***DWD 1 (Suwasini) :*** It was evolved at University of Agricultural Sciences, Dharwad. It is a clone of single plant root suckers.The leaves are dark-green, shiny and highly aromatic. It is sensitive to low temperature in winter season and hence the bud burst is poor. The laves have an oil content of 5.22 % and can be dehydrated at 50°C without loss of quality and made into powder. The chlorophyll content is 0.1629 mg of per gram fresh leaf.

***DW 2*** **:** It was evolved at UAS, Dharwad. It is a seedling progeny of unknown source. The leaves are slightly pale-green and less aromatic. It is not very sensitive to low temperature and much superior in number of bud burst, internodal length and 8 times higher in growth of shoot than DWD 1.

# FENUGREEK

Fenugreek (*Trigenella foenum-graecum* L.), 2n = 16, family-Leguminosae (Fabaceae) sub family Papilionaceae; syn. Methi, Menti, Mentulu, Vendayam, Venthayam, Menthya, Hulabaha, Shamlita, Fenugreek, Sag, Menthai, Uluva, Vendia, Keerai; is an important leaf vegetable cum condiment having several medicinal properties. Commonly grown throughout the country especially in areas having cool and dry climate. Fenugreek (methi) is extensively cultivated in Rajasthan, Madhya Pradesh, Gujarat, Maharashtra and Punjab for its dry seeds for use of condiment.

There are two types of methi, grown in the country.

I. **Common methi (*Trigonella foenum-graecum* L.) :** This is quick growing herb producing upright shoots to a height of 40-70 cm. It produces dark green leaves with or without pink margins. Two-to-three white flowers are produced in axils of each leaf. Pods are straw coloured and boat shaped. This is harvested either by clipping young plants from base or by uprooting. It is cultivated almost in all parts of the country.

II. **Kasuri methi (*Trigonella corniculata*)** : This is also known as Marwari methi and Champa methi and it gives better performance in cooler parts. This is generally scented types. Growth of plant is slow initially and remains in rosette condition during vegetative growth period. It produces bright orange to yellow flowers. Pods are sickle-shaped and smaller than common methi. Its green leaves are used as vegetable while its seeds are used as spice and as pulse. Harvesting is done by clipping leaf growth several times. Harvesting by uprooting entire plants is not practiced.

Fenugreek is an annual erect herb with tri-foliate leaves, papilionaceous flowers and leguminous pods. The plants growing to height of 90 cm, flower in 30-37 days after sowing. Anthesis take place between 9.0 am and 6.0 pm and plants are typically self-polinated. Cultivated forms are diploid.

## Varieties and Hybrids

In certain parts, fenugreek also grown as a green fodder, but distinct types of methi are used for fodder purpose and for leaf vegetable purpose. Both scented and non-scented cultivars are also grown traditionally.

**Detail characteristics of improved varieties and hybrids are given below:**

Some improved varieties are Lam Sel-1, Pusa Early Bunching, Kasuri selection, Col, Rajendra Kranti, RMT-1, Methi No.14, Methi No.47, UM-112, Barbara, Kasuri Improved.

Lam selection 1, CO-1, Rajendra Kranti and RMT-1 are dual purpose varieties suitable for seed purpose as well as for green purpose.

***Pusa Early Bunching*** : An early variety developed at IARI, New Delhi is a quick growing, producing upright shoots in bunches and having white flowers, yellow and bold seeds. Its pods are 6 to 8 cm long, green in colour having 10-12 seeds per pod. First cutting can be made after 55-60 days of sowing. It is a heavy yielder.

***Lam Selection-1*** : It is a selection from the Regional Agricultural Research Station at Lam, Guntur of N.G.Ranga Agricultural University. Plants are bushy in nature with medium height, it is a high yielding variety with better leaf quality. It is a capable of yielding 740 kg of grain per hectare. The green yield is around 10-12 t/ha under irrigated conditions. It contains 5.3% of protein in seed.

***Prabha (NLM)*** : It is released by College of Agriculture, Jobner, Rajasthan, suitable for cultivation in Rajasthan and Gujarat. It is a pure line selection from a local collection of Nagpur district. Semi erect growth, leaves soft, sweet with dark green colour, can tolerate many pests and diseases.

***Rajendra Kranti (RMT-16)*** : This is a selection made at the Dholi Station of the Rajendra Agriculture University. The plants are medium tall, bushy and are suited for raising as a pure crop or as an intercrop. It yields around 1,200-1,400 kg/ha in a crop duration of 120 days. Seeds contains around 9.5% protein.

***RMT-1*** : It is a selection from the SKM College of Agriculture, Jobner, Rajasthan. The selection yields around 1500 kg of seed from a hectare over a crop duration of 140-150 days. The variety is tolerant to root rot disease and is moderately resistant to powdery mildew disease. The seeds contain 21.0 % protein.

***RMT-143*** : Plants are medium sized, seeds are typical yellow colour. Tolerant to powdery mildew. Maturity period is 140-150 days with an average yield fo 16 q/ha. Suitable to grow in heavier soils of Chittor, Bhilwara and Jodhpur area (Rajasthan). It was released in 1997.

***CO 1*** : It is a dual purpose variety released by the Tamil Nadu Agricultural University, Coimbatore. The crop matures in 85-90 days with a grain yield of 680 kg/hectare. When grown for leaf, the yield is around 9 tonnes/ha. Because the plants are dwarf and has a short duration, it suits well for intercropping and relay cropping. The seeds contain 20-22 % protein while, the leaf contains 15.9 % protein.

***CO-2*** : Plants are medium sized, high yielding, less susceptible to root-rot, suitable to grow in coimbatore, Dharmapuri, Guddalore, Ramanathapuram, Karur, Tiruchirapalli, Perambalur and Tirunelveli districts, maturity period is 95-100 days with an average yield of 140.0 q/ha.

***Pusa Kasuri (Kasuri selection)*** : It is a slow growing variety and remains in a rosette condition during most of its vegetative growth period. Its seeds are small and dark brown in colour and its flowers are yellow. Its pods are smaller in size than the former variety. It is a heavy yielder. Leaves of this variety have special fragrance. It becomes ready for first harvest after 60 to 65 days of sowing.

***Methi No.14*** : It is developed by Maharashtra Agriculture Department. It is high yielding and contain more Vitamin C in the leaves.

***Methi No.47*** : It is an improved variety of Maharashtra, high yielding and containing more vitamin C in the leaves.

***HM-10*** : Plants are bushy, bold, yellow, attractive seeds moderate tolerant to leaf spot disease, maturity period is about 140-150 days with an average yield of 20.1 q/ha.

***HM-57*** : It is suitable variety for Haryana and it is capable of yielding 1000 kg of grain per hectare.

***UM-50*** : It is superior cultivar for the greens and seed yield.

***UM-112*** : The plants of this variety are very long and cultivated for both seed and leaf purpose. It contains more amount of diosgenin, which is used for preparation of medicines related to embryo distrrson.

***IC-74*** : It is local cultivar having erect greens, more branching type plants, good as greens as well as seed.

***Barbara*** : It is having erect plants, yellow seeded and high trigonellin content in seeds.

***Lam Selection*** : The plants are medium in height and bushy with better yield potential.

***Lam Selection 1*** : It is developed by APAU, Andhra Pradesh through selection. Plants are medium in height and bushy with more branches having better yield potential of greens, 30 per cent more than local cultivars. Seeds are golden yellow in colour. It matures in 90 days and produces an average yield of 7.40 q/ha. It is tolerant to root-rot, powdery mildew, caterpillars and aphids.

***Hissar Sonali*** : Plants are bushy, semi erect with bold yellow attractive seed, moderate tolerant to leaf spot and root rot diseases. maturity period is 140-150 days with an average yield 19.0 q/ha.

***Hissar Suvarna (HM 103)*** **:** A quick growing dual purpose high yielding variety suitable for cultivation throughout the country.The plant is tall and erect with more number of branches. It takes 130-140 days maturity period and yields 16 to 19 q/ha bold seeds with typical yellow colour. This variety is tolerant to cercospora leaf spot and powdery mildew diseases. It was developed through pureline selection from local germplasm collected from Haryana Agricultrual University, Hissar.

***Hissar Madhavi (HM-350)*** **:** It was developed by HAU Hissar through single plant selection from the line PME 46-1. It is a quick growing dual purpose, plant erect, tall with more branches. Pods are long, dark green leaves, typical yellow colour. It yields 19 to 20 t/ha seeds. It is suitable to grow in Haryana, Rajasthan and Gujarat. It is tolerant to powdery and downy mildews.

***Hissar Mukta (HM-346)*** **:** This variety was developed by Haryana Agricultural University. Hissar through a pure line mutant from an indigenous collection from U.P. This is a quick growing seed type suitable to grow in Haryana, Rajasthan and Gujarat under both irrigated and rainfed conditions. Seeds are bold with tan green seed coat colour. It yields 20-30 q/ha seeds. It is tolerant to both powdery and downy mildew.

## KANG KONG

Kang Kong (*Ipomoea aquatica* syn. *I. Repens*); 2n = 28, 30 for annuals & perennial; also known as water convolvulus, belongs to the family convolvulaceae; is an aquatic trailing and floating herbaceous perennial or annual with hollow stem. Leaves are ovate-oblong and flowers white or pale purple. Young terminal shoots and leaves are used after boiling, steaming or frying. The plant is also a nutritious fodder and is ideally suited to areas where water stagnation is a problem. Plant produce root at nodes. Seeds and herbaceous cuttings can be used for propagation.

### Varieties and Hybrids

Kang Kong is an underutilized leafy vegetable which do not have much area under cultivation. Kang Kong is having local plant types which are used for cultivation of the crop. It is mainly growing wild in and around ponds particularly in heavy soil.

## KURRAT

Kurrat (*Allium ampeloprasum* L. var. *kurrat* Schweinf. ex Krause; 2n = 24; belongs to the family alliaceae; resembles a small leek with which it is interfertile, and is an autotetraploid. It is propagated by seed. The kurrat is not known in a wild state. It is of ancient cultivation in the near East, and is grown by Arab peoples for its edible green tops, which are used for fresh consumption and for seasoning foods. In Egypt, where there has been some increase in production, about 37 tonnes of tops per hectare, per annum are produced. The first cutting is taken 3-4 months after sowing, with further cutting at 3-6 weeks for 15-17 months, after which yields decline and the crop is again sown.

### Varieties and Hybrids

A number of named local materials are grown in different countries. Many clones have named on the name of area of production.

## LETTUCE

Lettuce (*Lactuca sativa* L.) ; 2n = 18, belongs to the family Asteraceae/ Compositae, is the most popular salad crop, generally grown for its tender crisp leaves/heads which are used as salad. It is highly self-pollinated crop. There are several wild species of *Lactuca* like; *L. sativa, L. sarrida,L. salinga, L. virosa.*

### Varieties and Hybrids

There are four types of lettuce : (a) Head type (var. *capitata*). (b) Leaf or cutting type (var. *crispa.*). (c) Cos or romian type (var. *longifolia*). (d) Asparagus type (var. *asparagina*).

Thompson (1951) and Rodenburg *et al.* (1960) recognized six morphological types of lettuce, crisp head, butter head, cos or romaine, leaf or bunching, stem and latin.

***Crisp head*** : The cultivars are heading types with wrinkled non-wrapper leaves, round large head (upto 1 kg), brittle-textured and tightly folded with green outer and white or yellowish inner leaves, having excellent shipping and handling abilities, var. New York, 515, Imperial 44, Great Lake and Alamo-1.

***Butter head*** : It forms relatively small, loose heads, with broad, oily, crimpled, soft- textured leaves vary in colour, size and appearance.

Var. Big Boston, White Boston, May King, Salamander and Waya head. These are most popular in Northern Europe and includes summer and winter type.

*Cos or Romaine* : The cultivars have elongated leaves forming a leaf shaped head; outer leaves are dark or light green with heavy ribs, inner leaves are finer and lighter in colour; coarse in appearance but eating quality is quite good. Var. Effel Tower, Pairs Island, Paris White and Dark Green Valmaine.

*Leaf or Bunching* : The cultivars are non-heading or leaf type; colour, heat sensitivity and quality vary but all produce a rosette or leaves. Var. Black Seeded Simpson, Prize Head, Australian and Salad Bowl for out door growing and Grand Rapids for greenhouse production.

*Stem* : It is also called 'celery lettuce' and is grown for its thick stem which is pealed and eaten either or cooked as vegetable. The leaves are edible but inferior in quality to the leaves of the other types var. Celtuce.

*Latin* : It resembles butterhead in forming loose head but the leaves are elongated and similar to Cos lettuce. It is mainly grown in the Mediterranean countries and in South America. Var. Gallega, Coriolla, Verde, Criolla Blenca and Madrilene (Ryder, 1979).

*Varieties distributed by IARI, New Delhi* : Chinese yellow (Loose leaf), Imperial 847 (Cabbage head), May king, Paris White (Cos), Slowbolt (Leaf lettuce).

**Varieties and hybrids from seed dealers :**

*Suttons* : Golden Boll, A.I. Unrivalled, Little Gem.

*Pocha's* : Big Boston, Burpeana, Waya-head, Great lakes, New York or Wonderful, Black seeded, Simpson

*Sada Shiv's* : Ice Berg, Imperial, Paris Cos.

*Verma's* : Big Boston, Wonderful.

*Nunhems* : Revolution, Bergamo, Dublin.

*Tropica seeds* : Minetto, X-01, Panama, Red sun, Fast Fall.

**Detail characteristics of varieties and hybrids are described below:**

***Great Lakes*** : It is crisp head type and has large head with firm green leaves and outer leaves being blistered. It is fairly resistant to tip burn but susceptible to powdery mildew.

***Chinese yellow*** : Early, loose leaf variety, light green crisp and tender leaves.

***Revolution*** : It has sensational red colour, very slow bolting, resistant to Bremia NL-1-16 and good vigour are important features of this Lollo Rosa type variety. Good for outdoor and indoor cultivation.

***Bergamo*** : This Lollo bionda type variety is bigger and slow in bolting. Fine serated leaves are very uniform in shape and have a high quality appearance. Resistant to bremia NL1-4,6,7,10,13-15 and suitable for glass house and outdoor cultivation.

***Fast Fall*** : This black-seeded loose-leaf lettuce tolerant to diseases. This variety performs large and vigorous plant, produce attractive highly filled crisp, light green leaves, popular all around Asia for fresh leaves consumption and can be harvested after 40-45 days.

***Dublin*** : This is a new medium sized variety with a fresh medium green colour. The head is filled quickly and it generates a firm fare sized ball, which is well closed at the base. Resistant to bremia NL 1-16.

***Punjab Lettuce No.l*** : It is a non-heading type. The leaves are light green shining. crisp and bears loose leaves. It takes 45 days from sowing to first harvest of fully developed leaves. Average yield is 8.75 tonnes per hectare. It gives high yield under suitable package of practice and in favourable climatic condition.

***Red Sun*** : A very attractive and fine testing leaf lettuce. Loose leaf type, with maturity 45-50 days after sowing. An all-year round growing type, with distinct glossy red colour leaves. Good resistance to bolting. Displays excellent development at high temperatures.

***Panama*** : Panama is a loose leaf type, very resistant to diseases, light green coloured leaves. The variety is slow to bolt.

***Minetto*** : Crisp head, batavia type, medium-small size. Leaves are crispy, deeply notched and medium-green. Black seeded. Maturity 25 days; nursery and 34 days from transplanting to harvesting. Minetto is resistant to heat and humidity (it can form heads even in tropical lowlands) and show very high tolerance to tip burn.

***X-01*** : It is a butterhead type. The head is medium size; leaves are slightly wavy and medium-green colour. Black seeded. Maturity 55 to 60 days after sowing. It is tolerant to heat and is slow bolting.

***Mansang*** : It is a new romaine type cultivar developed from a cross between Late Texagreen and Ithaca by the lettuce breeding team of National Institute of Highland Agriculture (NIHA), RDA in 2002. This new cultivar has brown seed colour and grey-green leaves. Bolting is a little later than 'Chaser green'. Average market yield is 33.9 t/ha. It shows a good taste, resistant to lettuce mosaic virus and tolerant to soft rot caused by *Erwinia carotovora* sub sp. *carotovora*.

***Adam*** : It is crisp head lettuce cultivar of NIHA, Korea Republic. It is a cross of 'Montello' and 'Salad Crisp'. It has good growth, compactness, head weight and high yield of 45.2 t/ha.

## MUSTARD

Mustard (*Brassica* spp. L.); 2n = 20; known as sarson, belongs to the family Brassicaceae, an important oil seeds and condiment, is also used as green leaf vegetable in plains and hills of north India. The mustard leaves and tender stems form the most popular winter greens in the rural areas of the Punjab, Delhi and western Uttar Pradesh. Mustard leaves are ready for harvest 20-25 days after sowing. Flower stalks appear after 30-35 days and are cut before opening of flowers. Flower stalks produced at base of stem are harvested subsequently.

Number of species are grown in different parts of the country. Specific varieties/types are available for vegetable purpose. These are:

*Brassica campestris* var. *dichotoma* – Brown sarson of Punjab.

*Brassica campestris* var. *sarson* : Yellow sarson of U.P., Bihar and Bengal.

*Brassica juncea* : Banga sarson of Delhi.

*Brassica campestris* var. *Campestris* : Lai of Assam.

*Brassica campestris* var. *rugosa* : Dried in summer and stored.

*Brassica napus* : Late type free from attack of aphids and is also grown for tender stalks.

### Varieties and Hybrids

There are very few varieties for leaf purpose. These are Pusa Sag 1, MR 702, Pusa Bold.

***Pusa Sag 1*** : It is the first ever released improved variety of leaf mustard. Plants fast and vigorous growing with good rejuenation habit, large, glabrous, green, pedurate, slightly blistered, ovate to broad, elliptical leaves having high density of double dentate margin and greenish white, round, tender, fleshy and long petiole. Leaf number per plant varies from 25-30 depending upon sowing time.It takes about 35-45 days from sowing to first harvesting, gives 2-3 cuttings in normal time sown crops and has late bolting habit. It is high yielding (75-80 tonnes/ha leaves).

*MR 702* : Its leaves are also used as a green. It gives 33.33 tonnes of green leaves in one hectare crop. Leaves are highly nutritious.

## ORACH

Orach (*Atriplex hortense*); belongs to the family Chenopodiaceae, also known as chakwat, is a winter vegetable grown in South (Karnataka) during winter season and in North (Kashmir) during warm season. In Kashmir, it is known as 'Ushtahak'. The tender stem and attractive green leaves are harvested by cutting or by uprooting young plants and picking young leaves frequently. Chakwat is propagated through seeds and are sown in 60-90 cm distanced rows at 45-60 cm apart. A full grown plant gives around 900 g leaves. It produces four types of seeds which are either black or yellowish brown in colour and with or without bracts. Yellowish brown seeds are of good quality. Black seeds require nitric acid treatment for germination.

### Varieties and Hybrids

There is no any improved variety in this crop. Only local genotypes are used for cultivation of this crop.

## PARSLEY

Parsley (*Petroselinum crispum* mill); 2n = 22; belongs to the family Umbelliferae (Apiaceae), is the most popular garden herb and cool season crop. It is one of the most familiar and widely cultivated of the garden herbs. It is a biennial or short lived perennial and a cool season crop. Parsley is a fragile plant with a white tap root and branched striated cylindrical stem. The leaves are three pinnate with oval lobes, basal ones being narrower than the apical and with a dentated or incised margin. The flowers are grouped in compound umbels of ray florets. The fruit is ovoid, small ribbed diachemium.

Parsley leaves, in general, are used in a fresh, green condition. It is used as a garnish and for flavouring sauces, soups, omlets and stuffing. It is minced and stew over potatoes, carrot or other boiled vegetables. The characteristics flavour and green colour can be retained if leaves are dried rapidly at 120°C in oven.

As parsley is a temperate crop; it requires cool climate with higher humidity and well distributed more in rainfall during the vegetative growth period. It needs a little shaded, moist, deep loamy fertile soil for its cultivation. Fresh seeds are to be sown outdoors in early summer in shallow hills, 25 cm apart under 1 mm soil. The bed must be watered regularly to ensure proper germination. The seed rate is 3-5 kg per hectare. The plants are to be thinned to 15-20 cm apart. The flowering stem arising in the second season is to be removed for more vegetative growth. Major pest of the crop is Septoriosis (*Septoria petroselini*), the carrot fly and the celery fly. Leaf yellowing due to nematodes is also noticed. Spraying of dimethoate or methyldemeton and Endosulfan or fenvalerate @ 2 ml/litre of water alternatively check the pest incidence.

## Varieties and Hybrids

Parsley cultivars can be grouped into different five types. These are (a) plainleaf, (b) celery leaf, (c) curled leaf, (d) fern leaf and (e) turnip rooted. The most common and attractive type is curled leaf, which contain sub-type (i) double curled, (ii) ever green and (iii) triple curled.

The best known varieties are Moss Curled, Extra Double Curled, Fern Leaved, Curled Dwarf, "Extra Double Curled Paramount". Imperial Curled, Hamburg, Parsley Petra.

**Detail characteristics of some commercial varieties are given below:**

***Hamburg*** : It is a variety of turnip-rooted parsley. It has tough stem, deeply cut plain leaves and moss curled. It is dwarf with finely cut serrated and deeply curled leaves and grown for its edible root. It is grown in the vicinity of the large cities, where it is utilized mainly by the foreign population.

***Parsley Petra*** : It matures in 75 days after transplanting. It can be sown from August to January. It is very uniform with very attractive curled leaves. High dry matter content and tolerant to bolting.

***Hybrids*** : Festival 68, Hybrid No.41 and Hybrid No.42 are used for cultivation in several parts of the parsley growing area to obtain higher yield.

## PARSNIP

Parsnip (*Pastinaca sativa* Linn.); 2n = 22; belongs to the family Apiaceae is an underutilized vegetables. Its roots are used for the preparation of multiple items. It is considered as a winter vegetable because its flavour is not fully developed until the roots have been exposed to near-freezing temperatures for 2-4 weeks in the fall and early winter. The starch in the parsnip root changes into sugar, resulting in a strong, sweet, unique state.

Parsnip is biennial in habit but it is cultivated as annual.It is similar to carrot in anatomy and development. It is hardy, white roots store well in the ground all winter. The plant develops rosette of leaves and large fleshy tap root which is economic parts. Flowers in terminal compound umbels; fruit (commonly called head) broad, elliptic and ridged, made of two maricarps.

The parsnip roots are consumed raw and after boiling as vegetable and salad. They are also used for Jam, cakes, wine and in fritters or pan cakes. They make an excellent soup and wines. Roots contain more sugar than sugar beet which reduces to about three quarters when cooked. The roots are also fed to live stock more than 20 per cent of the total fodder. It possesses diuretic and carminative properties. The fruit is based as a folk medicine for stomach, bladder and stone troubles.

### Varieties and Hybrids

Only a few varieties are available. These are Cobham Improved Marrow, Tender, Harris Model, Andover, Lancer, Alba. Unicorn, White Gem. All American, Fullback and Hollow Crown *etc.*

## PURSLANE

Purslane (*Portulaca oleracea* L. var. *sativa*, DC.); syn. Portulaca, 2n = 54; belongs to the family portulaceae, known as kulfa in Hindi;

has wild form and cultivated form, grown during hot summer. Both forms are spreading in nature. Cultivated forms have broad leaves and grow erect when sown thick. It occur as weeds in all tropical and subtropical areas and a popular winter vegetable in North India, it is also cultivated to a limited extent in some subtropical regions and propagated from seed. Seeds @ 2-3 kg/ha are sown by broadcasting and raked just like in amaranth. Irrigate field frequently. Crop becomes ready for harvest in a month. Tender shoots of about 10-15 cm long are harvested by cutting near the base and 3-4 cuttings at fortnightly interval are possible. Care should be taken in collecting seeds, as capsules break open during maturation. Plants retained for seed purpose are cut when early formed capsules are fully developed. Plants are partially dried in a gunny cloth and seperated seeds are dried and stored.

### Varieties and Hybrids

There is no any improved variety in this crop. Only local materials are used for cultivation of crop.

## RAKKYO

Rakkyo *Allium Chinensis*, G.Don (Syn. *A bakeri* Regel); 2n = 16, 32; belongs to the family Alliaceae, known in Japan as 'rakkyo' and in China as 'chiaot', is a native of central and eastern China and has spread in cultivation to much of eastern Asia. It is used almost exclusively for pickles, some of which are exported by Japan. it forms clumps like chives, from which it can be distinguished by its more sharply angled hallow leaves, the production of well-developed bulbs and the solid scape. The bulbs, consisting of concentric sheaths of foliage leaf bases divide rapidly to form clumps. The leaves die down in mid summer and the new leaves of daughter lateral buds are produced with inflorescence in the autumn. The lavender flowers, 15-20 in number, on long pedicels, are produced in flattened umbels; stamens and style are long exserted. Most cvs are tetraploid and do not produce seeds. It is propagated by small bulbs planted 60 x 7.5 cm distance.

### Varieties and Hybrids

There is no improved varieties in this crop. A number of local materials (genotypes) are grown in different countries. Clones are named on the locality of production.

## SPINACH

Spinach *Spinacea oleracea* L. (2n = 12), belongs to the family Chenopodiaceae; also known as 'Vilayati palak' in Hindi, is an important leafy vegetable, of which the leaves and tender shoots are consumed fresh or processed. The name spinach is derived from the Spanish word *espinaca*. Spinach is annual for leaf production and biennial for seed production. Unlike palak, spinach has leaves with lobed margin.After production of a rosette of leaves of ground level on a shorter stem, stem elongates and forms flower stalks. Spinach plants are usually dioecious.

The male and female plants are identified on the basis of carbohydrates and pigment contents in the plants sugars, Chlorophyll and carotenoid contents are higher in female than male plants. Seeds are of two types ; round seeded (summer spinach) or prickly-seeded (winter spinach).

### Varieties and Hybrids

Spinach varieties have been classified on the basis of seeds and leaves as : (i) prickly- seeded and (ii) round seeded and according to leaves characters (i) smooth leaves, (ii) savoy-leaved.

According to Mac Gillivray (1961) spinach is of two types, prickly-seeded and (ii) smooth seeded. The prickly seeded spinach is an old one and was used in Europe in 1931, where as the smooth seeded was known in 1952.

Spinach cultivars have also been classified in different ways.

According to Kinney (1896) spinach cultivars, classified into 4 groups : (1) Norfolk or Bloomsdale (2) Round leaved (3) Thick leaved (4) Prickly seeded.

Geise and Farley (1929), Classified spinach cultivars into 2 groups

(a)flat-leaved type and (b) savoy-leaved type. Each class was further sub-divided into 3 sub-classes : (i) short-season group (ii) long standing group (iii) slow-growing group.

Magruder *et al.* (1938) classified spinach as follow :

#### I) Leaves savoyed

(A) Leaves heavily savoyed :

i) Plants late seeding

(a) leaves blue green

(b) leaves dark green

ii) Plants not late seeding

(a) Plants very early seeding

(b) Plants medium-early seeding.

(B) Plants medium - savoyed.

II. Leaves not savoyed (or slightly savoyed in cool weather)

(A) Plants early seeding

(i) seeds prickly

(ii) seeds not prickly.

(B) Plants not early seeding

(i) Seeds prickly

(ii) Seeds not prickly

a) leaves blue-green

b) leaves medium green

According to Drewes (1943) spinach cultivars can be classified as follows :

I. Round-seeded

(A) Smooth-leaved (or slightly savoyed in cool weather)

i) Early bolting or not long –standing

ii) Slow bolting or long-standing

(B) Savoy leaved

i) Early bolting or not long-standing

a) mosaic resistant

(b) not mosaic resistant

ii) slow bolting or long standing.

II. Prickly seeded

(A) Smooth leaved (or slightly savoyed in cool weather)

i) early bolting or not long-standing

(a) mosaic resistant

(b) not mosaic resistant

ii) Slow bolting or longs-standing

(B) Savoy-leaved
   i) early bolting or not long standing
      (a) mosaic resistant
      (b) not mosaic resistant
   ii) slow bolting or long standing.

Ryder (1979) classified spinach cultivars on the basis of leaf type, into 3 groups (1) savoy (wrinkled leaf) (2) semi-savoy (3) smooth.

A good variety should have dark green leaves without stem, crisp, succulent and savoy leaves are preferred for bresh market while flat and smooth leaves for processing.

**Details characteristics of cultivars and hybrids are described below:**

***Virginia savoy*** **:** This is prickly seeded cultivars having large dark green leaves with round tip. Plants are upright and vigorous growth.

***Early smooth leaf*** **:** Smooth seeded cultivars and produced small light green leaves with a pointed apex tip.

***Sporter*** **:** It is developed by Bejo Sheetal Seeds and high yielding $F_1$ hybrid. It has broad leaves without fibres and becomes very soft on cooking. It matures in 45 days and suitable for cultivation in both kharif and rabi season. It is tolerant to mildew races, 1, 2 and 3.

***Spinach early*** **:** It is developed by Tropica Green Co. Ltd. Plant is upright with deep green leaves. It is early type and is harvested all the round year in tropical high lands. Spinach early is resistant to downy mildew races 1 and 3, heat and cold and also medium bolter.

**Varieties/Genotypes/Hybrids resistant to different diseases**

***Downy mildew*** : Several genotypes including varieties, hybrids or species have develop as resistant against different races of downy mildew are Early Giants, Hybrids-320, 308, Vikimum, Dorema, Matares, Vital, Fall Green, Noord Brabant, West Mass, Triathlon, Triade, Hermes, Triangel, Trisolde, Solar, Tritan, Tristar, Correnta, Mazurka, Valeta, Achilles, Atorasu, Aubade Hybrid, Capella Hybrid, Califlay Febris, Gurobaru, G-one, Nores, Orai, Sevenadehybrid, Soloman, Samson, Sanraito, EH7, EH-10, EH 424, EH 425.

***Fusarium wilt*** **:** (Moderate level of resistance). The sources hybrids lines EH 7, EH 10, Hybrid 50, EH 424, EH 425. Ozarka, Green Valley and WRG 70-5 are sources of resistance to wilt.

***White rust*** : Varieties/hybrids/lines exhibiting resistance to white rust caused by *Albugo occidentalis* Wilson, are Fall Green, Green Stand, Hybrid 178, WRG 70, Green Valley 11 x PH 1452, 4C x 119, Coho.

***Mosaic (CMV-1)*** : Resistance source - Dixie Market and Fall Green and in Roga, Lavires and Mona Lisa.

***Beet yellows*** : Transmitted through aphids (*Myzus persicae* Sulz), but may be transmitted by other aphids and mechanically also. Dutch varieties, least susceptible - Roga, and Ravires.

***Beet mosaic*** : Least susceptible lines-Novera, Estivato and Mona Lisa. Roga and Lavires are resistant to beet yellow.

## WATER LEAF

Water leaf [*Talinum triangulare* (Jacq) wild]; 2n = 78, 72; belongs to the family Portulaceae; is a soft mucilaginous leafy vegetable grown in the tropics. Being a shade-loving species, it is suitable for cultivation under other crop canopy.The leaves and tender shoots are used as vegetable. The plant is a herbaceous perennial, erect or decumbent and 35-90 cm in height. The stem is stout, succulent and fleshy. The leaves are light green, shining, triangular in shape and fleshy. The crop is highly self-pollinated.

Tender shoots and leaves are harvested 6-8 weeks after planting. Seeds mature rapidly after flowering and make plant unfit for consumption.

The genus *Talianum* which has nearly 50 other species of annual or perennial herbs are more or less succulent, sometimes woody at base.The related species in which leaves are eaten as vegetable are *T. arnotti, T. caffrum, T. patens, T. portulecefolium, T. crassifolium, T. cuneifolium* and *T. indicum*.

### Varieties and Hybrids

There is no named varieties in this crop. Only local genotypes are grown for production of edible leaves and mother plants are maintained in nursery. Propagation is possible through seeds and cuttings. Cuttings are more common and used as seed materials.

□□□

# 9

# Perennial Vegetables

## AGATHI

Agathi (*Sesbania grandiflora*); 2n = 24; known as Humming bird, Agati, Aal and Aich in Hindi; belongs to the family Fabaceae is a perennial tropical tree valued for its leaves and flowers, rich in minerals and vitamins. The tender pinnatified leaves and large white flowers are used after cooking. Red flowered type is slightly bitter and is used in medicines.

Agathi originated in Malaysia, crop prefers humid and sub-humid tropical climate. In addition, to its use as a vegetable, the plant has many medicinal uses. Juice of leaves and flowers are used in nasal catarrh and head ache and for curing sour mouth.

Agathi is propagated through seeds by sowing in situ and by transplanting 1-2 months old seedlings at a distance of one metre. Plants usually flower during September-December and tender leaves along with flowers are harvested. A grown up tree yields 5-10 kg of leaves/year.

### Varieties and Hybrids

There is no any improved varieties of this underutilized vegetable crops. A local materials/cultures are used for production purposes.

## ASPARAGUS

Asparagus (*Asparagus officinalis* L.) , 2n = 20; family - Liliaceae, is also known as Halgun, Halyun, Hillua and Shatawar. It has at least 150 species native to Europe, Asia and Africa. In India nearly 17 species have been reported. The tender shoot called spears are used as vegetables. Being a dioecious crop with male and female plants, cross pollination is the rule and the cultivars should be grown in isolation to maintain the genetic purity.

### Varieties and Hybrids

*The asparagus varieties are broadly divided into two groups* : (a) Spears overcast with purple and (b) Spear tips almost free from purple overcast. The common varieties are : Perfection, Mary Washington, Martha - Washington, Colossal, Berr's Mammoth, Columbian Mammoth White, Donald's Elmira, Palmetta, Consvers Colossal, UC 72, Brock Imperial 84, Palmetto, Argentenil, Huchels Leistung Sausbese and Schewetzinger, Reading Giant, Gersey Queen, Jersey Giant, New Jersey, Improved Perfection, UC 66, UC 711, SL 831, Limbrass 22, Librass 126, Lucullas 1813, Paradise Meisterschuss, Bergamo, Violet Dutch and White German are grown in different countries.

### Hybrids

HD - INDA, Gijnlim, Revolution, Lara and Mira, Faribo, Aneto and Desta.

**Detail characteristics of improved varieties and hybrids are given below :**

***Perfection*** : It is an early variety , spears large, green, succulent, heavy cropper and uniform in maturity.

***Consver Colossal* and *Mammoth White*** : They produces light green or whitish spears. They are high yielder and commonly grown for marketing purpose.

***Palmetts, Argentenil* and *Mary Washington*** : These are most important varieties and produces very higher yield. The tips of spears before exposer to the light are purplish and the spears become dark green in sunlight.

***Violet Dutch*** : It is the oldest variety which may be regarded as the form from which all the modern types have been derived.

***White German*** : It is selected in Germany and widely grown by Germans. It has rounded tip and good quality spears.

***Palmetto*** : It has a little difference in appearance than other varieties and considerably resistant to rust and extensively grown throughout America.

***Schewetzinger*** : It is a commercial variety and has a greater vigour when planted by clones.

***Lara and Mira*** : These hybrids are early and high yielding evolved through breeding.

***Aneto and Desta*** : These are inter - clone hybrids, high yielding and less heterogeneous. Evolved through 'In - Vitro' tissue culture.

***Faribo*** : It is a high yielder and produce good quality spears.

***UC 711*** : It is a better hybrid than other and give maximum yield and high return.

***Gijnlim*** : This asparagus hybrid is very early in production with medium thick, straight spears of excellent quality. A very good total yield by means of 100% male plants gives 20% higher production compared to female plants and gives the advantage of no production of seed berries and seedlings under normal conditions.

***Paradise*** : It is a selection of Mary Washington developed in California are said to be superior to the standard strain in some regions.

***Jersey Queen*** : A clone cultivar selected from Mary Washington. It produce high quality spear and maximum yield.

***Jersy Giant*** : It produce maximum thick spears and give maximum yield.

***Brock Imperial 84*** : It is highly suitable cultivar for commercial cultivation.

## BREAD FRUIT

Bread fruit (*Artocarpus altilis* Postperg); 2n = 2x = 56; known as Barhal or Khadel belongs to the family Moraceae, is a perennial tree grown for its seedless and whitis meat pulped fruits rich in carbohydrate (27.98 %). The name 'bread fruit' came after serving it as a staple food, as bread for British. It grows to a height of 10-15 m with an equal spread.Stem is woody and leaves are simple and palmately lobed. It has a long pre-bearing period of around 10 years, if grown from seedlings. Root suckers start yielding $5^{th}$ to $6^{th}$ year after planting. Plant is monoecious and bears separate male and female inflorescence on specialized laterals. Fruit is developed through stimulative parthenocarpy.

All the three species of Artocarpus, viz., *A. altilis* (Bread fruit), *A. integar* (Chempedak) and *A. haterophyllus* (Jack) are basically diploids (2n = 2x = 56). In addition, triploidy is also common among seedless bread fruit clones.

### Varieties and Hybrids

There are two distinct types, i.e., seeded and seedless and they do not differ much morphologically. The seeded type is not used in coolinary preparation, but seeds are roasted or boiled and consumed.

The seedless type is popularly cultivated and is believed to be a 'sport' of the seeded variety perpetuated vegetatively through man's effort. It has no seeds but a mass of whitish-meaty pulp is present.

***Yellow Heart*** **:** This variety is grown in Jamaica, which is considered to be the best among all other cultures.

## CHEKKURMANIS

Chekkurmanis (*Sauropus androgynus* meeril syn. *S. albicans* Blume, *S.gardnerianes* Wight., *S.zeylanicus* Wight, *S.indicus* Wight., *S. Sumatrans* mig.); is also known as Madhura Cheera in Malayalam and Thavarai Murungai in Tamil Nadu; belongs to the family Euphorbiaceae, is a small shrub becoming popular as a leaf vegetable. It is grown as a live fence or hedge for kitchen garden. The succulent shoots and tender leaves used for culinary purposes are noted for its high nutritive value and is popularly known as 'Multi-vitamin green' and 'multi-mineral packed' leaf vegetable.

The genus *Sauropus* consists of several other species. Of them some species are related to *S. androgynous*. Four related species have been identified, these are : *S. retroversus*, Wight, *S. assimilis* Thw, *S. rigidus* Thw, and *S. quadrangularis* Muell.

The crop is highly cross-pollinated and entomophilous, the reasons for cross pollination being monoecy, alongwith protogynous nature of plant. Crop is raised through semi hard wood cuttings of length 20-30 cm. Cuttings are planted 10-15 cm apart during May-June. First harvest is possible after 2-3 months of planting, when plants reach around 60 cm height. Even though the crop is drought tolerant and with better management, number and frequency of cutting can be increased. Plants are trimmed to one metre height to encourage fleshy growth.

### Varieties and Hybrids

There is no distinct varieties or types available of this crop. It is often mistaken for some broad-leaved types and small leaved typies. The difference in the leaf size is due to the shade effect, when the plants are raised under shade, it produces broader leaves, where as under full sunlight it produces narrow leaves.

## DRUM STICK

Drumstick (*Moringa oleifera* Lamk); 2n = 28; known as Seeng, Sahgan in Hindi, belong to the family Moringaceae, a monogeneric family, is distinguished by parietal placentation, 3 valved fruit, elongated, non-dehiscent berry and winged seeds. There are 2 common species *M. oleifera* and *M. concanensis* Nimmo, the former being the vegetable species. The *M. oleifera* species is being distinguished by leaves usually tripinnate, leaflets 12-18 mm long; petioles yellow or white without red streaks and the tree being medium sized. *M. concanensis* is distinguished by leaves bipinnate, leaflet 15-30 mm long, petioles with red streaks or reddish at base and the tree being large sized. A few of the species of moringa related to *M. oleifera* are *M. pterygosperma* Gaertn and *M. longefolia*.

The flowers are bisexual, oblique, stalked, axillary and heteromorphic. The plant is highly cross-pollinated due to heteromorphism and is entomophilous, bees being the pollinators. The tree can be propagated from seed or from cuttings. The stem cuttings are usually preferred which root very easily. Plants raised from seeds produce fruits of unpredicted quality.

Several types are broadly classified on the basis of branching habit of the tree and the pigmentation of the floral parts and the fruits. With regard to the branching habit, tree types could be broadly classified into 2 groups, viz., types having profuse branching habit in which the branches are straggly and drooping and types having only a few erect branches. Of these the former is preferred by growers on account of their larger yields.

Based on the pigmentation of the plant, floral parts and the fruit, 3 broad groups could be distinguished.

***Group I:*** The tender twigs are dirty brown, leaf stalk and peduncle of panicles tinged purple; stamens and ovary completely purple; fruits in tender stages reddish, turning green with reddish streaks running from base to tip at full maturity.

***Group II:*** Twigs dirty brown, leaf stalk, nodes, peduncle, pedicel and their point of attachment deep purple; basal portion of sepals and petals tinged purple; stamens and ovary completely purple; fruits at tenders stages deep reddish turning to black purplish on full maturity.

***Group III:*** The plant parts are green except the basal portions of the leaf stalk which are sometimes tinged purple; sepals whitish, petals either white or light yellow; stamens and pistil light yellow; fruits at tender stages light green with or without purple tinge turning to darker green on maturity.

The types which are under cultivation generally falls into the first or the third group. The second group is not found under large-scale cultivation presumably because of the comparatively lower yields and the poor quality fruits.

## Varieties and Hybrids

There are few locally grown cultivars/varieties known by the place of their cultivation in drumstick. These are : Chavakacheri Murunga, Chem Murunga, Kadu Murunga, Yazhpanam Murunga, Jaffna, Kodikal Murunga, Annual Drumstick, PKM 1, Dwarf Drumstick, Pal Murungai, Poona Murungai, Moolanoor Murungai, Palamedu Murungai, KM 1, PKM 1 and PKM 2 drumstick cultivated in different parts of the country.

**Detail characteristics of varieties and hybrids are given below:**

*Jaffana* : It is popular drumstick in south India and said to be introduced from Jaffna. It is highly suited for coastal tracts of Tamil Nadu, Kerala and Karnataka. This type bears long pods (60-90 cm in length) and with a soft flesh of good taste. The type of moringa can yield 400 pods per tree per year from the second year of planting which increases to 600 pods from the third year of planting.

*Moolanoor Murungai* : It is a local commercially grown in Moolanoor of Erode district. The frutis are 30-35 cm in length with very soft flesh. One tree yields about 500-600 fruits per year.

*Chavakacheri Murungai* : It is reported to have been introduced from a place called Chavakacheri near Jaffna. It is an ecotype of Jaffna moringa, which bears pods as long as 90-120 cm. Due to long size of pods, this type of moringa is highly damaged during transportation and is grown mostly in home garden. It yields 500 - 600 pods per tree per year.

*Chem Murungai* : It is another type of Jaffna muringa. It is said to flower and fruit throughout the year and also yields heavy crops of 400-500 fruits per tree per year. The tip of the pod is red in colour.

*Pal Murungai* : It is also another ecotype of Jaffna moringa. It is preferred for its thicker pulp and better taste of the pods. It yields 400-500 pods per tree per year.

*Puna Murungai* : It is another ecotype grown in the home garden of Tirunelveli and Kanyakumari districts in Tamil Nadu. It yields 300-400 pods per tree per year.

*Kodikal Murungai* : This is cultivated predominantly in the betelvine gardens of Trichirapalli, Tirunelveli, Tanjavur and Dharmapuri districts of Tamil Nadu.It is an annual type of moringa. The seeds of the variety are sown in the betelvine field just three months before planting of betelvine cuttings. The betelvine cuttings are planted near the moringa. This moringa tree is highly useful for training of betelvine and also gives shade. The pods of this type are shorter (20-25 cm in length) and thick flesh. The pods and leaves are very tastier. The trees are short statured and the leaves smaller. This is a distinct type propagated by seeds.

*Kattu Murungai* : This type of moringa are seen in Chidambaram areas in South Arcot district of Tamil Nadu. This is a wild type producing small and inferior quality pods. The trees are larger and leaves are bigger. The pods are 30-60 cm in length and fleshy but used as cattle feed.

***Palamedu Murungai*** : It is cultivated in Madurai district in Tamil Nadu. The pods are long (60-75 cm) having good taste.

***Yazhpanam Muringa***: It is same as Jaffna type.

***Annual Drumstick*** : It is short duration variety raised as a border and mixed crop in chilli and brinjal fields. It is claimed that a single tree gives a yield of 800 fruits. It is precocious in flowering and bearing without any definite peaks, the tree becomes exhausted and appears to senesce at the end of one year.

***KM (Kudumianmalai 1)*** : This variety was developed at Anna farm, Kudumianmalai of Pudukkottai district in Tamil Nadu. It is a selection from an annual type propagated by seeds. The plants do not grow very tall and harvest of fruit becomes easier. Pods are short (32-37 cm) and thick (5.5-6.0 cm). The pod weight ranges from 65-82 g. The number of pods per tree/year is 226-328 weighing 14.69 -26.90 kg. The number of seeds per pod is 10-13. The plants come to bearing 6 months after planting. After each harvest, the plants can be ratooned for 2-3 years by cutting the trunk at a height of one metre. Fresh planting of seedlings can be taken after three years.

***PKM 1*** : It is an annual type of murungai, evolved at Horticultural College and Research Institute, Periyakulam of Tamil Nadu Agricultural University. The plants grow to a height of 4-6 metres and come to flowering in 100-125 days after planting. Each tree bears on an average of 200-250 pods/year. The pods are 60-75 cm long with 6.0 cm girth and 150 g weight. They are very pulpy containing 70 % of edible portion. Every year after the harvest is completed, the trees have to be cut back to about one metre from ground level during September and three ratoon crops can be taken in a period of three years.

***PKM 2:*** This variety was released during January 2000 from Horticultural College and Research Institute, Periyakulam of Tamil Nadu Agricultural University. It is a hybrid derivative of the cross between MP 31 and MP 28 of the types maintained in the germplasm. The plants are quick growing reaching a height of 4.8 m in six months. Each tree has 12 branches and bears flowers in clusters. Three to four pods are obtained per cluster. Plant starts flowering in 100-110 days and pods can be harvested in 170-180 days. Each pod weighs 280 g with a length of 125 cm and girth of 8 cm . Seed content is less with more flesh. Each tree yields on an average 220 pods weighing 62 kg with an hectare yield of 98 tonnes.

***Dwarf Drumstick*** : Plants of this variety are medium in size and bearing 480 numbers pods (fruits) per plant from 7-9 months. The leaves are falling gradually away from the plants from 10 - 12 months and again bearing to yield 350 - 450 numbers of fruits from 13 months continuously.

***Dhanraj*** : This is also an annual drumstick propagated through seeds and is evolved at KRC College of Agriculture, UAS, Arabhavi, Karnataka.

Three promising perennial drumstick clones viz., MO 70, MO 95 and MO 44 and one annual seed drumstick, AD 4 have been developed by Department of Olericulture, Kerala Agricultural University, Vellanikkara.

## GLOBE ARTICHOKE

Globe artichoke (*Cynara scolymus* L.); 2n = 34; belongs to the family compositae (Asteraceae); is generally known as artichoke. It is said to be derived from the two Arabic words ard (earth) and shoke (thorny). It is a herbaceous perennial plant grown for its flower and head or flower bud, are used as vegetable. The edible portion consists of immature tender bases of scale like leaves as well as receptacle or fleshy base upon which the flower buds are borne. The heads range from 2.5 - 10.0 cm in diameter. The flower head is large and composed of many florets which are fertile. After flowering, the above ground portion (the rosette) dries and dies every year during winter and the crown of individual plant decays after a years growth. New growth emerges in spring renewed by off-shoots from the root stocks. The development of axillary buds of the rosette leaves produces a cluster of new rosettes.

The flowers of a globe artichoke are hermaphrodite. The pollen viability is decreases very quickly after 2 days and stops after 4-5 days. The duration of one plant cycle, seed to seed, is about 12 months.

Globe artichoke is propagated through seeds, sucker or off - shoots. Since plants raised from seeds are highly variable and take longer time, suckers or off-shoots are generally used for propagation. Seedlings are usually transplanted. Plants grown from seeds take 8 months for harvest while from suckers, it is only 6-7 months. Terminal flower buds are the most desirable economic product. Harvesting continues during entire winter. Inflorescence heads are harvested alongwith 2.0-2.5 cm long stem. Average yield is 10-12 t/ha.

## Varieties and Hybrids

There are very limited cultivars in this crop. They are: Green globe and Purple globe very popular as commercial varieties. Some fresh market cultivars like Violet de Provence, Catanese and Spinosa Sarda are under cultivation. The processing varieties - Bianco Tarantino and Precoce di Jesi are also common for cultivation. Other cultivars include Bull, Tudella, E15, Brindisino and Romanesco. Some clonal selections are also used for cultivation in different countries. These are - 'Blanca de Espana' in Spain, 'Violetta di Sicilia' in Italy and 'INRA VP 45' in France.

**Some vegetatively propagated cultivars are :**

(a) ***Terom*** is a clone selected in the progeny of an open - pollinated plant of Violetta di Toscana - In Italy

(b) **'Camerys' and 'Caribou'** are two clones selected in the progeny of the hybrid between 'Camus de Bretagne' and Romanesco. They give an earlier production of green globular heads than does the standard cultivar 'Blanc Hyerois'.

(c) **'Salanquet and Cacique'** are clones produced from vegetative propagation of two hybrids between inbred lines. They give a high yield of violet heads, but their harvest period is late.

(d) **'Jaja'** produced from 'Precoce di Jesi'-became a commercial cultivar.

(e) **'Talpiot'** is an improvement on Jaja and grown commercially in Israel. It gives a late production suitable both for fresh marketing and processing.

# JERUSALEM ARTICHOKE

Jerusalem artichoke (*Helianthus tuberosus* L.); 2n = 102; belongs to the family Asteraceae (compositae); known as Hathichuk in Hindi; is a cool season perennial vegetable grown for its edible tubers in Europe, parts of Asia and throughout Southern hemisphere in India, the crop is not popular and grown in a limited scale in hill stations only. It is a tuber-bearing type plant cultivated for its underground oblong or extremely elongated tubers, weigh 80-120g each. Tubers are eaten raw or boiled and used for making pickles, chips or powder. It is equal to potato in food value and has relevance in the diet of persons suffering from diabetes since it contains carbohydrate insulides. Concentration of insulides increases during storage at expense of inulin through action of inulase. Tubers are commercial source of levulose and used as sweetening agent by diabetic patients.

Jerusalem artichoke is an erect perennial plant grown as an annual for tuber production. It is propagated by vegetative means using small sets of tubers. Tubers are harvested when leaves start withering. As quality is not deteriorated in soil, tubers are usually lifted as and when required. Yield is as high as 370-390 q/ha have been recorded under favourable conditions.

## Improved Varieties

The varieties of Jerusalem artichoke have been devided into three groups on the basis of branching type. They are, monopodial, sympodial and combined (monopodial plus sympodial). Branching type is associated with earliness and seed production, with monopodial branching being typical of late varieties (growth period 179-227 days), sympodial branching typical of early varieties (140-175 days). Mixed branching types are of mid-season and exceptionally, some late varieties (176-239 days).

Some improved forms of Jerusalem artichoke has been developed by hybridization of different species and few others hybrids have also been developed these are :

***'Vostorg'*** ('Rapture') is the most promising, hybrid. ***'Tijumen'*** is a improved variety. *'V 71'*, *'D'$_1$*, *'D'$_2$* and ***'Patate Vilmorin'*** (Vilmorin Potato) are resistant to *Sclerotinia minor*, and *S.sclerotiorum*.

***'Comber'*** showed high resistance to *Puccinia helianthi*. ***'White Early'***, ***'Kiev white'*** and ***'Ukrainskii, 108'***, are mid-season varieties having hardiness to cold (Winter).

***'Leningradskii'***, ***'Nakhodka'***, ***'White*** - productive' and ***'Ivanova Red'*** are late varieties less hardness to winter.

***Seyanets*** and ***G120*** are the most productive varieties tested in the field experiment in Grondo province.

***'Moder'*** (M6) produced the best combination of tuber yield, high reducing sugar concentration (16.3%) and high concentration of fructose in the reducing sugars (77.4%).

***'Challenger'*** is dual - purpose cultivar producing a large, dense mass of top growth (which could be used as animal feed for protein production, or for ethanol production) and tan-coloured, white-fleshed tubers. Tuber carbohydrate content ranged from 14.5 to 20%; of this, 65-82% is fructose.

***'Columbia'*** is selected from a mixed/intercrossed population of accessions *H.tuberosus, H.annuus* and several other Helianthus species. Plants are upright (135-170 cm) and produce white-fleshed tubers and yields, up to 76.2 tonnes per hectare.

***'Minnesota 1'***, has the highest crude protein and iron concentration.

***'Iowa 1'***, genotype has the highest calcium and magnesium concentrations.

***Illinois*** : This genotype has the highest potassium concentration.

## JHUMPURI

Jhumpuri (*Phyllochlamys spinosa* Bueav; syn. *Streblsu, taxoides* Kurz., *Trophis spinosa* Roxb., *Taxotrophis roxburghii* Blume , *Ericarpusus spinosus* wight, *Trophis taxiflormis* spring, *Trophis taxoides* Heyne, *Trophis heyneana* Wall) ; also known as Sheora in Bengali, Kurrera in Marathi, Sukali in Telugu, Kattari in Tamil, Jhumpuri in Oriya and Phutkuli by tribal people in Mayurbhanja district of Orissa.

Jhumpuri an under-exploited tree vegetable belongs to the family Moraceae, native of India is an evergreen small tree grown up to the height of five meters arranged with woody straight, sharp glabrous spines about 1-3 cm. long irregularly arranged which exudes a milky, sap or juice when a scratch is made, its branchlets are also thorny and often bear leaves.

Jhumpuri flowers almost round the year leaving 3-4 months during winter. However, the maximum flowering takes place during the rainy season. Culinary preparation with flowers and immature fruits in form of curry and fry are very delicious. Besides the use of its flower as vegetable, the plant has also got medicinal value. The bark is boiled to make a poultice for ulcers; it is burnt and the smoke is inhaled to cure cold in the head (Anonymous, 1962).

Kurz (1877) reported that this plant was mostly confined to the Indo-Burmese regions starting from India, Burma, Ceylon to Indo China. It is also reported to be found in Thailand, Hainau, Malayasia, Java, Timor, South East Celebes, Mindoro and Palawan. This plant is present in Indo-China region (Vidal, 1967). This plant occurs in Himalayas, from Himachal Pradesh to West Bengal, in hills of Assam and Tripura upto 450m, Peninsular India upto 600 m altitude and also in Andaman. According to corner (1975) this plant was distributed in the regions like Srilanka, India, Burma, Indo-China, Hainan, Thailand and Malaya.

Jhumpuri is widely distributed in various parts of Orissa, Bihar and South India particularly in the tropical forests regions and the coastal belts. It has not been reported to be used as vegetable except in parts of Orissa where it is grown in a very limited scale and its cultivation is confined only to backyard plantation.

Jumpuri belongs to the family Moraceae, subtrib-Strebleae; Tribe morae, order-urticaceae and genus-phyllochlamys; is an ever green small tree or shrub which grows upto 5 metre height, if unprunned , armed with woody, straight, sharp glabrous spines (1-3 cm long), irregularly armed, which exudes a milky or green colour sap or juice and later when a scratch is made. Branchlets are thorny or often terminating in a sharp thorn and often bear leaves, trunk is dark, gray. Leaves, alternate, very rarely opposite, simple, stipules (2-5 mm long) are lateral and deciduous and often sheathing and leaving a permanent annular scar after fall,. Leaves are mostly oblongelliptic to lanceolate, entire with wavy margins often with sharply pointed apex. Lamina is 5-10 x 0.7 - 6.5 cm elliptic, ovate to obovate very often asymmetric, lateral veins 8-12 pairs, basal veins are very short without any glands, petioles are 2-5 mm long.

It is a dioecious plant but rarely found to be monoecious. The male inflorescence, very shortly racemose, flowers are pedicellate, 2-5 mm wide, axillary. Male inflorescence is capitate with 8-10 flowers. Sessile except the central ones. Pedicellate, 1.5 mm long bracts-ovate, concave 3mm, tepals 4, obovate, concave stamens 4, filaments-stout 3mm, anthers 1.5mm. Female flower appear solitarily in the leaf axils, pedicellate sometimes two together with bracts at the base of the short pedicel. Bracts and bracteoles lanceolate, unequal, tepals 4, erect and conniving over the fruit, oblanceolate 4.5 mm style branching from below 1 mm long stigmatic arms 0.5-1.5 mm long, style is bipartite at length, lateral. Ovary compressed ovate 2 mm, superior. Female tepals foliaceous upto 1 inch in case of genus phyllochlamys whereas in streblus female tepals small enclosing the ovary. Flowering in *Streblus taxoides* starts from May and also between April to June.

Fruits 5-10 mm wide, ellipsoid, asymmetric, very fleshy at the base, covered and exceeded by the foliaceous, narrow or broad, much enlarged tepales, 20 mm long. Berry-obliquely oval, size of a cherry, half immersed in the increased receptacle, pulpy, fruit splitting in to two halves when ripe, one seeded. Seed oblique, albumen, scanty, embryo globose, one, cotyledon large and bifid enclosing the other, radicle ascending. Seeds are oblique, albumin scanty, embryo globose having two cotyledon large and bifid enclosing the other and radical ascending.

Jhumpuri is propagated by seed as well as by vegetative part. With respect to sexual propagation, maximum percentage of seed germination (92 %) was recorded with medium containing sand, leaf mould (1:1) followed by soil + leaf mould (89 %). Maximum survival of seedlings (93 %) was found in the medium containing soil and leaf mould followed by soil + leaf mould + well decomposed cowdung manure.

As regard to vegetative method of propagation, application of Rootex-3 was found to be most suitable which enhanced rooting in 68.34 % of the layers in 72 days by producing maximum number of thicker and larger roots followed by IBA-50 ppm treatment. Propagation through cutting was proved unsuccessful. The layered twigs detached from the mother plant after 98 days were proved to be the most successful layers for propagation. Sphagnum moss proved its superiority over soil as the favourable medium in providing better condition for rooting.

Jhumpuri requires a tropical climate with high humidity and a lot of sunshine. Climatic conditions typical to forests of tropical to subtropical India and foothills of high mountain ranges are suitable for better growth of the plant.

## Varieties and Hybrids

As stated by Kurz (1877) *Streblus microphylla* might be a botanical variety of *Streblus taxoides*. He mentioned that in *Streblus microphylla*, young shoots were puberulous unlike *Streblus taxoides* where it was glabrous. Besides in *Streblus microphylla* spines were also puberulous and leaf bearing, leaves 1-15 long, oval blunt, flowering and fruiting unknown. According to him the habitat of variety-1 viz., *streblus taxoides* was the rocky coasts of Andamans in tropical forests, whereas variety-2 viz., *Streblus microphylla* was common in the swampy forests of the *Irrawaddi alluvium*. Haines (1925) reported that *Trophis spinosa* was a synonym of *Phyllochlamys spinosa*. However, Roxburghi (1832) stated that *trophis spinosa* seemed to be a new species uniting *Trophis* and *Texus* in itself. According to him it was a small rare tree native of tops of mountains flowering during cold season.

## PISONIA

Pisonia (*Pisonia grandis* syn. *P. alba*); n = 68, octaploid, multiple of 17; 2n = 136 also known as Lettuce tree, Porush (Hindi); belongs to the family Nyctaginaceae; is a perennial tree growing to a height of 9-12 m. Leaves are pale green or yellowish green. Plant flowers rarely and are dioecious in habit. Leaves of plant resemble lettuce in taste and are used as cooked vegetables as well as salad. Leaves of male trees are darker and are less appetizing. It is propagated through cuttings.

### Varieties and Hybrids

There is no named variety in this crop. Only locally available genotypes/culture are used for cultivation.

□□□

# 10

# Other Vegetables

## BABY CORN

Baby corn (*Zea mays* var. *rugosa*); 2n = 20, belongs to the family Poaceae, is the stage of sweet corn, harvested at very tender, immature, young de-husked female inflorescence before fertigation, is sweet in taste and crisp. Baby corn is an ingredient for soup, vegetable, biryani, mixed vegetables and Chinese food preparations. It is used in crispy vegetable salad. Very tasty pickle is prepared from baby corn kernels.

The plants are stout annual, usually with a single stem, 1-4 m high, but one or more tillers may produce in some forms. The plant is monoecious and diclinous with male and female inflorescences borne separately on the same plant.Male inflorescences, called a tassel, is a terminal panicle and female inflorescence is called the ear is a modified spike. Fruit is known as the kernel or grains, contain a glossy sweetish endosperm which is translucent when immature and picked for boiling as corn on the cob. It is also used for canning and freezing and has become a highly popular vegetable.

## Varieties and Hybrids

A number of good varieties and hybrids have been developed for cultivation in Indian climatic conditions.

***VL-42*** **:** It is an open pollinated variety, developed at Vivekananda Parvatiya Ansundhan Sansthan, Almora.

***COBC-1:*** It is a high yielding and synthetic variety for commercial cultivation.

***NHRDF-1:*** It is an open pollinated variety developed at National Horticultural Research and Development Foundation, has been found very promising for baby corn production.

***MEH-114 ($F_1$)*** : It is developed by Maharashtra Hybrid Seed Company, Jalna.

***Prakash Hybrid (G.H.-3189):*** It is recommended for baby corn production in different region of the country.

***Golden Baby*** : It is a hybrid, 2 to 2.5 m talls, growth is fast, there are 2-3 uniform sized (8-10 cm) earheads per plant. Earheads are ready for harvesting 45 days after sowing.

Punjab Agricultural University, Ludhiana has developed two varieties for this purpose. They are Composite Kesri and Prakash Hybrids.

# OKRA

Okra (*Abelmoschus esculentus* (L.) Moench.); 2n = 130; syn. Lady's finger, Bhindi, Bhendi, Gumbo, Dheras, Venda, Vendi, Bhinda, Bhida, Bendekayi, Ramturai, Tindisha, Gandhmula, Babniyah, Bandaka; belong to the family Malvaceae, is predominantly crop of tropics and subtropics. Okra is a warm-season fruit vegetable, grown in summer as well as in rainy season. It occupies fifth position, next to tomato, in area under vegetables in the country. The crop is cultivated for its young tender fruits (pods), used in curry and soups after cooking. The dehydrated bhindi is a processed product for preservation and export. Okra seeds form a nutritious ingredient of cattle feed and is a source of vegetable oil.

The order Malvales has only one family Malvaceae (Mallow family). The family has 82 genera and 1,400 species. In India, about 111 species belonging to more than 20 genera are reported. The geneus

*Hibischus* (syn. *Abelmoschus*) has the distinguished feature,the caducous calyx where calyx, corolla and the stamen are fused together at the base and falls off as one piece after anthesis. Hochruitner (1924) listed 14 species under *Abelmoschus* later many species were added. The Index Kewensis lists about 30 species under *Abelmoschus* in the old world and 4 in the New World. Okra is considered both as a self pollinated and as an often cross-pollinated vegetable due to protogyny and male sterility in a few varieties.

## Varieties and Hybrids

Okra Varieties and hybrids differ in growth habit, height of the plants, presence of purple pigmentation on plant parts, length, colour and number of ridges of fruits etc.

The important varieties are : Pusa Sawani, Pusa Makhmali, Perkins Long Green, Pusa A-4, Arka Anamika, Arka Abhay, Parbhani Kranti, Punjab Padmini, Lucknow Dwarf, Long Green, Red Wonder, Violet Green, Smooth Velvet, Long White, Samalkota, Satdhari, Vaishali Badhu, Utkal Gaurav (BO-2), BO-1, Varsha Uphar(for rainy), Hisar Barsati, Sel.2-2, Kiran, Salkeerthi, Aruna, Susthira, MDU 1, CO1, Hisar Unnat, Azad Kranthi, Punjab-8, Harbhajan Bhindi.

## Hybrids

**Mahyco** : NO-6, NO-7, NO-8, NO –10, NO –12, MHOK-417.

***Century Seeds*** : Panchaali, Aadhunik.

***Ralish india*** : Tara.

***Nath Seeds*** : No.-1, Nath kranti-20, Nath sobha-110/111.

***Nuzividu Seeds*** : NBH-225, Sabour selection (Branching tendency),

***St John Bush*** : (Enormous size), long life.

***Modi Agritech*** : Lahari, Remani, M.A.O.H.-40, MADH-7/08/10.

***Heritage Seeds*** : H.S. 101, Geetanjali.

***N.F.C.L.*** : D.S.H.B-1.

***Royal Slouis*** : Priya, Somaya, Najuka, Seria, Sukomal.

***Syngenta*** : Ashwarya, Mistik, Perpiyat, Reshma.

***Pro Agro Seeds*** : Uphar.

***Sungro seeds*** : Supriya, S-45, S-51, S-222, Priya.

*Beejo, Sheetal Seeds* : Anokhee, Karishma, Saloni, Praful, Snehal, Tara, RIL-56.

*Unicorn Seeds (Choice)* : Kirti, Kirti Plus, OKT-072.

*Golden Seeds* : Hybrid- 99, Harishan.

*Ankur Seeds* : AROH-9, AROH-10, AROH-85, AROH-113.

*Nunhem Seeds* : S-40, S-008, Uphar, Makhmali, S-90, Tulsi (NP 1104), Lady's Pride (NP 1103), Shaina (NP 1105).

*Namdhari Seeds* : NS 502, NS 503, NS 504, NS 531, NS 532.

*Malav Seeds* : Malav Shakti, Malav-31 Spreme, Malav-27 Prime.

*Sreemaa Seeds* : Sarathi, Kalyani (OP).

*Tokita Seeds* : Kamini.

*IIVR Varanasi* : DVR-1, DVR-2, VRO3, VRO4, VRO-5, VRO-6, IIVR-10.

*Annapurna beej* : Pragati, Sel. No.10, Prava, Pankaj.

*Indo-American Hybrid Seeds* : Varsha, Vijaya, Vishal.

**Detail characteristics of varieties and hybrids are given below:**

*Pusa Sawani* : This variety has smooth, dark green, 5-ridged fruits, 12-15 cm long and takes 40-50 days to first harvest. First fruit is born on $6^{th}$ to $8^{th}$ node.It is suitable for cultivation in both spring-summer or well as kharif seasons as it is day neutral. It is a cross between 'IC-542' and Pusa Mahmali. It is a very high yielding variety.

*Pusa Makhmali* : It is has 15-20 cm long, straight, smooth and attractive green fruits. During summer the first harvest is taken in 50 days whereas in monsoon crop it takes about 10 days more.

*Red Bhindi* : The fruits are red with longer size and are slender fleshy and have fewer seeds than Pusa Sawani. This is local type bhindi, grown almost all parts of the country during rainy season particularly.

*Parbhani Kranti* : Pusa Sawani was crossed with *A. Manihot* a related wild sps. having resistant Y.V.M. Two back cross were made to Pusa Sawani followed by selection in subsequent selfing generation. Plant tall, single stemmed and with dark green foliage. Plants flower in 45 days. First harvest in 55 days. Fruit size is 8-9 cm. The fruits are smooth, dark green tender, slender, 5 ridged with long back. Yield is varies from 90-115 q/ha during summer and rainy season.

***Punjab Padmini*** : It is a cross between *A. esculentus* x *A. manihot* a wild sp. Plants are taller than Pusa Sawani.The fruits are quick growing, dark green in colour, thin, long, 5 ridged and remain tender for a long time. Produce 50 % more yield than Pusa Sawani and Presistent to Yellow Vein Mosaic.

***Pusa A-4*** : The plants have single stem with short internodes (2-4 cm)and the fruits are 5-ridged, attractive dark green, 10-12 cm long having low mucilage and excellent shelf life. The variety is resistant to YVM virus and tolerant to aphid and jassids and least preferred by shoot and fruit borer.Tender green fruit yield 100-120 q/ha in summer and up to 175 q/ha in kharif.

***CO-1:*** Plants are 90-120 cm tall with 6-8 branches. The fruiting starts from 5$^{th}$ node. The fruits are glossy, slender 5 ridged, scarlet red (colour non persistent on cooking) and bear about 20 fruits per plant. It has resistance to YVM virus at field level but susceptible to fruit borer and powdery mildew.

***CO-2:*** It is a cross of AE 180 x Pusa Sawani. Fruits very long (22-25 cm), thick, 7-8 ridged and light green in colour. Productivity 15-16 t/ha. Susceptible to YVMV.

***CO-3:*** Parbhani Kranthi x MDU1, Fruits very long (22-25 cm), thick, 7-8 ridged, light green in colour. Productivity 16 to 18 t/ha. Moderately resistant to YVMV. These two $F_1$ hybrids ($CO_2$, $CO_3$) developed by Tamil Nadu Agril. University.

***Punjab 7 (P7)*** : Developed through back cross method from *A. esculentus*, cv. Pusa Sawani x *A. manihot* ssp. manihot cv. Ghana. Fruits 5-ridged, dark green, medium long, petiole base is deeply pigmented. Productivity 9.5 t/ha.

***Punjab 8 (EMS 8):*** Induced mutant from Pusa Sawani using EMS (1 %). Fruits 5-ridged, green and 15-16 cm long. Field tolerance to YVMV and resistance to fruit borer.

***Punjab No.13:*** It produces light green and 5-ridged fruit of medium length. It is susceptible to mosaic and hence only suitable for spring-summer season. All above three varieties developed by Punjab Agril. University, Ludhiana.

***VRO-5:*** It is bushy in nature, heigh up to 75 cm with 2-3 effective branching, five ridges, more than 20 fruits/plant, fruit length 10-12 cm, 1.5 cm diameter, dark green colour having good marketable attributes,. Flowering starts on $4^{th}$ to $5^{th}$ node after 40 days of sowing. It is the first high yielding bushy and dwarf variety resistant to yellow vein mosaic virus and Okra Enation leaf curl virus and producing an average yield of more than 150 q/ha in rainy season and 115 q/ha in summer season.

***VRO-6***: Plants are tall (17 cm) resistant to YVMV and ELCOM, fruit 12 cm long with 1.40 cm diameter at marketable stage. First flower appears after 37 days of sowing on $4^{th}$ nodes during rainy and $3^{rd}$ nodes during summer. Average yield recorded 201.5 q/ha.

***IIVR-10:*** The plants are 7 ridged, 90-120 cm tall, branched (2-3), internodal length 5-7 cm, flowering, starts from $3^{rd}$ to $4^{th}$ nodes at 43 days after sowing, fruit length 13-15 cm at marketable stage, 1.3-1.45 cm diameter and fruits having dark green colour. On an average it bears 18 fruits per plant. The average yield of fruit is 138 q/ha in rainy season and 110 q/ha in summer season. It is highly resistant to TVMV.

***Perkins long Green***: Plants very tall, thick and prolific leaves with rough surface, bright green long fruits with 8-ridges. Suitable for hilly areas only.

***DVR-1*** : It is a combination of PDIB-2 (female) and PDIB-1 (male) parent. ON an average green fruit yield was 266 q/ha. This hybrid is resistant to yellow vein mosaic as well as enation leaf curl virus.

***DVR-2*** : It is recommended through AIC VIP for Arid Western Plain for okra hybrid cultivation. Plants are resistant to YVMV and ELCOM, pods length medium and dark green in colour. Average yield is 220 q/ha.

***BO-1:*** Plant medium height, fruit long and bold, deep green, 7-8 ridges per fruit, very good yielder (105-110 q/ha). Highly tolerant to Y.M.V. diseases.

***Utkal Gaurav (BO-2):*** Plant tall, fruit long green, tender, 5-ridges, very good yielder (95 q/ha), resistant to YMV disease.

***Kalyani*** : It is resistant to yellow vein mosaic and having high yield potential. Pods are thin, long and dark green in colour. It is suitable for cultivation throughout year.

***Satsira (seven ridges) :*** It has small plants with branching habit. Pods are bold. Small and deep green in colour, it can be cultivated throughout the year.

***Panchali :*** It is a popular hybrid, it come to marketing stage within 40-45dasys after seed sowing. It has high consumer acceptability than other types.

***Sarathi (F1) :*** Sarathi is vigorously growing tall plants hybrid with early maturity. Fruits are not ridged, slender, dark green in colours and 16-18 cm in length.Harvesting can be done for 18-20 times, first harvest start after 50 days of sowing. The average yield is about 7-8 tonnes per acre. It is ressisant to yellow vein mosaic.

***Varsha Uphar :*** It is a cross between ' lam selection 1' and ' Parbhani Kranti', the plants are medium-tall (90-120 cm) with internodes, producing 2-3 branches each. The fruit smooth dark green with 5 ridges, mature fruits 18-20 cm long with long tapering tips. It is ready for first picking 47 days after sowing. On an average yield 9.8 tonnes of vegetable and 1.5 to 2.0 tonnes of seeds from one hectare. Since it is resistant to yellow-vein mosaic-a dreaded viral disease, it is recommended for cultivation in disease-prone areas.

***Arka Abhay (IIHR-4) :*** It is a yellow vein mosaic virus resistant, has been eveloped through inter specific hybridization, using *Abelmoschus manihot* spp. *tetraphyllus* var. *tetraphyllus* as a source of resistance. The plants are tall, erect, well branched, fruits are medium long, with 5-6 ridges, spineless, tender and lush green in colour. Fruit stalk is long and easy to snap. It set fruit in two flushes and can be grown commercially throughout the year. Yield: 175-250 q/ha in 120 -135 days.

***Arka Anamika (IIHR-10) :*** It is a sister line of 'IIHR-4'. The plants are tall, erect, well branched, fruits are medium long to long with 5-6 ridges, spineless, tender and lush green in colour, fruits stalk is long and easy to snap. It set fruits in two flushes and can be grown commercially throughout the year in South. Yield 175-250 q/ha in 120–135 days.

***Hisar Unnat :*** It is a cross between selection 2-2 X 'Parbhani Kranti'. Its plants are medium tall (100- 110 cm) with short inter nodes, foliage colour is green, lower leaves shallow lobes, whereas subsequent upper leaves are narrow and deeply lobed; producing 3-4 branches each. Fruit being start from 4th to 5th node, taking 45-46 days to first picking, fruits beaked,5- ridged, smooth, green, 8-10 cm long at marketable size, where as average length and diameter of fully grown fruits are 16 cm and 2-3 respectively. Since 'Hisar Unnat' is resistant to yellow vein mosaic virus, it is recommended for cultivation in the rainy season. It performs equally good during hot summer season. Its average yield is 10.2 tonnes /ha.

***Gujarat Bhindi-1*** : It was developed at the GAU Ahmedabad in 1983. It is pure line selection from an unknown bulk seed sample received from IARI, New Delhi. Plants grow to about 60 cm height in summer season and 90 cm in rainy season and are erect with purple tinge on stem. Leaves broad, dark green with purple tinge on veins. It takes 40 days to first flowering and first picking is ready in 55-60 days of sowing. Fruiting begins from 4-5 node. Fruits 5-ridged, tender, 14-15 cm long and 6-7 cm girth in the centre at marketable stage. It yields about 70 q/ha of green pods.

***Anokhee*** : Plant height 4-5 ft in summer and winter; 5-6 ft. in rainy season. Days of 1st harvest ; 50-60 days from sowing. Colour of fruit; dark green in colour. Tolerant to yellow vein mosaic virus. It is very high yielding having short internode distance. It has 3 to 4 side branches.

***Karishma*** : This is a dark green coloured hybrid. It is high yielding and having thick fruits. This hybrid has very good virus tolerance. It can be grown round the year, except cold winter.

***Saloni*** : This is a green coloured hybrid. It is single branches hybrid with good fruit quality. It is tolerant to Yellow Vein Mosaic Virus.

***Praful*** : This is high yielding green coloured hybrid. The fruits are having good tenderness. It has got branching habit and very good, virus tolerance.

***Snehal*** : Plant height :4-5 ft. in summer and winter, 5-6 ft. in rainy season. First harvest started after 45-55 days from sowing. Fruits are green in colour. Tolerant to yellow vein mosaic virus. It has good fruiting, which starts from bottom and has short internode distance. It has 2-3 side branches and gives sort marketable fruits.

***S-040*** : This vigorous growing hybrid has good resistance to yellow vein mosaic virus. First harvest is possible 47-50 days after sowing; five ridged fruits are long (12-13cm), slim and attractive dark green in colour.

***Uphar*** : This early maturing (42-43 days) and widely adapted hybrid with vigorous plant growth has good resistance to yellow vein mosaic virus. Five ridged, long fruits (10-12cm) are smooth, tapered and light green in colour.

***S-008*** : Plants of this hybrid are vigorous in growth and has high tolerance to yellow vein mosaic virus. Five to six ridged fruits are 10-12cm in length, tapered and green in colour with smooth surface. Edible pods can be harvested 50-55 days after sowing.

***Makhmali (9777)*** : This newly developed hybrid has vigorous plant growth and good tolerance to yellow vein mosaic virus. The pods are tender, glossy, deep green in colour, silky, five ridged and 11-12 cm in length.

***S-090*** : This high yielding hybrid is vigorous in growth and recommended for areas with max. temperature below 36° C medium long thick walled pods, glossy deep green fruit colour and good shelf life makes this hybrid an excellent choice for processing and export.

***Tulsi (NP-1104)*** : This medium tall hybrid is high yielding, early in maturity (45 days) and well adapted to a very wide range of okra growing areas. Five ridges fruits are 12 cm long, glossy green in colour and attractive, Partial resistance to YYMV.

***Lady's Pride (NP1103):*** The plants of this short slimmed hybrid are compact and have highly prolific branches. Attractive dark green pentagonal fruits are straight, thick walled sweet in taste and very ideal for fresh export and processing. Partial resistance to hot and humid weather condition and yellow vein mosaic virus.

***Shaina (NP1105):*** This hybrid has a vigorous plant growth. Pods are beautiful, glossy green in colour, five ridged, smooth, 15 cm long, suitable for all okra growing areas.

***Kirti:*** Plants are medium tall and erect. This heavy yielding variety produces long, straight, tender, green and fibreless fruits of length 12 -16 cm with 5 ridges.

***Kirtiplus*** : These hybrid plants are medium tall and semi spreading. Fruits with 5 ridges are 12 -16cm long with medium dark green colour.

***OKT -072:*** New hybrid variety with tall plants produces medium thick dark five ridged fruits. Yield is extremely good.

***Aishwarya***: Plant is tall with strong and sturdy stem. Leaves are palmate-shaped, dark green and spineless. Fruits are slender, dark green spineless and pentagon in shape with five ridges. Each fruit weighs 14-15 g and is tender and uniform. Fruit bearing starts 40-45 days after sowing.

***Mistic***: Plant is tall with strong and sturdy stem. Leaves are green and palmate-shaped with deep libations. Pods are green, tender, five-ridged with high degree of uniformity. Fruits are 12-14 cm long and ready in 40-45 days after sowing.

***Purpiat***: Compact plant with strong, sturdy purple stem. Leaves are purple green and palmate-shaped with deep libations. Pods are purple, tender, five ridged with high degree of uniformity. Fruits are 10-12cm long and ready in 40-45 days after sowing.

***Reshma***: Tall and compact plant with strong, sturdy stem, leaves are green and palmate-shaped with deep libation. Pods are green tender, five-ridged with high degree of uniformity. Fruits are 10-12cm long and ready in 40-45days after sowing.

***Supriya*** : Plant tall and erect, first picking starts 45 days after sowing, fruit long and straight with 5–6 ridges and green in colour, fruit stalk long and facilitates picking, highly tolerant to yellow vein mosaic.

***S–45*** : Plants tall with late side shoots, first picking starts 40-45 days after sowing, fruit long with ridges and dark green in colour, suitable for late sowing and tolerant to yellow vein mosaic, recommended for areas prune to yellow vein mosaic.

***S–51*** : Plant tall with branches first picking start 45 days after planting, fruit long with 5–6 ridges and dark green in colour, tolerant to yellow vein mosaic.

***Selection 2-2***: Fruits green, tender and long (16-20 cm) and 5 ridged. Developed by IARI, New Delhi.

***VRO 3***: Fruits 5-ridged, green, medium long. Productivity 11.2-16.5 t/ha. Resistant to ELCV. Released by IIVR, Varanasi.

***VRO 4***: Resistant to YVMV and ELCV.Fruits 40 ridged, green medium long, productivity 160 t/ha. Developed by IIVR, Varanasi.

***Kiran***: Shy tranching plant with light green and long fruits, length 20-30 g. Productivity 17.21. t/ha.

***Salkeerthi*:** Attractive and long light green fruits, fruit length 24-30 cm, number of ridges on fruit 5, days to first harvest 44; Productivity 16.2 t/ha.

***Aruna*:** Attractive red coloured long fruits; fruit length, 25-30 cm, number of ridges on fruit 7; productivity 15.8 t/ha.

***Susthira (A caillei):*** Fruits green coloured, fruit length 22 cm; number of ridges on fruit 5; productivity 18 t/ha. All above 4 varieties developed by Kerala Agril. University.

***MDU 1*:** Developed through gamma irradiation of seeds of Pusa Sawani. Plants compact with close arrangement of nodes.Fruits 5 ridged, light green, medium long, productivity 10-11 t/ha. Susceptible to YVMV. Released by Tamil Nadu Agril. University.

***Azad Kranthi*:** Fruits 5 ridged, shining green with long beak, productivity 12.5 t/ha. It is developed by CSAUA &T, Kanpur.

***Harbhajan Bhindi*:** Perkins long Green was released as Harbhajan Bhindi. Fruits long tapered, green and 8 ridged. Released by YSPUH&F, Solan, H.P.

***F1 99*:** Tall and erect plants with good lateral branches. Foliage deeply palmate and spineless, first fruiting after 45-50 days of sowing. Fruits are smooth, medium long 10-15 cm, straight and slender, lustrous dark green, spineless, five ridged and very tender; yields 18-20 mt/ha. The hybrid has very high tolerance to yellow vein mosaic virus diseases.

***Harishan* :** Vigorous, tall, erect plants with good branching habit. Leaves deeply palmate and spineless. First fruiting after 38-40 days of sowing. Fruits are long, more slender, straight, green to lush green, 10-12 cm long, spineless with five ridges and very tender and possess very good keeping quality. The hybrid has strong resistance to yellow vein mosaic virus.

***NS 502 (F1)* :** Medium vigorous plants, erect and well branched, bearing in two flushes. During first flush fruits are borne on the main stem 40-50 days after sowing. In second flush the fruits are borne on shoot branches. Fruits are medium to long (15-20 cm), five ribbed, smooth, dark green in colour and tender. Resistant to yellow vein mosaic virus.

***NS 503(F1)* :** Tall vigorous plants bearing two flushes and well branched. Fruits are thick, smooth, 15-20 cm long, bush green, five ribbed. It is an excellent yielder and resistant to YVMV. Begins to yield from 45th day.

*NS 504 (F1):* Vigorous, well branched plant and bears in two flushes. Fruits are very dark green, shiny, five ribbed, smooth, 15-20 cm long. The fruits are easy to harvest and has good keeping quality. This hybrid is an excellent yielder and resistant to YVMV.

*NS 531(F1) :* Plants dwarf and well branched with short internodes (5 cm). Fruits are dark green and attractive with five ribs. They have very thick smooth well and are short (8-9 cm) and wide. This hybrid is a prolific yielder.

*NS 532 (F1):* Plants are dwarf, well branched and early.Fruits are lush green, short (8 cm), attractive and thick. Fruit quality is excellent. The walls are also shick. It bears very high yields.

*AROH-9 (F1):* Fruits cylindrical, uniform slender, 5-ridged, medium sized green coloured. Length of fruit is 10-12 cm. Plant habit is erect with 2-3 branches. First harvest can be done in 49-50 days. Good quality picking, 10-12. High yield potential.

*AROH-10 (F1):* Vigorous and rapid growth habit. Plants are medium tall, with earth growth habit. Less internodal distance. Fruits are elongated smooth attractive green in colour, 5-ridged, 12-13 cm fruit length. First picking in 50-55 days. 12-15 excellent quality picking. Good bearing and high yield.

*AROH-85 (F1):* Erect growth habit red pigmented stem. Attractive green slender fruits, thin red pigment at fruit tip. Good fruit weight. Less mucilage content. Excellent fruit bearing potentiality.

*AROH-113 (F1):* Erect growth habit with 2-4 branches. 50 % flowering in 44-45 days. Fruits 5-ridged, slender dark green, smooth. Good fruit weight and keeping quality. First harvest can be done in 48-50 days. Approximately 15 quality pickings. High yielding potentiality.

*Kamini (F1):* Strong, vigorous, medium tall with short internode with 2-3 branches. Dark lush green, smooth without hair, five ribbed, short tip. First picking after 45-50 days of sowing. Tolerant to yellow vein mosaic. Wide adaptable and suitable for round the year sowing.

*Aadhunik* : A front line hybrid where no compromise has been made on fruit quality and low incidence of yellow vein mosaic (YVM).

The pods are tender, glossy, deep green, silky and five angled. Low crude fibre and high total sugars, iron and phosphorus content makes, this hybrid an excellent choice for nutrition conscious. A good yielder.

***Panchali***: A vigorous and early hybrid has good resistance to yellow vein mosaic (YVM). Fruiting begins from $4^{th}$-$5^{th}$ node and first picking starts in 40-45 days. Pods are glossy, green and five angled. Harvesting of pods at immature stage (usually forefinger size) would fetch good market premium as also better harvest.

***No10*** : It is suitable for fresh market and home garden use. Fruits tender, dark green, 5 ribbed, length 12-14 cm. First harvest short from 48-50 days after sowing. Numebr of pickings 25-30. Yield potential 20-25 t/ha.

***No.12*** : First picking starts 48-50 days after sowing.Fruits dark green and attractive, length 10-12 cm, number of pickings 25-30. It gives continuous fruiting and high yield potential. Yield potential 25-28 t/ha.

***MHOK-417***: First harvest starts 45-50 days after sowing and it gives 20-28 pickings. Fruits thin, tender, green 5-ribbed, fruit length 12-14 cm. It is suitable for fresh market and home garden. Yield potential is 28-30 t/ha.

***Varsha***: It is a vigorous, toll growing hybrid, mature in 45-50 days after sowing. Fruits green, 5-ridged, length, 9 cm long, thick walled, tender, highly tolerant to yellow vein mosaic. Optimum temperature requirement is 30°C.

***Vijaya*** : Plants are dwarf, mature in 45-50 days after sowing. Fruits 13 cm long, smooth green, 5-ridged lengthy. Medium thick and tender fruits, highly tolerant to yellow vein mosaic. Optimum temperature requirement is 30°C.

***Vishal***: Plants taller with few branches, mature in 45-50 days after sowing. Fruits smooth green, 11 cm long, five ridged, medium sized and tender. Highly tolerant to yellow vein mosaic. Optimum temperature requirement is 30°C.

❑❑❑

# Colour Plates

Amaranthus (Arka Suguna)

Bean (Arka Anup)

Bean (Arka Suvidha)

Bitter gourd (Arka Harit)

Brinjal

Plate-1

Chilli (Arka Neelachal Agni)

French bean (Arka Bold)

Ivy gourd (Arka Neelachal Kunkhi)

Ivy gourd (Arka Neelachal Sabuja)

Onion (Arka Keerthiman)

Plate-4

Onion (Arka Pitambar)

Pointed gourd (Arka Neelachal Kirti)

Pointed gourd (Arka Neelachal Kirti)

Ridge gourd (Arka Sujat)

Spine gourd (Arka Neelachal Shree)

Plate-5

Sweet gourd (Arka Neelachal Arakta)

Teasel gourd (Arka neelachal Gaurav)

Tomato (Arka Ananya)

Tomato (Arka Shreshta)

Tomato (Arka Abhijit)

Tomato (Arka Alok)

Plate-6

# References and Suggested Readings


Adesuyi, S.A. (1973). Trop. Root Crops. Proc. 3rd Int. Symp. Ibadan, Nigeria.

Ankur Seeds (2001). Success of Growing Crops. Ankur Seeds Ltd., 27, Cotton Market Layout, Nagpur.

Annapurna Beej (2002). Superb Quality Vegetable Seeds. Annapurna Beej Bhandar, I. Municipal Market, P.O. Contai, Dist., Midnapur, West Bengal.

Anonymous (1943). Food values of Common Indian Food Stuffs, Published by Calcutta Corporation.

Anonymous (1952). Wealth of India, C.S.I.R., Vol. 6, New Delhi.

Anonymous (1952). Wealth of India. Publ. C.S.I.R., New Delhi.

Anonymous (1962). Wealth of India (Raw Materials) Vol. III, ph-Re, Council of Scientific Industrial Research, New Delhi, pp. 36.

Anonymous (1965). Vegetables - A mine of Nutrition. Intensive Agriculture. III(9) : 25-26.

Anonymous (1970). New Vegetable Varieties. Deutsch. Gartenbau. 7, 247.

Anonymous (1980). Achievement at a glance 1969-1980. All India co-ordinated Research project on Tuber Crops (Other than Potato), ICAR, Tirhut College of Agriculture, Dholi (Bihar).

Anonymous (1980). Nutritive Value of Indian Food. National Institute of Nutrition. ICMR., Hyderabad.

Anonymous (1981). CO-3, Amaranthus. Faculty of Horticulture. T.N.A.U., Coimbatore.

Anonymous (1982-83). Package of Practices for Rabi Crops. Haryana Agric. Univ., Hisar.

Anonymous (1984). Amaranthus. *In* : Seed Technology Scientific Workers Conference. T.N.A.U., Coimbatore, p. 68.

Anonymous (1989). Silver Jubilee, Commemoration Volume (1963-88). Central Tuber Crops Research Institute, Trivandrum, Kerala.

Anonymous (1990). Improved Varieties of Tuber Crops. Central Tuber Crops Research Institute, Sreekariyam, Thiruvananthapuram, Kerala.

Anonymous (1990). New Varieties. Division of Vegetable Crops. I.I.H.R., Bangalore.

Anonymous (1997). Vegetable Varieties. Vivekananda Parvatiya Krishi Anusandhan Sansthan (ICAR), Almora, U.P.

Anonymous (1998). Achievements of All India Co-ordinated Vegetable Improvement Project, Crop Improvement (ICAR), OUAT, Bhubaneswar.

Anonymous (1998). Varieties and Hybrids, Division of Vegetable Crops. I.A.R.I., New Delhi.

Anonymous (1998). Varieties and Hybrids. Division of Vegetable Crops. I.A.R.I., New Delhi.

Anonymous (1998). Vegetable Hybrids. Project Directorate of Vegetable Research, (ICAR), Varanasi.

Anonymous (2007). Achievements on Vegetable Research, AIC, V.I.P. (ICAR). OUAT, Bhubaneswar.

Aoba, T. (1983). Root Crops : 'Radish', in Japanese Vegetables. Green and Roots, Yasakashobou, Tokyo, 106.

Arya, P.S. and Sant Prakash (2002). Vegetable Growing in India. Kalyani Publishers, New Delhi.

Aykroyd, W.R. (1941). Nutritive Value of Indian Foods. Pub. by Nut. Res. Lab. Koonoor.

Aykroyd, W.R. (1963). The nutritive value of Indian foods and the planning of satisfactory diets. ICMR Special Rept., 42, New Delhi.

Bailey, L.H. (1949). Manual of cultivated plants. The MacMillan Co. New York.

Bailey, L.H. (1973). Manual of Cultivated plants. The Mac-Millan Company, New Delhi.

Bailey, L.W. (1930). The cultivated Brassicas. Genetes herbarium.

Banga, O. (1962). Main Types of the Western Carotene Carrot and their Origin, W.E., Tjeenk Willink Zwolle, Netherlands.

Bassett, M.J. (1986). Breeding Vegetable Crops. AVI Publications.

Bejo Sheetal (2000). Hybrid Seeds. Bejo Sheetal Seeds Pvt Ltd., A-3, Old M.I.D.C. Jalna, Maharashtra.

Bhattacharjee, S.K. (2001). Handbook of Medicinal Plants. Pointer Publications, Jaipur, CSIR, New Delhi.

BIOSTADTHI Seeds (2003). BIOSTADT, Agri Sciences. Wockhardt Tower, Bandra, Kurla Complex, Bandra (East), Mumbai.

Bohn, G.W. and Whitaker, T.W. (1951). Recently introduced varieties of head lettuce and methods of their development. U.S. Dept. Agr. Circ. 881.

Bose, T.K. and Som, M.G. (1986). Vegetable Crops in India, Noya Prakash, Calcutta.

Boss, G.V. (1959). On classification of Cucumber. Trud Pricklad Bot Genet Seleke, 32(3) : 80-6.

Century Seeds (2002). Century Seeds Pvt. Ltd., B.A. 22-24, Phase-II, Mangolpuri Industrial Area, Delhi, India.

Chadha, K.L. and Kalloo, G. (1993). Advances in horticulture, Vol. V. Malhotra Publishing House, New Delhi.

Chandola, R.P. (2002). A vegetable boon in Uttaranchal ... Popularizing cultivation of Horse Radish. Indian Horticulture. 47(2) : 10.

Chandola, R.P. (2002). A vegetable boon in Uttaranchal : Popularizing Cultivation of horse radish. Indian Horticulture 47(2) : 10.

Chatterjee, S.S. and Swarup, V. (1972). Indian Cauliflower has a still Greater Future. Indian Horticulture. Oct-Dec. Issue : 1-3.

Chauhan, D.V.S. (1972). Vegetable Production in India. Ram Prasad and Sons, Agra.

Chhabra Neelum (1992). The Tomat. Indian Horticulture, 37(2) : 11-16.

Choice (2003). Quality Seeds. Bumper Yields. Unicorn Seeds Limited Regd. Off. 1-7, 139/3, S.D. Road, Secunderabad, Andhra Pradesh.

Choudhury, B. (1976). Vegetables. National Book Trust, India, New Delhi.

Choudhury, B. and Tewari, R.N. (1993). Tomato 'Pusa Gaurav", an all Purpose Variety, Indian Horticulture, 37(4) : 14-15.

Corner, E.J.H. (1975). The Evolution of *Streblus lour* (Moraceae).

Coursey, D.G. (1967). Yams. Tropical Products Institute, Longmans, Green & Co. Ltd., London.

Dabas, B.S., Thomas, T.A., Mital, S.P. and Chopra, D.P. (1981). Sharad-bahar : A New Cluster bean. Indian Horticulture, 25(4) : 17-18.

Della, A. (1981). Report on the exploration and collection of broad beans (*Vicia faba*) in Ciprus, Plant Genetic Resources Newsl. FAO, Rome.

Dhankar, B.S., Saharan, B.S. and Pandita, M.L. (1997). 'Hisar Unnat' : new YVMV resistant Okra. Indian Horticulture, 41(4) : 50-51.

Dhankar, B.S., Saharan, B.S. and Sharma, N.K. (1999). 'Hisar Unnat' : new YVMV resistant Okra. Indian Horticulutre. 44(1) : Cover II & Page 6.

Dhar, S., Mani, V.P., Dube, S.D., Lal, S. and Prakash, V. (1997). V.L. Ageti Matar 7 : New Garden Pea. *Indian Horticulture*. 41(4) : 47-49.

Dhar, S., Mani, V.P., Prakash, V. and Lal, S. (1999). 'Vivek 6" A New garden pea. Indian Horticulutre, 44(2) : 19-21.

Dhillon, G.S. and Singh, M. (1993). 'Matar Ageta 6' : an early and high yielding garden pea. Indian Horticulture, 37(4) : 16-17.

Drewes, H. (1943). Spinach Varieties, Mich. Spl. Bull., 225.

Filippetti, A. and Ricciardi, L. (1993). Faba bean (*Vicia faba* L.). In Genetic Improvement of Vegetable Crops, edn. 1, pp. 355-79. Kalloo, G. and Bergh, B.O. (Ed.) Pergamon Press, New York.

Fortune Seeds (2002). Fortune Seeds, Gopalpur Junction, Berhampur, Ganjam, Orissa.

Geise, F.W. and Farley, H.B. (1929). Variety Studies of Spinach. Md. Bull., 312.

George, R.A.T. and Evans, D.R.A. 91981). A classification of winter Radish cultivars. Euphytica, 30, 483.

Ghosh, S.P. (1992). Tropical Tuber Crops. Indian Horticulture, 37(3) : 24-25 & 27.

Gill, H.S. (1993). Improvement of Cole Crops. *In* : Advances in Horticulture Vol. 5, (Eds. K.L. Chadha and G. Kalloo, 1993). Vegetable Crops : Part-1. Malhotra Publishing House, New Delhi, pp. 296-298.

Golden Seeds (2002). Buy with Confidence. Golden Seeds, Gigro, Golden Seeds Pvt. Ltd., B-22, Block-B, Bridge M.M. Building, K.R. Road Cross, Yediyur, Bangalore.

Gopalakrishnan, T.R. (2007). Vegetable Crops-Horticulture Science Series-4. New India Publishing Agency, Pitam Pura. New Delhi.

Gopalkrishnan, R. (1957). Medicinal Ash gourd (a new variety). Madras Agric. J. 44 : 345.

Haines, H.H. (1925). Botany of Bihar and Orissa. Vol. III. 819.

Hazra, P., Chattopadhaya, A., Dasgupta, T. and Som, M.G. (2001). 'Bidhan Barbati 1' and Bidhan Barbati 2; New high yielding and non-viny vegetable Cowpeas. Indian Horticulture. 46(1) : 13-15.

Hazra, P., Ghosh R., Sahoo, P. and Som, M.G. (1998). Characterization of the pointed gourd (*Trichoranthes Dioica* Roxb.) Clones. *Vegetable Science.* 25(2) : 162-65.

Hazra, P., Ghosh, R., Sahoo, P. and Som, M.G. (1980). Characterization of the Pointed gourd (*Trichosanthes Dioica* Roxb.) Clones. Vegetable Science, 25(2) : 162-65.

Heritage Seeds (1999). Brinjal Hybrids. Heritage Seeds, Pvt. Ltd., 201, Lusa Tower, Azadpur, Delhi.

Herkolts, G.A.C. (1972). Vegetables in South-East-Asia (1st edn.), George Allen and University, London.

Honma, S. and Heechat, O. (1971). Green Ball- A New type of Cauliflower Michigan Agric. Exp. Sta. Res. Rep. 135.

Indira, P. and Peter, K.V. (1987). Underexploited Tropical Vegetable. Kerala Agricultural University, Mannuthy.

Indo-American Hybrid Seeds (2001). Salient Fruit and Plant Characters of Hybrid Tomato. P. Box. No. 7099. 17th Cross, 2nd 'A' main, BSK 2nd stage, Bangalore.

Indo-American Hybrid Seeds (2001). Salient Fruit and Plant Characters of Hyrbid Capsicum, P. Box. No. 7099, 17th Cross, 2nd 'A' Main BSK 2nd Stage, Bangalore.

Indo-American Hybrid Seeds (2001). Salient Fruits and Plants Characters of Hybrid Brinjal. P.B. No. 7099. 17th Cross, 2nd 'A' Main, BSK 2nd stage, Bangalore.

Iwasa, S. (1971). Root Crops : 'Radish', in Vegetables in Tropics, Youkendo, Tokyo, 293.

Jones, H.A. and Mann, L.K. (1963). Onions and their allies. Leonard Hill (Books), London.

Jones, H.A. and Mann, L.K. (1963). Onions and their allies. Leonard HWH (Books), London.

Joseph, Philip, Peter, K.V. and Gopalakrishan (1981). Curry leaf - a mineral packed leafy vegetable, Indian Horticulture, 25(4) : 1-2.

Kalia, Pritam (2002). Palam Samridhi : a new variety of sprouting broccoli, Indian Horticulture, 7(1): 24-25.

Kalia, Pritam and Singh, Yudhvir (2004). Palam Kanchan : a new nutritious broccoli. Indian Horticulture, 49(2) : pp. 36.

Kalloo, G. (1989). Tomato. Indian Horticulture. 33 & 34 (4&1) : 12-15.

Kalloo, G. (1997). Research and Development of the 50 years. Solanaceous Vegetables; Indian Horticulture. 42(20) : 56-59.

Kalloo, G. and Bergh, B.O. (1993). Genetic Improvement of Vegetable Crops (Eds.) Pergamon Press, Oxford, U.K.

Kalloo, G., Dudi, B.S., Singh, Amir and Balyan, D.S. (1993). 'Hisar Harit'. A New pea. *Indian Horticulture*. 38(2) : 1.

Kalloo, G., Rai, Mathura, Kumar, R., Prasanna, H.C., Singh, M., Kumar, S., Singh, B., Ram, D., Pandey, S., Lal, H., Pandey, P.K., Rai, S., Pandey, K.K. and Satpathy, S. (2006). New Vegetable Varieties from IIVR, Varanasi. Indian Horticulture, 51(3) : 22.

Kalloo, G., Singh, Amir, Balyan, D.S., Baswana, K.S. and Pratap, P.S. (1993). 'Hissar Gairic' : Carrot bred for nutrition. Indian Horticulture, 37(4) : 21.

Kalloo, G., Singh, Major and Kumar Sanjeet (1996). Vegetable Hybrids (eds.) Project Directorate of Vegetable Research, Varanasi.

Kamalanathan, S., Sundarajan, S. and Shanmugham, A. (1973). CO-1 Amaranthus - A high yielding delicious strains, Madras Agric. J., 60 : 355.

Kamalanathan, S., Sundarajan, S., Thamburaj, S. and Subbiah, R. (1971). A high yielding Ash gourd strain. CO-1, Madras Agric. J., 58 : 25.

Kashyap, R. and Singh, J. (1988). 'Bilaspuri Arum' : A delicious local variety of Madhya Pradesh, Indian Horticulture, 32(4) : 14.

Kathiresan, C., Manoharan, M.L. and Velusamy, R. (1999). Package of practices for drumstick 'PKM-1'. Indian Horticulture, 44(2) ; 32.

Katyal, S.L. (1977). Vegetable growing in India. Oxford & IBH Publishing Co., New Delhi.

Kaul, B.L. and Kanwar, H.S. (1997). Chinese radish is a promising Vegetable for hills, Indian Horticulture, 42(1) : 16-17.

Key, D.E. (1973). Root Crops. Tropical Products Institute, London, 149-59.

Kinney, L.F. (1896), R.I. Bull., 41.

Knott, J.E. (1988). Handbook of Vegetable Growers. John Wiley and Sons, New York.

Kolhe, A.K. (1961). Poona Agri. Coll. Mag., 52 : No. I & II.

Kolthe, A.K. (1971). 'Jampa' : A New high yielding variety of French bean. Indian Horticulture. 16(1) : 19-20.

Koul, B.L. and Kanwar, H.S. (1997). Chinese radish is a promising vegetable for hills. Indian Horticulture, 42(1) : 16-17.

Kulkarni, C.K. (1962). How to get bumper crop of Snakegourd Crop. Indian Farming (March, 62) : 13-14.

Kumar, Sanjeet (2001). Production Technology of Babycorn. In : National Training Course on Vegetable Production Technology, I.I.V.R., Varanasi, India.

Kunjamma Hirshi, V.K. and Rajashekaran, V.P.R. (1969). CO2, CO3, CO4 and CO5 the four economic strains of Garden lablab (*Lab Lab niger* var. *typicus*). Madras Agric. J. 56(7) : 475-79.

Kurz, S. (1877). Forest Flora of British Burma. Vol. II : 463. M/s Bishen Singh, Mahendra Pal Singh, 23-A, New Connaught Place Dehradun.

Lal, Tarsem and Singh, Surjan (1993). 'Cowpea 263' is duel season vegetable. Indian Horticulture, 37(4) : 18-19.

Leeper, P.V. and Scully, B. (1991). 'Valcas' : A Cos lettuce cultivars of Southern Texax. Hort. Sci., 26(7) : 928-929.

Levanddovskaja, L. (1965). Leafy forms of Celery and Parsley. Kartofel i. Ovosei, 10, 1:30.

Mac Gillivary, J.H. (1961). Vegetable Production, McGraw Hill Book Company, New York. pp. 171-9.

Mac Gillivary, J.H. (1961). Vegetable Production. Mc Graw-Hill Book Company, Inc. New York.

Madzharova, D. (1971). The new parsley varieties 'Festival 68 B'. Igarski Plodve, Zelenchustii i Konservi, 4, 13.

Madzharova, D., Benbasat, E., Baburova, M. and Chavdarov, I. (1973). New forms of leaf and root celery. Gradinarskai lozarska nauka. 10(3) : 83.

Magruder, R., *et al.*, (1938). U.S. Dep. Agri. Misc., Pub., 316.

Mahyco (2002). Quality Vegetable Seeds. Mahyco Vegetable Seeds Ltd., B-4, Industrial Estate, Jalna, Maharashtra.

Malav Seeds (2003). Prime Quality Seeds Malav Seeds Pvt. Ltd. 5, Sahu Bavdi, Manek Chowk, Ratlam, Madhya Pradesh.

Mandal, R.C. (1978). Amorphophallus in Agronomy of Root and Tuber Crops. CTCRI, Trivandrum, 27-32.

Marshall, Cavendish (1981). The Fruit and Vegetable Gardeners Handbook. Marshall Cavendish Books Ltd., London.

Martin, F.W. (1982). A second edible Okra species, and its hybrids with common Okra. *Ann. Bot.*, 50 : 277-83.

Mehta, Y.R. (1959). Vegetable Growing in Uttar Pradesh. Bureau of Agriculture Information. Uttar Pradesh.

Modi Seeds (1999). Brinjal Hybrids. Modi Agritech (P) Ltd., No. 8, Jalavayu Vihar, Kukatpally, Hyderabad.

More, T.A., Sawant, S.V. and Jadhav, S.B. (2002). 'Phule Surekha' : A New French bean variety. Indian Horticulture. 47(2) : 21 & 25.

Mudaliar, V.T.S. (1950). Sweet Potato, Madras Agri. J., 37 (17) : 431-439.

Nadkarni, K.M. (1927). Indian Materia Medica, Published by Nadkarni & Co., Bombay.

Nagarjuna Seeds (1999). Bhindi Hybrid. Nagarjuna Fertilizers and Chemicals. Ltd., Bidadi, Bangalore.

Nagarjuna Seeds (1999). Brinjal Hybrids. Nagarjuna Fertilizers and Chemicals Limited, Nagarjuna Hills, Hyderabad.

Namdhari Seeds (2003). Seeds for a Better Future. Namdhari Seeds Pvt. Ltd., Bidadi, Bangalore.

Narasinghani, V.S. and Bedi Ramatala (1989). Pea. Indian Horticulture. 33 & 34 (4&1) : 20-23.

Natarajan, S., Shanmugavelu, K.G., Nambisan, K.M.P. and Krishnan, M. (1985). A New varieties of Horticultural Crop released by Tamil Nadu Agricultural University, Coimbatore during 1985.4. MDU-1, Snake gourd. South Indian Horticulture, 27 : 133.

Nath Prem, Sundari Velayudhan and Singh, D.P. (1987). Vegetables for the Tropical Region, ICAR, New Delhi.

Nath, P. (1971). Tinda - A Popular Vegetable. Indian Horticulture, 76 : 16-17.

Nath, Prem (1975). Coccinia for Commercial Cultivation. Finance Agric., July-Sept., 1975 : 23-24.

Nath, Prem and Dutta, O.P. (1971). New High-yielding Varieties in Cucurbits. Indian Horticulture. 15(4) : 11-13.

Nath, Prem and Seenappa, M. (1973). The Snake gourd. Indian Horticulture, 18(1) : 19 & 21.

NIAB (1989). Recommended Varieties of butterhead and crisphead lettuce for field production. Cambridge National Institute of Agricultural Botany.

Nieuwof, M. (1969). Cole Crops : Botany, Cultivation and Utilization. Leonard Hill Books, London.

Novartis Seeds (1990). Novartis India Limited; Seed Sector 'Seeds House', 117027, Revenue Colony, Shivajinagar, Pune.

Nunhems (2003). Vegetable Seeds. Nunhems Seeds Pvt. Ltd. Regd. Office : 10-1-107/1, Masab Tank, Hyderabad, India.

Onwueme, I.C. (1978). The Tropical Tuber Crops. John Willey & Sons, New Delhi.

Oomen, H.A.P.C. and Grubben, G.J.H. (1978). Tropical Leaf Vegetables in Human Nutrition. Communication 69. Koninklijk Instituut Voor de Tropen, Amesterdam.

Panda, J.M. (1994). Production Technology of Spices. Dwijabar Das, Brahmeswar Patna, Bhubaneswar.

Pandey, U.B. (1993). Onions in India. Bulletin No. 5. National Horticultural Research and Development Foundations, Nasik.

Pandey, U.B. (1994). Garlic Cultivation in India. Bulletin No. 7. National Horticultural Research and Development Foundation, Nasik.

Paroda, R.S. and Chadha, K.L. (1996). 50 Years of Crops Science Research in India. ICAR, New Delhi.

Patnaik, S. (1961). Arum. Indian Farming, Feb., 1961 issue.

Perrino, P. (1985). Conservazione delle risorse genetiche vegetali in Italia : Realizzazioni e programmi, in Risorse Genetiche Vegetali : Autonomia di Svilupp, Centro Internazionale CROCEVIA - Roma.

Peter, K.V. (1998). Genetics and Breeding of Vegetables. ICAR, New Delhi.

Pillai, A.A.O., Velukutty, B. and Jayapal, R. (1979). PKM-1. A New Snakegourd (*Trichosanthes anguina* L.), South Indian Horticulture. 27 : 133.

Proterfield, W.M. (1955). Loofah - The Sponge Gourd. *Ecom. Bot.*, 9 : 211-223.

Purewal, S.S. (1957). Vegetable Cultivation in Northern India. Tech. Bull. 39. Indian Council of Agricultural Research, New Delhi.

Purseglove, J.W. (1974). Tropical Crops : Dicotyledons, edn. 1, 2, pp. 114-115, English Language Book Society, London.

Purseglove, J.W. (1977). Tropical Crops : Dicotyledons, Longman, London.

Purseglove, J.W. (1978). Tropical Crops : Monocotyledons, Vol. 1 and 2. The English Language Book Society and Longman, London.

Purthi, J.S. (1987). Spices and Condiments, National Book Trust, New Delhi.

Rai, M. Vidyadhar, Sisodia, G.S., Yadav, T.P.S., and Singh, R.M. (1992). In Arunachal Pradesh Replace taro with 'Kandyam C7'. Indian Horticulture, 37(1) : 28-29.

Rai, Mathura and Pandey, A.K. (2008). Vegetable Sector : Growing with the nation. Agriculture Today. The National Agriculture Magazine; Agriculture Year Book, 2008 : 66-72.

Rai, N. and Yadav, D.S. (2005). Advances in Vegetable Production. Researchco Book Centre, New Delhi.

Rajendran, R. and Satyanarayana, A. (1988). Grow New French Beans. Indian Horticulture. 33(2) : 6-7.

Rajput, J.C., Palve, S.B., Kanade, V.M. and Wagh, R.G. (1994). Konkan Bhushan : A dwarf hyacinth bean. Indian Horticulture. 38(4) : 7.

Rajput, J.C., Patil B.P. and Palve, S.B. (1991). 'Sheetal' Cucumber may be raised as Kharif and Summer Crop. Indian Horticulture. 36(2) : 36 & Cover III.

Rallis Seeds (1999). Hybrid Brinjal. Rallis India Limited Agrochemicals Division, 21, D.S. Marg, Bombay.

Rallis Seeds (1999). Hybrid Carrot. Rallis India. Ltd., Agro-Chemicals Division, 21, D.S. Marg, Bombay - 400 001.

Rallis Seeds (1999). Tomato Hybrids. Rallis India Limited. Agro-Chemicals Division, 21, D.S. Marg Bombay.

Rallis Seeds. (1999). Watermelon Hybrids. Rallis India Ltd., Agro-Chemical Division, 21, D.S. Marg, Bombay.

Ram, D., Kalloo, G. and Banerjee, M.K. (2002). Popularizing Kakrol and Kartoli : The Indigenous Nutritious Vegetables. *Indian Horticulture*, 47(3) : 6-9.

Ram, H.H. (1998). Vegetable Breeding, Principles and Practices. Kalyani Publishers, New Delhi.

Ram, Harihar and Singh, B.P. (1984). Pant Anupama : A French bean variety for Indian bean variety for U.P. hills. Indian Farmers' Digest, 17(4) : 19-20.

Rana, M.K. and Kalloo, G. (1989). High Temperature Tolerance in Tomato : Evaluation of genotypes. Vegetable Sci., 16:156.

Rick, C.M. (1976) Tomato (in) Evolution of Crop Plants, edn. 1, pp. 268-73. Simmonds, N.W. (Ed.) Longman, London.

Rodenburg, C.M., Basse, H., Glaschke, E.B., Drouzy, J., Trebouchet, G., Haigh, J.C., Watts, L.E. and Huyskes, J.A. (1960). Varieties of lettuce, Instituunt Voor de Veredeling van, Tuinbouwgewassen, Wageningen.

Roxburghi, W. (1832). *Flora Indica*, pp. 715.

Roy, A.K., Medhi, G. and Rathiah, Y. (1992). 'VC 48-1' Tomato for Disease Prone Assam. Indian Horticulture, 37(1): 6-7.

Roychawdhury and Verma, S.P. (2003). Coleus. In : Handbook of Horticulture (eds. by K.L. Chadha). Indian Council of Agricultural Research, New Delhi.

Ryder, E.J. (1968). Evaluation of lettuce varieties and breeding lines for resistance to common lettuce mosaic. U.S. Dept., Agri. Tech., Bull., 1391.

Ryder, E.J. (1979). Leafy Salad Vegetables. AVI, Publishing Co., Westport Connecticut.

Saimbhi, M.S. (1993). Agrotechniques for cucurbits. Advances in Horticulture, 5: 401-423.

Salekha Abu and Dutta, O.P. (1995). 'IIHR8' : High Yielding ridge-gourd. Indian Horticulture, 40(3) : 8-9.

Salikutty, Joseph and Peter, K.V. (1984). Curry Leaf. Economic Botany. 39: 68-73.

Sandhu, A.S. (1963). Asparagus. Intensive Agriculture, 1. X.

Sarnaik, D.A., Sharma, G.L. and Verma, S.K. (2001). Ivy gourd : a perennial remunerative vegetable. Indian Horticulture. 46(3) : 22-23.

Sarnaik, D.A., Sharma, G.L. and Verma, S.K. (2001). Ivy gourd : A Perennial remunerative vegetable. Indian Horticulture, 46(3) : 22-23.

Sarnaik, D.A., Verma, S.K. and Sharma, G.L. (1999). Evaluation of Ivy gourd (*Coccinia grandis*) germplasm. Vegetable Science, 26(1) : 58-60.

Seelig, R.A. (1971). Broccoli. Fruits & Vegetables Facts and Pointer. United Fresh Fruit and Vegetable Association, Alexandria, Virginia.

Seemanthini, B. (1962). Quadruple your yields with three Improved Sweet Potatoes. Indian Farming, August, 1962.

Seshadri, V.S. (1989). Cucurbits. Indian Horticulture. 33 & 34 (4&1) : 28-31.

Shanmugavelu, K.G. (1989). Production Technology of Vegetable Crops. Oxford and IBH Publishing Co. Pvt. Ltd., New Delhi.

Sharma, D.K. and Sharma, P.P. (1994). Performance of Asparagus cultivars in Himachal Pradesh, Vegetable Science, 21(2) : 118-121.

Sharma, R.K., Tewari, R.N. and Pachauri, D.C. (1997). 'Pusa Sem 2' and 'Pusa Sem 3', Promising lab lab beans. Indian Horticulture, 41(4) : 45-46.

Sharma, R.R. and Verma, V.K. (1997). 'Pusa Supriya' : New Sponge-Gourd. *Indian Horticulture*. 42(3) : 12-13.

Sharma, S.R., Gill, H.S., Pachauri, D.C. and Sharma, R.K. (1994). 'Pusa Himlata': A new climbing French bean. Indian Horticulture. 38(4) : 12-13.

Shimotsuma, M. (1965). Watermelons collected in Afganistan and Iran, in Cultivated plants and their Relatives, Yamashita, K. ed., Results of the Kyoto Univ. Sci. Expedition to Karakhoram and Hindukush, Kyoto.

Shivashankar, G. and Kulkarni, R.S. (1989). Field bean. Indian Horticulture, 33 & 34 (4&1) : 24-27.

Shoemaker, J.S. (1953). Vegetable Growing. Chapman and Hall Ltd., London.

Singh, D., Atwal, A.K. and Bajaj, K.L. (2000). Evaluation of some promising type lettuce cultivars for leaf and quality characters. Vegetable Science, 27(1) : 92-93.

Singh, D.N. (2002). Performance of Colocasia (*Colocasia esculenta*) var. Antiquorum Genotype in Kharif season on the Entisol of Orissa. Indian Journal of Environment and Ecoplaning. 6(2) : 241-243.

Singh, D.N. and Singh, S.P. (2007). Production Management of Underutilized Vegetables. Agrihortica Publications Junagadh, Gujarat.

Singh, D.P., Swarup, V. and Chatterjee, S.S. (1981). Development of Synthetic Varieties of Indian Cauliflowers. Indian J. Hort., 38 (1&2) : 79-82.

Singh, Dharam and Patel, R.N. (1990). 'Pusa Komal' : A Cowpea for Vegetable. Indian Horticulture, 35 (3) : 23-25.

Singh, H. (1962). Choosing Varieties for the Home Garden. Delhi Garden Magazine I(1) : Oct.-Dec., Delhi Agri-Hort. Soc. New Delhi.

Singh, H. (1999). An emerging potential vegetable. Baby corn hybrid. "Golden Baby". Indian Horticulture. 43(4):35.

Singh, H.B. (1963). 'Pusa Kesar' and 'Pusa Kanchan' Carrot and Turnip varieties. Indian Horticulture, 7 (4) : 21.

Singh, H.B. and Bhagchandani, P.M. (1967). Bhindi Cultivatrion in India. ICAR, New Delhi.

Singh, H.B. and Joshi, S.A. (1960). Our leafy vegetables. Farm Bull. 56, ICAR, New Delhi.

Singh, H.B. and Sikka, S.M. (1955). Cowpeas that do better. Indian Farming, 4:16.

Singh, H.B., Mital, S.P. and Mansoor, Kazim (1968). 'Pusa Dofasali' : A New Cowpea variety. Indian Horticulture. 12(3) : 13.

Singh, H.B., Swarup, V. and Singh, B. (1975). Three Decades of Vegetable Research in India, ICAR, Tech., Bull. New Delhi.

Singh, J. and Kashyap, R. (1983). Some Promising Varieties of Chilli. Indian Horticulture, 28(1) : 7-9.

Singh, J.P., Gill, H.S., Swarup, V. and Arora, R.S. (1971). 'Pusa Himani' Radish. Indian Horticulture, 16(1) : 13-14.

Singh, K.P., Singh, D.K. and Singh, B.K. (1982). All about Sweet Potato of Bihar. Indian Horticulture, 27(2) : 9-10.

Singh, K.P., Singh, J.R.P. and Rai, B. (1982). Rajendra Shakarkand-5, A New Sweet Potato. Indian Horticulture, 26(4) : 5-6.

Singh, Kirti (1989). Pointed-gourd. Indian Horticulture, 33 & 34 (4&1) : 33-37.

Singh, Kirti (1989). Pointed-gourd. Indian Horticulture. 33 & 34 (4&1) : 33-37.

Singh, N.D., Srivastava, B.P. and Srivastava, J.P. (1994). Brinjal 'Azad Hybrid'. Indian Horticulture, 40(1) : 13-14.

Singh, N.P., Bharadwaj, A.K., Kumar, A. and Singh, K.M. (2004). Modern Technology on Vegetable Production. International Book Distributing Co., Lucknow, India.

Singh, S.J.P., Rajput, C.B.S. and Singh, K.P. (1988). Cluster bean : for vegetable, forage and gum. Indian Horticulture, 32(4) : 18-19.

Singh, S.P. (1989). Production Technology of Vegetable Crops. Agric. Res. Comm. Centre, Kernal, Hisar, Haryana.

Singh, S.P. (2001). Seed Production of Commercial Vegetables. Agrotech Publishing Academy, Udaipur.

Sirohi, P.S. (1997). Improvement in Cucurbit Vegetables. Indian Horticulture, 42(2) : 64-67.

Sirohi, P.S. (1997). Pumpkin 'Pusa Vikas'. Indian Horticulture, 41(4) : 41-42.

Sirohi, P.S. (1998). New Indian Spinach : Pusa Bharati, Indian Horticulture, 43(3) : 19.

Sirohi, P.S. (2000). 'Pusa Hybrid 1" : New Pumpkin Hybrid. Indian Horticulture, 44(1) : 32 & 34.

Sirohi, P.S. and Sivakami, N. (1997). Amaranth, Pusa Lal Chanlai. Indian Horticulture, 41(4) : 43-44.

Sirohi, P.S., Choudhury, B.C. and Kalda, T.T. (1991). Pumpkin Horticulture. 36(1) : 24-26.

Smart, J. (1976). Tropical Pulses. Trop. Agric. Ser. Longman Group Ltd., London.

Sreemaa Seeds (2001-02). Krushi Darpan. Sreemaa Seeds Pvt. Ltd., N.H.-5, Naya Chowk, Madhupatna, Cuttack, Orissa.

Srivastava, B.P., Srivastava, J.P. and Singh, K.P. (2000). 'Azad B3' : New Brinjal. Indian Horticulture. 44(4) : 36.

Sundarajan, S. and Muthukrishnan, C.R. (1979). CO-1. A Promising Snakegourd selection. Indian Horticulture, 24 : 21.

Sundarajan, S. and Muthukrishnan, C.R. (1982). A high yielding Ash gourd. Indian Horticulture. 27 : 15.

Sundarajan, S. and Veeraraghavathan, D. (1986). New Varieties of horticultural crops released by the T.N.A.U., Coimbatore during 1986. 2. CO-2 Snakegourd. South Indian Horticulture, 24 : 21.

Sungro Seeds. 2000. Sungro Seeds Ltd., Head Office : 207, Aradhana Bhawan, Azadpur Commercial Complex, Delhi.

Swarup, V. and Chatterjee, S.S. (1972). Origin and Genetic Improvement of Indian Cauliflower. Economic Botany Vol. 26 : 381-393.

Swarup, Vishnu (1998). Genetic Resources of Vegetable Crops in India. National Symposium on Emerging Scenario in Vegetable Research and Development. Indian Society of Vegetable Science, Silver Jubilee Memorial Lecture. 12-14 December, 1998 at PDVR, Varanasi.

Swiader, J.M., Ware, G.W. and Mc Collum, J.P. (1994). Producing Vegetable Crops. International Book Distributing Co., Charbagh, Lucknow.

Symmonds, N.W. (1976). Evolution of Crop Plants. Longman, London and New York.

Thakur, M.R. (1989). Okra. Indian Horticulture. 33 & 34 (4&1) : 49-51.

Thakur, P.C., Joshi, S., Verma, T.S. and Verma, H.C. (1998). 'Pusa Deepti' : New Capsicum Hybrid. Indian Horticulture 43(1) : 6 & Cover III.

Thamburaj, S. and Singh, Narendra (2003). Vegetables, Tuber Crops and Spices (Eds.) I.C.A.R., New Delhi.

Thamburaj, S., Ramaswamy, N., Muthukrishnan, C.R. and Shanmugavelu, K.G. (1981). CO-2 A short duration variety of sweet potato. Indian Horticulture, 25(3) : 17-19.

Thompson, H.C. and Kelly, W.C. (1983). Vegetable Crops. Tata McGraw-Hill Publishing Company Ltd., New Delhi.

Thompson, R.C. (1951). Lettuce Varieties and Culture, U.S. Dept. Agr. Farmer's Bull. 1953.

Thompson, R.D. and Ryder, E.J. (1961). Description and Pedigrees of nine varieties of lettuce. USDA, Techno. Bull., 1244.

Tokita Seeds (2002). Seeding in Believing. Tokita Seed India (P) Ltd.

Tracy, W.W. Jr. (1904). American Varieties of lettuce. U.S. Dept. Agri. B.P.I., Bull. 69.

Tropica (2000). Brassica, Green Co., Ltd. 81/10B, Ho Van Hue St., Ward 9, Phu Nhuan, Dist., HCMC. Tropical Seeds Com.

Tropica (2000). Gourd, Pumpkin, Tomato, Eggplant. Green Co. Ltd., 81/10B Ho Van Hue St., Ward 9, Phu Nhuan Dist., HCMC.

Tropica (2000). Hort. & Sweet Pepper Co. Ltd., 81/10B Ho Van Hue. St. Ward 9. Phu 10 Huan Dist., HCMC. Tropical-Seed Com.

Tropica (2000). Leaf Vegetable, Bean & Allium. Green Co., Ltd., 81/10B Ho Van St., Ward 9, Phu Nhuan Dist., HCMC, Tropical-Seeds Com.

Tropica (2000). Melon & Watermelon, Green Co., Ltd. 81/10B, H O Van Hue St., Ward 9, Phu Nhuan Dist., HCMC. Tropical Seeds Com.

Varalakshmi, B., Naik, G., Reddy, P.V.V. and Pandey, S.C. (1998). 'Arka Suguna' : New multient amaranth. Indian Horticulture 42(4) : 14-15.

Varalakshmi, B., Reddy, P.V.V., Naik, G., Chalam Celia V. and Pal, D.K. (2001). 'Arka Anupama' : A New Spinach variety. Indian Horticulture, 46(1) : 28 & 30.

Varalakshmi, B., Reddy, P.V.V., Naik, G., Chalam Celia, V. and Pal, D.K. (2001). 'Arka Arunima' : A New, multient amaranth. Indian Horticulture, 46(1) : 29-30.

Verma, S.P. and Vimala, B. (1993). All India Co-ordinated Research Project on Tuber crops. Indian Horticulture, 30(3) : 7-9.

Verma, V.K., Murthi, A.D. and Sharma, R.R. (2003). 'Pusa Sneha' : A New Sponge Gourd from IARI, *Indian Horticulture*, 48 (4) : 28-29.

Vidal, J.E. (1967). Presence in Indochina De *streblus indicus* (Bur) Corner Adonsonia 7 : 365-369.

Vijay, O.P. (1987). Vegetable Varieties for Tribal of Panchmahal District of Gujarat. Indian Horticulture. 32(2) : 14-16.

Vijay, O.P. and Vani, A. (1990). Dwarf bushy Dolichos-bean gives early high yield at low cost. Indian Horticulture, 35(2) : 14, 15 & 17.

Vijay, O.P., Nath Prem and Dutta O.P. (1972). Three New Varieties in Brinjal. Indian Horticulture, 16(3) : 21-22.

Vuurmans, J. and Grubben, G.J.H. (1977). Vegetable for Hot Humid Tropics. 2: 40-43.

Walters, T.W. and Walters, D.S. 91988). Balsam Pear (*Momordica Charantia, Cucurbitaceae*), *Econ. Bot.* 47 : 286.

Whitaker, T.W. and Davis, G.N. (1962). Cucurbits : Botany, Cultivation and Utilisation. Leonard Hill and Interscience, Publications, INC, New York.

Whitaker, T.W. and Davis, G.N. (1962). Cucurbits, edn. 1, pp. 1-136. Inter Science Publishers Inc., New York.

Work, P. and Carew, J. (1955). Vegetable Production and Marketing. John Wiley and Sons. Inc., New York.

Yamaguchi, M. (1983). World Vegetables. AVI, Publishing Co., West Port, Connecticut.

Yawalkar, K.S. (1980). Vegetable Crops of India. Agro-Horticultural Publishing Horse, Nagpur, India.

Zandstra, B.H. and Marshed, D.E. (1982). A growers Guide to Rhubarb Production. American Vegetable Growers: Dec. 1982.

Zhang, S. (1989). An extremely long variety of *Luffa Cylindrica*, Zuowu Pin Zhong Ziynam, No. 4, 7.

□□□

## About the Authors

**Dr. Dina Nath Singh** is Professor of Vegetable Science, Orissa University of Agriculture and Technology, Bhubaneswar. He received B.Sc. (Ag.) degree from Udai Pratap College, Varanasi under Gorakhpur University, Gorakhpur (UP) in 1974 and M. Sc.(Ag.) Horticulture and Ph.D., Horticulture from Banaras Hindu University in 1976 and 1982, respectively. Dr. Singh has 30 years of research experience in the field of Vegetable Science along with 12 years experience of teaching of Graduate, Post Graduate and Ph.D. students in the field of vegetable science. He has 80 research papers published in National and International Journals. Dr. Singh has guided several Post Graduate and Ph.D. students in the field of vegetable science. He is a author of a text book entitled "Production Management of Underutilized vegetables". He has developed a number of technologies for increasing production of vegetables in Orissa. Dr. Singh is a fellow of Society of Environmental Sciences. He is the life members of 4 Scientific Societies. He has attended several workshops, seminars and symposia of national and international repute during his service careers related with Horticultural crops. Dr. Singh is also Officer-in-Charge of All India Co-ordinated Research Project on Medicinal & Aromatic Plant and Betelvine, OUAT, Bhubaneswar Centre.

**Prof. Vishal Nath, Ph.D.** has completed his Master's degree in Horticulture from Narendra Dev University of Agriculture and Technology, Faizabad and Ph.D. (Hort.) from Kanpur. He started his career as Scientist (Hort.) from Central Institute for Sub-tropical Horticulture, Lucknow in 1993 and worked as Scientist at NRC for Arid Horticulture, Bikaner, Senior Scientist (Hort.) at Central Horticultural Experiment Station (IIHR), Ranchi and Horticulture and Agro-forestry Research Programme, Ranchi and moved as Professor (Horticulture) at Birsa Agricultural University, Ranchi. Prof. Vishal Nath is presently working as Principal Scientist & Head, Central Horticultural Experiment Station (IIHR), Bhubaneswar since 01.07.2006.